Smart Electric Vehicle Charging Approaches for Demand Response

Smart Electric Vehicle Charging Approaches for Demand Response

Guest Editors

Cesar Diaz-Londono
Yang Li

Basel • Beijing • Wuhan • Barcelona • Belgrade • Novi Sad • Cluj • Manchester

Guest Editors

Cesar Diaz-Londono
Politecnico di Milano
Milan
Italy

Yang Li
Chalmers University
of Technology
Gothenburg
Sweden

Editorial Office
MDPI AG
Grosspeteranlage 5
4052 Basel, Switzerland

This is a reprint of the Special Issue, published open access by the journal *Energies* (ISSN 1996-1073), freely accessible at: https://www.mdpi.com/journal/energies/special_issues/2PWHQ9TK2Y.

For citation purposes, cite each article independently as indicated on the article page online and as indicated below:

Lastname, A.A.; Lastname, B.B. Article Title. *Journal Name* **Year**, *Volume Number*, Page Range.

ISBN 978-3-7258-2929-3 (Hbk)
ISBN 978-3-7258-2930-9 (PDF)
https://doi.org/10.3390/books978-3-7258-2930-9

© 2025 by the authors. Articles in this book are Open Access and distributed under the Creative Commons Attribution (CC BY) license. The book as a whole is distributed by MDPI under the terms and conditions of the Creative Commons Attribution-NonCommercial-NoDerivs (CC BY-NC-ND) license (https://creativecommons.org/licenses/by-nc-nd/4.0/).

Contents

About the Editors vii

Cesar Diaz-Londono and Yang Li
Smart Electric Vehicle Charging Approaches for Demand Response
Reprinted from: *Energies* **2024**, *17*, 6273, https://doi.org/10.3390/en17246273 1

Fatma Gülşen Erdinç, Alper Çiçek and Ozan Erdinç
Resiliency-Sensitive Decision Making Mechanism for a Residential Community Enhanced with Bi-Directional Operation of Fuel Cell Electric Vehicles
Reprinted from: *Energies* **2022**, *15*, 8729, https://doi.org/10.3390/en15228729 4

Esmaeil Valipour, Ramin Nourollahi, Kamran Taghizad-Tavana, Sayyad Nojavan and As'ad Alizadeh
Risk Assessment of Industrial Energy Hubs and Peer-to-Peer Heat and Power Transaction in the Presence of Electric Vehicles
Reprinted from: *Energies* **2022**, *15*, 8920, https://doi.org/10.3390/en15238920 21

Lisa Calearo, Charalampos Ziras, Andreas Thingvad and Mattia Marinelli
Agnostic Battery Management System Capacity Estimation for Electric Vehicles
Reprinted from: *Energies* **2022**, *15*, 9656, https://doi.org/10.3390/en15249656 45

Kiran Bathala, Dharavath Kishan and Nagendrappa Harischandrappa
Soft Switched Current Fed Dual Active Bridge Isolated Bidirectional Series Resonant DC-DC Converter for Energy Storage Applications
Reprinted from: *Energies* **2023**, *16*, 258, https://doi.org/10.3390/en16010258 62

Yesid Bello, Juan Sebastian Roncancio, Toufik Azib, Diego Patino, Cherif Larouci, Moussa Boukhnifer, et al.
Practical Nonlinear Model Predictive Control for Improving Two-Wheel Vehicle Energy Consumption
Reprinted from: *Energies* **2023**, *16*, 1950, https://doi.org/10.3390/en16041950 82

Andrea Mazza, Giorgio Benedetto, Ettore Bompard, Claudia Nobile, Enrico Pons, Paolo Tosco, et al.
Interaction among Multiple Electric Vehicle Chargers: Measurements on Harmonics and Power Quality Issues
Reprinted from: *Energies* **2023**, *16*, 7051, https://doi.org/10.3390/en16207051 108

Zahid Ullah, Kaleem Ullah, Cesar Diaz-Londono, Giambattista Gruosso and Abdul Basit
Enhancing Grid Operation with Electric Vehicle Integration in Automatic Generation Control
Reprinted from: *Energies* **2023**, *16*, 7118, https://doi.org/10.3390/en16207118 125

Bingkun Song, Udaya K. Madawala and Craig A. Baguley
Optimal Planning Strategy for Reconfigurable Electric Vehicle Chargers in Car Parks
Reprinted from: *Energies* **2023**, *16*, 7204, https://doi.org/10.3390/en16207204 143

Josef Meiers, and Georg Frey
A Case Study of the Use of Smart EV Charging for Peak Shaving in Local Area Grids
Reprinted from: *Energies* **2024**, *17*, 47, https://doi.org/10.3390/en17010047 164

Andrea Mazza, Angela Russo, Gianfranco Chicco, Andrea Di Martino, Cristian Giovanni Colombo, Michela Longo, et al.
Categorization of Attributes and Features for the Location of Electric Vehicle Charging Stations
Reprinted from: *Energies* **2024**, *17*, 3920, https://doi.org/10.3390/en17163920 189

About the Editors

Cesar Diaz-Londono

Cesar Diaz-Londono is currently serving as a researcher in the Microgrid and Renewable Energy Research Center at the Huanjiang Laboratory, China. He earned his B.Sc. and M.Sc. degrees in Electronics Engineering from Pontificia Universidad Javeriana, Colombia, in 2014 and 2016, respectively. Cesar holds a double doctoral degree, having completed a Ph.D. in Engineering at Javeriana and a Ph.D. in Electrical, Electronics, and Communications Engineering from Politecnico di Torino, Italy, in 2020. He graduated with Cum Laude honors for his master's degree, and his doctoral thesis received the highest honors. Following his academic achievements, Cesar served as a postdoctoral researcher at Politecnico di Torino in 2021. He was also an Assistant Professor at Politecnico di Milano, Italy, between 2022 and 2024. Additionally, he undertook two visiting research positions: one at Delft University of Technology, Netherlands, in early 2024 and another at Chalmers University of Technology, Sweden, later that year. His research expertise centers around the integration of electric vehicles into the electrical grid. He actively participates in collaborative research projects with industry partners specialized in electric vehicle chargers. Cesar also contributes to the development of real-time controllers for electrical grid studies. His diverse experiences have honed his skills in research, teaching, and managing both independent and team-based projects. He has also established valuable collaborative relationships with peers across various universities. Cesar is committed to advancing his career and making significant contributions to the fields of energy management and electric vehicle integration. His research has been disseminated in over 35 scientific publications, including conference papers and journal articles.

Yang Li

Yang Li is a Researcher at the Automatic Control research group, Division of Systems and Control, Department of Electrical Engineering, Chalmers University of Technology, Gothenburg, Sweden. His main research focuses on developing efficient energy conversion processes using energy storage systems for modern power grids and electrified transportation systems. He has authored over 80 relevant peer-reviewed papers published in top-tier journals and prestigious conferences. All the papers are in the areas of modelling, management, and control of energy storage and energy conversion systems.

Editorial

Smart Electric Vehicle Charging Approaches for Demand Response

Cesar Diaz-Londono [1,*] and Yang Li [2]

1. Dipartimento di Elettronica, Informazione e Bioingegneria, Politecnico di Milano, Piazza Leonardo da Vinci, 32, 20133 Milano, Italy
2. Department of Electrical Engineering, Chalmers University of Technology, 412 96 Gothenburg, Sweden; yangli@ieee.org
* Correspondence: cesar.diaz@ieee.org

This editorial explores the recent advancements in the field of smart Electric Vehicle (EV) charging approaches, particularly in the context of demand response. As EVs become increasingly integrated into the power grid, significant challenges arise in maintaining the balance between supply and demand. These challenges require the development of innovative charging strategies that not only offer energy flexibility but also predict and optimize the charging behaviors of EVs connected to charging infrastructure.

Smart charging strategies, including unidirectional (V1G) and bidirectional (V2G) approaches, offer a means to better manage demand peaks, enhance grid reliability, and improve power quality. According to [1], the strategic placement of EV charging stations is crucial for optimizing grid operations, urban planning, and customer convenience. This study underscores the importance of location analysis in maximizing accessibility while minimizing the stress on local grids.

Incorporating renewable energy sources into EV charging infrastructure is equally critical. Solar photovoltaic (PV) systems, for instance, can support local grid stability while reducing reliance on fossil fuels. Reference [2] illustrates the effectiveness of smart charging technologies in achieving peak shaving and cost savings when integrated with PV systems. Bidirectional charging, in particular, enhances these benefits, achieving up to 8.1% additional cost efficiency compared to unidirectional systems. These findings emphasize the role of advanced energy management strategies in fostering sustainable urban mobility.

Moreover, innovative planning approaches are required to address the growing demand for EV charging infrastructure. Reference [3] presents an optimization model for reconfigurable EV chargers, which reduces both investment and operational costs in large car parks while meeting diverse energy needs. Such strategies highlight the potential of flexible and scalable solutions to adapt to varying energy demands across different regions.

Citation: Diaz-Londono, C.; Li, Y. Smart Electric Vehicle Charging Approaches for Demand Response. *Energies* **2024**, *17*, 6273. https://doi.org/10.3390/en17246273

Received: 29 November 2024
Accepted: 9 December 2024
Published: 12 December 2024

Copyright: © 2024 by the authors. Licensee MDPI, Basel, Switzerland. This article is an open access article distributed under the terms and conditions of the Creative Commons Attribution (CC BY) license (https://creativecommons.org/licenses/by/4.0/).

1. Advanced Control and Optimization Techniques

The role of advanced control strategies and optimization algorithms in EV integration cannot be overstated. These technologies ensure the efficient operation of EVs and their seamless interaction with the grid. Reference [4] explores how EVs can contribute to grid reliability by compensating for the inherent variability of renewable energy sources, such as wind and solar. By leveraging EV batteries as distributed energy resources, grids can achieve greater stability even under fluctuating conditions.

Another critical aspect is improving the energy efficiency of EVs through real-time control. Nonlinear Model Predictive Control (NMPC), as discussed in [5], provides a practical framework for optimizing energy consumption. By adapting to dynamic driving profiles and environmental conditions, NMPC extends EV range and reduces overall energy use. These advancements are instrumental in enhancing the practicality and affordability of electric mobility.

However, the increasing use of EVs also raises concerns about power quality. Harmonics generated by multiple chargers operating simultaneously can affect grid stability and equipment performance. Reference [6] investigates these issues and provides insights into mitigating the impact of harmonic distortion. By addressing these technical challenges, researchers are paving the way for a more resilient and reliable charging ecosystem.

2. Energy Markets and Peer-to-Peer Transactions

Beyond technical optimization, integrating EVs into energy markets presents an opportunity for innovative business models. Reference [7] examines advanced DC-DC converters, which enable efficient energy storage and transfer, enhancing the operational capabilities of EV charging systems. Meanwhile, accurate battery management systems, as highlighted in [8], are crucial for ensuring transparency in capacity estimation and overall system performance.

Peer-to-peer (P2P) energy trading further expands the potential of EVs in modern energy systems. Reference [9] highlights the benefits of P2P transactions in industrial multi-energy hubs, demonstrating how these systems can enhance flexibility and reduce risk. By allowing decentralized energy trading among EV users, renewable energy generators, and other stakeholders, these models foster a collaborative approach to energy management that benefits all participants.

3. Enhancing Community Resilience

The integration of EVs into local energy systems is not only about efficiency but also resilience. During grid disturbances or natural disasters, bi-directional EV operations can provide critical backup power to residential communities. Reference [10] proposes a resiliency-sensitive decision-making mechanism that incorporates fuel-cell EVs in V2G mode. This approach ensures energy stability and supports local grids during abnormal conditions, offering a lifeline for communities in crisis.

Moreover, distributed energy resources, such as solar panels and battery storage systems, can work synergistically with EVs to create self-sustaining energy ecosystems. By leveraging these technologies, communities can reduce their dependence on centralized grids and enhance their ability to recover from disruptions.

4. Remarks

The integration of EVs into power grids represents a pivotal moment in the transition toward a sustainable energy future. Through smart charging strategies, advanced optimization techniques, and innovative market models, EVs can transform energy systems to be more efficient, resilient, and environmentally friendly. While challenges remain—particularly in ensuring power quality and managing infrastructure demands—ongoing research and technological advancements continue to address these obstacles.

Ultimately, the widespread adoption of EVs and their seamless integration into the grid depend on a holistic approach that balances technical innovation, market mechanisms, and community resilience. As evidenced by the studies cited, progress in this field not only enhances the viability of electric mobility but also contributes to a cleaner and more sustainable world.

Funding: This research received no external funding.

Conflicts of Interest: The authors declare no conflicts of interest.

References

1. Mazza, A.; Russo, A.; Chicco, G.; Di Martino, A.; Colombo, C.; Longo, M.; Ciliento, P.; De Donno, M.; Mapelli, F.; Lamberti, F. Categorization of Attributes and Features for the Location of Electric Vehicle Charging Stations. *Energies* **2024**, *17*, 3920. [CrossRef]
2. Meiers, J.; Frey, G. A Case Study of the Use of Smart EV Charging for Peak Shaving in Local Area Grids. *Energies* **2024**, *17*, 47. [CrossRef]

3. Song, B.; Madawala, U.; Baguley, C. Optimal Planning Strategy for Reconfigurable Electric Vehicle Chargers in Car Parks. *Energies* **2023**, *16*, 7204. [CrossRef]
4. Ullah, Z.; Ullah, K.; Diaz-Londono, C.; Gruosso, G.; Basit, A. Enhancing Grid Operation with Electric Vehicle Integration in Automatic Generation Control. *Energies* **2023**, *16*, 7118. [CrossRef]
5. Bello, Y.; Roncancio, J.; Azib, T.; Patino, D.; Larouci, C.; Boukhnifer, M.; Rizoug, N.; Ruiz, F. Practical Nonlinear Model Predictive Control for Improving Two-Wheel Vehicle Energy Consumption. *Energies* **2023**, *16*, 1950. [CrossRef]
6. Mazza, A.; Benedetto, G.; Bompard, E.; Nobile, C.; Pons, E.; Tosco, P.; Zampolli, M.; Jaboeuf, R. Interaction Among Multiple Electric Vehicle Chargers: Measurements on Harmonics and Power Quality Issues. *Energies* **2023**, *16*, 7051. [CrossRef]
7. Bathala, K.; Kishan, D.; Harischandrappa, N. Soft Switched Current Fed Dual Active Bridge Isolated Bidirectional Series Resonant DC-DC Converter for Energy Storage Applications. *Energies* **2023**, *16*, 258. [CrossRef]
8. Calearo, L.; Ziras, C.; Thingvad, A.; Marinelli, M. Agnostic Battery Management System Capacity Estimation for Electric Vehicles. *Energies* **2022**, *15*, 9656. [CrossRef]
9. Valipour, E.; Nourollahi, R.; Taghizad-Tavana, K.; Nojavan, S.; Alizadeh, A. Risk Assessment of Industrial Energy Hubs and Peer-to-Peer Heat and Power Transaction in the Presence of Electric Vehicles. *Energies* **2022**, *15*, 8920. [CrossRef]
10. Erdinç, F.; Çiçek, A.; Erdinç, O. Resiliency-Sensitive Decision Making Mechanism for a Residential Community Enhanced with Bi-Directional Operation of Fuel Cell Electric Vehicles. *Energies* **2022**, *15*, 8729. [CrossRef]

Disclaimer/Publisher's Note: The statements, opinions and data contained in all publications are solely those of the individual author(s) and contributor(s) and not of MDPI and/or the editor(s). MDPI and/or the editor(s) disclaim responsibility for any injury to people or property resulting from any ideas, methods, instructions or products referred to in the content.

Article

Resiliency-Sensitive Decision Making Mechanism for a Residential Community Enhanced with Bi-Directional Operation of Fuel Cell Electric Vehicles

Fatma Gülşen Erdinç [1], Alper Çiçek [2] and Ozan Erdinç [3,*]

[1] Department of Electrical and Electronics Engineering, Faculty of Engineering and Architecture, Istanbul Gelisim University, Istanbul 34310, Turkey
[2] Department of Electrical and Electronics Engineering, Faculty of Engineering, Trakya University, Edirne 22030, Turkey
[3] Department of Electrical Engineering, Faculty of Electrical and Electronics, Yildiz Technical University, Istanbul 34220, Turkey
* Correspondence: oerdinc@yildiz.edu.tr or ozanerdinc@gmail.com; Tel.: +90-212-383-58-44

Abstract: The trend regarding providing more distributed solutions compared to a fully centralized operation has increased the research activities conducted on the improvement of active regional communities in the power system operation in the last decades. In this study, an energy management-oriented decision-making mechanism for residential end-users based local community is proposed in a mixed-integer linear programming context. The proposed concept normally includes inflexible resiliency-sensitive load–demand activated as flexible during abnormal operating conditions, fuel cell electric vehicles (FCEVs) fed via the hydrogen provided by an electrolyzer unit connected to the residential community and capable of acting in vehicle-to-grid (V2G) mode, common energy storage and photovoltaic (PV) based distributed generation units and dispersed PV based generating options at the end-user premises. The combination of the hydrogen–electricity chain with the V2G capability of FCEVs and the resiliency-sensitive loads together with common ESS and generation units provides the novelty the study brings to the existing literature. The concept was tested under different case studies also with different objective functions.

Keywords: common energy storage systems; distributed generation units; energy management; fuel cell electric vehicles; resiliency-sensitive loads; vehicle-to-grid

Citation: Erdinç, F.G.; Çiçek, A.; Erdinç, O. Resiliency-Sensitive Decision Making Mechanism for a Residential Community Enhanced with Bi-Directional Operation of Fuel Cell Electric Vehicles. *Energies* 2022, *15*, 8729. https://doi.org/10.3390/en15228729

Academic Editor: Abu-Siada Ahmed

Received: 2 November 2022
Accepted: 18 November 2022
Published: 20 November 2022

Publisher's Note: MDPI stays neutral with regard to jurisdictional claims in published maps and institutional affiliations.

Copyright: © 2022 by the authors. Licensee MDPI, Basel, Switzerland. This article is an open access article distributed under the terms and conditions of the Creative Commons Attribution (CC BY) license (https://creativecommons.org/licenses/by/4.0/).

1. Introduction

1.1. Motivation and Background

The environmental issues have already necessitated many changes in electric power system operation from different points of view. On the one hand, the introduction of new types of electric loads, such as electric vehicles (EVs), and the integration of non-dispatchable renewable generation units in the last decades have improved the challenge of sustaining the demand–supply balance in different operating conditions. On the other hand, abnormal climatic events due to dramatically changing environmental conditions may also lead to serious additional challenges for electric power system operation. Hurricanes, earthquakes, heavy snow, etc., have shown high impacts on the physical structure of the electric power system of different regions in the world. Thus, considering the resiliency of electric power systems to such events is an operational challenge for electric power systems with increasing importance [1].

Many solutions have been proposed in this manner to enhance the power system operation under the impacts of environmental concerns driven by new technologies as well as environmental impacts based on new operating requirements. These solutions are generally considered within the context of smart grid technologies, ensuring additional possible benefits, especially for enabling a more flexible grid structure [2]. The vehicle-to-X operation option enables large amounts of EVs to act as a large-scale source during long-term outages, uneconomical operating periods, etc. [3]. Additionally, demand-side flexibility and different energy storage options can also further be active grid-edge technologies as vital parts of a more flexible electric power system structure [4,5].

The mentioned attempts to enable a smarter grid also have improved the utilization of more distributed solutions instead of a centrally operated structure. Therefore, for smaller communities, each with separate distributed generation, energy storage, multi-type energy utilization, flexible load, etc., options have been implemented even in real-world pilot examples in different regions of the world. Here, especially providing residential communities with or without grid connection enhanced by distributed energy management structures has drawn significant attention in the aforementioned applications.

1.2. Literature Overview

The literature in this manner consists of a vast number of studies dealing with specific or combined parts of the residential communities' combined energy management problem. Coordinated management of residential end-users considering varying price signals as well as grid constraints was proposed in [6] with the opportunities of photovoltaic (PV) based distributed generation, an energy storage system (ESS) and load rescheduling based flexibility options at the end-user premises. A control strategy, also with the development of proper power electronics interfaces, was presented in [7] for a multi-household residential area, including sole battery-based EVs (SBEVs), PV and ESS units at each end-user's premise. A residential microgrid with multiple renewable common generation options as well as conventional generation and ESS units was considered both from planning and operational points of view in [8]. A centralized operation scheme for a residential community composed of end-users, each equipped with PV, ESS, vehicle-to-grid (V2G) capable EVs and flexible loads, was proposed in [9].

The energy management of different residential end-users with varying operational possibilities such as distributed generation, ESS, EV, etc., availability was considered this time in a hierarchical and distributed manner consecutively in [10,11]. Another hierarchical study was given in [12] considering the flexibility of thermostatically controllable loads of end-users combined with common ESS (CESS) units. A combined centralized–decentralized management strategy for a residential neighborhood was presented in [13], where each household was equipped with PV, ESS and EVs, and the general management scheme also included the inclusion of a market operator. The study of Ancona et al. [14] considered the optimum design and operation problem for a residential community, including a PV-based common generation option together with the availability of a dual energy storage option based on a combined electrolyzer–hydrogen tank-fuel cell (FC) structure and a battery-based CESS. A game-theoretic approach-based energy management concept was proposed in [15] for multiple residential communities, including common PV and wind turbine-based generation and ESS units as well as flexible loads.

There are numerous more studies on the centralized or decentralized energy management of residential communities. In order to direct for more compact evaluations, a detailed review of community-level coordination of residential end-users considering demand-side flexibility options in a recent study in [16] can be referred to. Additionally, the role of optimization-based decision-making mechanisms for residential communities' energy management problems was discussed in detail in another recent study [17].

The studies given in [6–15] and more cited in [16,17] neglected the simultaneous availability of CESS and distributed generation units as well as the vehicle-to-grid (V2G) supply of possible commonly connected EVs, while none of the mentioned studies considered the possibility of the resiliency-sensitive loads in the residential community.

The study in [18] considered the resiliency conditions as mimicking a grid-outage event for a residential community, including SBEVs, load flexibility and common PV unit. Another study in [19] considered a residential neighborhood operation in which each residential end-user was equipped with a PV-based distributed generation option, and the common area of the residential community was equipped with V2G capable common EV parking lot and a large-scale ESS unit. The study in [20] considered multiple types of storage options (heat and electricity) for a multi-energy community, including a common PV, common ESS and V2G enabled SBEVS, as well as demand side flexibility. However, the study in [18] did not consider a common ESS option, while the study in [19,20] neglected the flexibility from the demand side arising during resiliency-challenging operational conditions. Additionally, the studies in [18–20] did not consider the possible multi-energy flow options (electricity and hydrogen) in which FC-based EVs (FCEVs) could act in V2G mode instead of SBEVs ensuring a closed hydrogen chain from hydrogen production via an electrolyzer unit step to the last step of FCEVs acting also as a source. Even a more combined structure compared to [15,16,20] was proposed in [21], including common ESS and PV-based common generation units, pricing-based indirect demand flexibility, multiple types of EVs (FCEVs and SBEVs), and V2G availability for SBEVs. However, the resiliency-based conditions and the relevant energy management system behavior to utilize the relevant demand-side flexibility options via direct load control were not considered in [21].

1.3. Content and Contributions

In this study, a resiliency-sensitive decision-making tool for a residential community is proposed in a mixed-integer linear programming (MILP) framework. The mentioned residential community includes a common PV-based generation unit, a CESS, hydrogen production and storage-connected FC-based electrified transportation solutions in the common usage area. During outages caused by abnormal operating conditions, the normally inflexible residential loads become partially flexible by curtailment of defined resiliency-sensitive loads. Additionally, FCEVs may also act as generating units in both normal and abnormal operating conditions to enhance the supply capability and economic operation of the residential community.

The contributions of the proposed study are two-fold also compared to the detailed taxonomy given in Table 1:

Table 1. The taxonomy of the relevant literature.

Reference	Common Distributed Generation	Common ESS	Additional ESS	Demand Side Flexibility	Resiliency	EV Type	V2X
[6]	X	X	X	√	X	X	X
[7]	X	X	X	X	X	SBEV	X
[8]	√	√	X	X	X	X	X
[9]	X	X	X	√	X	SBEV	√
[10]	X	X	X	√	X	SBEV	√
[11]	X	X	X	X	X	X	X
[12]	X	√	X	√	X	X	X
[13]	X	X	X	X	X	SBEV	√
[14]	√	√	√	X	X	X	X
[15]	√	√	X	√	X	X	X
[18]	√	X	X	√	√	SBEV	X
[19]	X	√	X	X	X	SBEV	√
[20]	√	√	√	√	X	SBEV	√
[21]	√	√	√	√	X	SBEV and FCEV	√
This study	√	√	√	√	√	FCEV	√

- The combination of dual-side FCEVs integration together with common distributed generation, CESS and multi-energy chain (hydrogen and electricity) availability is considered in such a structure;
- The resiliency conditions are considered a sub-decision period by enabling a flexible portion in a normally inflexible residential load profile leading to a resiliency-sensitive decision-making mechanism.

1.4. Organization of the Paper

The rest of this paper is organized as follows. The methodology is described in Section 2. The obtained results are presented and discussed in Section 3. Finally, concluding remarks are presented in Section 4.

2. System Description and Methodology

In the proposed concept, the residential community includes a common PV-based generation unit and an electrolyzer system, as well as a common hydrogen storage option. Here, FCEVs of household owners are assumed to be parked in a common area in the community. The hydrogen needs of the mentioned FCEVs are supplied by the common hydrogen storage system. Additionally, each household in the residential community owns a PV-based distributed generation unit. Moreover, apart from the grid failure-based resiliency conditions, the load of each household is totally inflexible. However, only for abnormal operating conditions, each household has resiliency-sensitive loads to be curtailed if necessary. Moreover, FCEVs can also support the residential community power needs both in normal and abnormal operating conditions.

The objective function in Equation (1) consists of two parts that can be activated separately or simultaneously. The first part, defined as the cost, comprises the difference between the economic correspondence of the energy exchanges between the upstream grid and the residential community, as shown in (2). The second part, defined as the curtailment, represents the total energy value curtailed under abnormal operating conditions from the resiliency-sensitive loads of households, as depicted in (3).

$$\min A \cdot Cost + B \cdot Curtailment \quad (1)$$

$$Cost = \sum_{t} \left(P_{UG2RC,t} \cdot \Delta T \cdot \tau_{buy,t} - P_{RC2UG,t} \cdot \Delta T \cdot \tau_{sell,t} \right) \quad (2)$$

$$Curtailment = \sum_{t} \sum_{n} \sum_{h} \left(\left(P_{rs-load-profile,h,n,t} - P_{rs-load,h,n,t} \right) \cdot \Delta T \right) \quad (3)$$

The power balance in (4) corresponds to the contributions of the total reverse power flow from the households to the residential community, the power drawn from the upstream grid by the residential community, the power production of common PV unit, discharging power of CESS unit and the total discharging based power contribution of FCEVs on the one hand, and the total power transfer to households from the residential community, the possible reverse power flow from the residential community to the upstream grid, charging power of CESS unit and the power directed to the electrolyzer unit for hydrogen production on the other hand.

$$P_{H2RC-tot,t} + P_{UG2RC,t} + P_{CPVU,t} + P_{CESS-disc,t} + P_{FCEV-disch-tot,t} \\ = P_{RC2H-tot,t} + P_{RC2UG,t} + P_{CESS-ch,t} + P_{elec,t}, \ \forall t \quad (4)$$

The logical constraints in (5)–(12), respectively, hinder the simultaneous occurrence of the bidirectional power exchanges, respectively, for the upstream grid and residential community in CESS charging and discharging conditions, the households, residential community and the electrolyzer and FCEV discharging operation.

$$P_{UG2RC,t} \leq N \cdot u_{1,t} \cdot u_{grid,t}, \ \forall t \quad (5)$$

$$P_{RC2UG,t} \leq N \cdot (1 - u_{1,t}) \cdot u_{grid,t}, \quad \forall t \tag{6}$$

$$P_{CESS-disc,t} \leq N \cdot u_{2,t}, \quad \forall t \tag{7}$$

$$P_{CESS-ch,t} \leq N \cdot (1 - u_{2,t}), \quad \forall t \tag{8}$$

$$P_{H2RC-tot,t} \leq N \cdot u_{3,t}, \quad \forall t \tag{9}$$

$$P_{RC2H-tot,t} \leq N \cdot (1 - u_{3,t}), \quad \forall t \tag{10}$$

$$P_{FCEV-disch-tot,t} \leq N \cdot u_{4,t}, \quad \forall t \tag{11}$$

$$P_{elec,t} \leq N \cdot (1 - u_{4,t}), \quad \forall t \tag{12}$$

Equations (13)–(15) represent the model for the CESS unit. Here, the state-of-energy variation in the CESS is given by (13), considering the discharging and charging power variations as well as efficiencies and time granularity. Equation (14) initiates the state-of-energy value of the CESS at the starting time while the mentioned state-of-energy value is lower and upper bounded by (15).

$$SoE_{CESS,t} = SoE_{CESS,t-1} + P_{CESS-ch,t} \cdot \Delta T \cdot CE - \frac{P_{CESS-disc,t} \cdot \Delta T}{DE}, \quad \forall t > 1 \tag{13}$$

$$SoE_{CESS,t} = SoE_{CESS-init}, \quad if\ t = 1 \tag{14}$$

$$SoE_{CESS-min} \leq SoE_{CESS,t} \leq SoE_{CESS-max}, \quad \forall t \tag{15}$$

The local power balance in the residential community, together with the bi-directional power exchanges with the upstream grid, is ensured by (16). The power balance within each household is given in (17), while (18) presents the breakdown of the residential demands into totally inflexible and resiliency-sensitive loads. The mentioned resiliency-sensitive loads are also inflexible during normal operating conditions and are just activated during abnormal operating conditions. The change in power profile in such loads is represented by (19). The activation of these loads is only possible during abnormal operating conditions, ensured by (20).

$$P_{H2RC-tot,t} + \sum_{h} P_{buy,h,t} = P_{RC2H-tot,t} + \sum_{h} P_{sell,h,t}, \quad \forall t \tag{16}$$

$$P_{PV,h,t} + P_{buy,h,t} = P_{sell,h,t} + P_{totalload,h,t}, \quad \forall h, t \tag{17}$$

$$P_{totalload,h,t} = P_{inflexload,h,t} + \sum_{n} P_{rs-load,h,n,t}, \quad \forall t \tag{18}$$

$$P_{rs-load,h,n,t} = k_{h,n,t} \cdot P_{rs-load-profile,h,n,t}, \quad \forall h, n, t \tag{19}$$

$$k_{h,n,t} \leq u_{grid,t}, \quad \forall h, n, t \tag{20}$$

The calculation of hydrogen amount produced by the power value directed to the electrolyzer unit is calculated by (21). The hydrogen amount variation within the main hydrogen tank of the residential community is represented in (22), considering the produced hydrogen and the hydrogen demand of the FCEVs. The hydrogen amount in the main tank is initiated as in (23) and bounded by lower and upper limits as in (24).

$$m_{H2-CS-prod,t} = P_{elec,t} \cdot a_{H2-P}, \quad \forall t \tag{21}$$

$$m_{H2-CS,t} = m_{H2-CS,t-1} + m_{H2-CS-prod,t} - \sum_{k} m_{FC-inj,k,t}, \quad \forall t > 1 \tag{22}$$

$$m_{H2-CS,t} = m_{H2-CS-init}, \quad if\ t = 1 \tag{23}$$

$$m_{H2-CS-min} \leq m_{H2-CS,t} \leq m_{H2-CS-max}, \quad \forall t \tag{24}$$

The total discharging power gathered by FCEVs is calculated as in (25), while the hydrogen consumption at each FCEV regarding this discharging operation is presented as in (26). The supplied, as well as consumed (with VG2 operation), hydrogen values result in a hydrogen amount variation in each FCEV's hydrogen tank depicted in (27). The hydrogen value at the hydrogen tank of each FCEV is initiated by (28) and lower-upper bounded by (29). Finally, each FCEV is ensured to leave the residential community with a hydrogen level greater than a predefined desired value as in (30).

$$P_{FCEV-disch-tot,t} = \sum_k P_{FCEV-disch,k,t}, \forall t \quad (25)$$

$$m_{H2-cons,k,t} = P_{FCEV-disch,k,t} \cdot a_{H2-P}, \forall k,t \quad (26)$$

$$m_{H2,k,t} = m_{H2,k,t-1} + m_{FC-inj,k,t} - m_{H2-cons,k,t}, \forall k, t > T_{a,k} \quad (27)$$

$$m_{H2,k,t} = m_{H2-init,k}, \forall k, t = T_{a,k} \quad (28)$$

$$m_{H2-min,k} \leq m_{H2,k,t} \leq m_{H2-max,k}, \forall t \quad (29)$$

$$m_{H2,k,t} \geq m_{H2-des,k}, \forall k, t = T_{d,k} \quad (30)$$

3. Test and Results

The problem of the resiliency-sensitive energy management strategy of a residential community with FCEVs is created through the MILP approach. The proposed structure is tested with the GAMS v.24.1.3 software and CPLEX v.12 solver. Input data and simulation results are presented in the subsections of this section, respectively.

3.1. Input Data

In this study, a community consisting of 40 individual dwellings with different numbers of residents was considered. It was assumed that only one person lives in 5 of these dwellings, 2 people live in 9 of the dwellings, 3 people live in 15 of the dwellings and 4 people live in 11 of the dwellings. It contains two group loads, inflexible and resiliency-sensitive in dwellings. While inelastic loads are priority loads that always need energy, flexible loads are in the category of interruptable loads. It should be underlined that the loads are different for each house. The inelastic load data for dwelling3 (single person), dwelling9 (two residents), dwelling24 (three residents) and dwelling39 (four residents) are presented in Figure 1. Resiliency-sensitive loads in dwellings are iron, kettle, TV, washing machine, vacuum cleaner and tumble dryer. Data on expected resiliency-sensitive load usage for some of the selected dwellings (dwelling15—3 residents and dwelling39—4 residents) are presented in Figure 2. In addition, the expected total load–demand data of the residential community are presented in Figure 3 when there is no power outage.

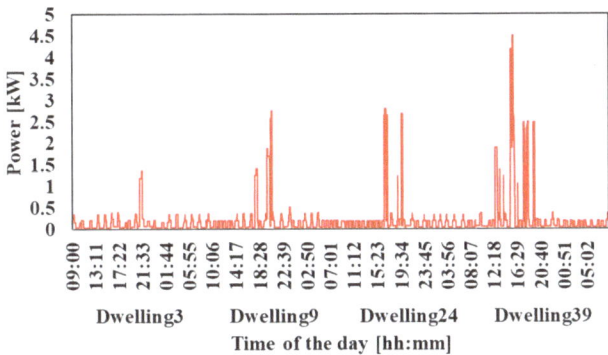

Figure 1. Inflexible load–demand of dwelling3, dwelling9, dwelling24 and dwelling39.

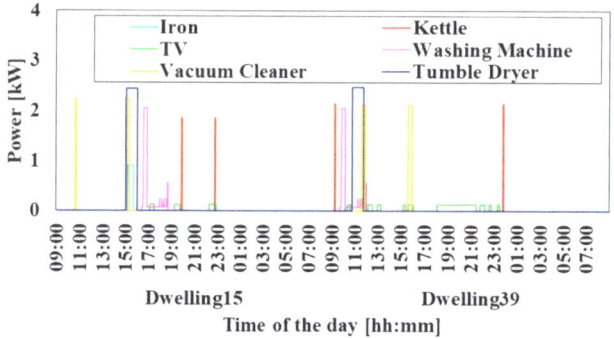

Figure 2. Resiliency-sensitive load–demand of dwelling15 and dwelling39.

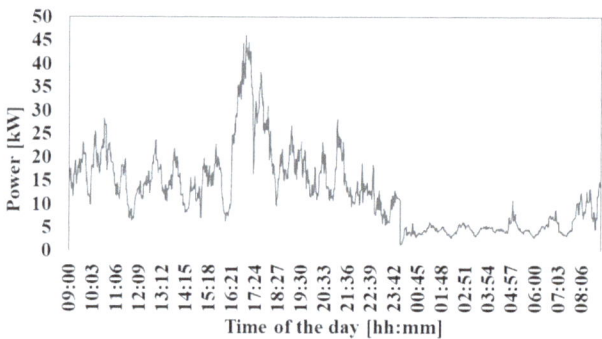

Figure 3. Total expected load–demand of residential community without power outage.

It is assumed that the residential community purchases energy from the electric power system and sells electrical energy to the grid. For electricity purchasing and electricity selling, the actual data of the Turkish electricity market dated 22 May 2022 in Figure 4 are used in Turkish Liras (TL)/kWh [22]. The community is considered to have a common PV power generation system. The power data produced using the real global radiation data of the same day are presented in Figure 5. When the power generation data were examined, it could be observed that the global radiation data belonged to a cloudy day. Additionally, in each house, there is a PV system that produces 1/25 of the common PV system.

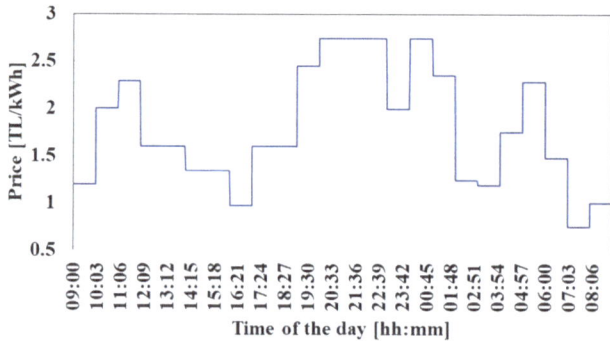

Figure 4. Electricity price data.

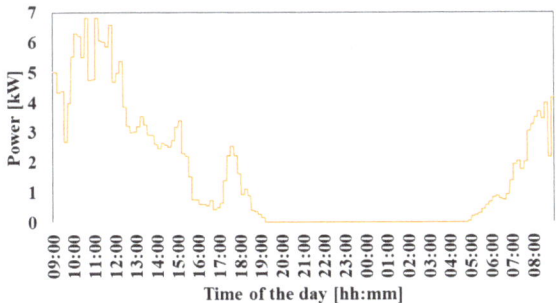

Figure 5. Power generated from common PV system in the residential community.

It is assumed that there are 40 FCEVs in the community. Data on the hydrogen molar amount in the hydrogen tanks at the time each FCEV arrives at the residential community and desired hydrogen molar amounts of FCEVs are shown in Figure 6. Additionally, it should be stated that the tank volume of FCEVs is 5.9 kg, while a_{H2-P} is taken as 1.25×10^{-4} (considering time resolution). The hydrogen molar amount for hydrogen exchange in FCEVs is determined as 2 kg for one minute period.

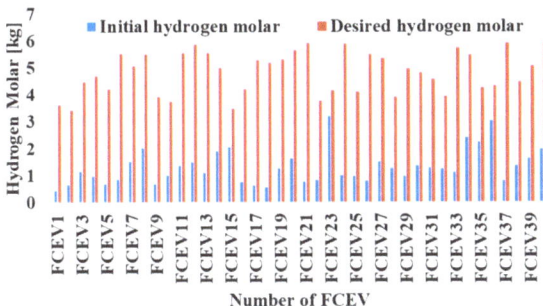

Figure 6. Initial and desired hydrogen molar amounts of FCEVs.

Data on the arrival and departure times of 40 FCEVs are presented in Table 2. Additionally, the technical characteristics of CESS owned by the community are presented in Table 3, while the features of the common hydrogen storage unit are given in Table 4. It should also be stated that the common hydrogen storage system is equal to its initial value in the final period. The maximum power consumption of the electrolyzer is 100 kW. The time period in the study is determined as one minute.

Table 2. Data of FCEV behaviors.

EV No.	* DT–AT	EV No.	* DT–AT	FV No.	* DT–AT	EV No.	* DT–AT	EV No.	* DT–AT
FCEV1	07:47–17:32	FCEV9	07:40–17:28	FCEV17	07:42–16:39	FCEV25	08:08–15:51	FCEV33	08:53–18:03
FCEV2	07:13–15:29	FCEV10	09:00–17:36	FCEV18	07:21–17:12	FCEV26	06:21–16:39	FCEV34	08:05–18:34
FCEV3	09:58–17:29	FCEV11	08:43–16:07	FCEV19	08:34–18:15	FCEV27	07:15–15:30	FCEV35	08:11–17:15
FCEV4	08:32–17:49	FCEV12	09:25–17:59	FCEV20	08:21–18:23	FCEV28	08:32–16:36	FCEV36	97:10–17:04
FCEV5	09:11–18:32	FCEV13	06:55–17:04	FCEV21	07:18–15:55	FCEV29	09:34–17:06	FCEV37	08:45–17:02
FCEV6	08:51–17:19	FCEV14	08:36–17:20	FCEV22	07:19–16:28	FCEV30	08:06–16:33	FCEV38	07:54–18:30
FCEV7	08:24–17:43	FCEV15	07:10–16:59	FCEV23	09:06–17:15	FCEV31	08:21–17:21	FCEV39	07:15–15:47
FCEV8	08:51–17:09	FCEV16	06:29–17:01	FCEV24	07:55–15:25	FCEV32	10:22–18:31	FCEV40	08:24–16:10

* DT–AT: Departure time from the residential community–arrival time to the residential community.

Table 3. Data of CESS.

CE [%]	DE [%]	$SoE_{CESS-init}$ [kWh]	$SoE_{CESS-min}$ [kWh]	$SoE_{CESS-max}$ [kWh]	Maximum Value of $P_{CESS-ch,t}$ [kW]	Maximum Value of $P_{CESS-disc,t}$ [kW]
0.95	0.95	500	100	500	250	250

Table 4. Data of common hydrogen storage unit.

$m_{H2-CS-init}$ [kg]	$m_{H2-CS-min}$ [kg]	$m_{H2-CS-max}$ [kg]
80	5	80

3.2. Simulation Results and Comparison

In order to demonstrate the effectiveness of the proposed optimization model, ten case studies were realized. The data relating to the test studies carried out are given in Table 5. These test studies are operated by changing the value of binary parameters A and B, availability of PV, CESS and from FCEVs to the residential community mode. It is thought that the power grid was not available between 17:30 and 19:30 in all test studies.

Table 5. Case Studies.

Cases	Parameter A (Cost Minimization)	Parameter B (Curtailment Minimization)	PV (Common and Dwellings)	Power from FCEVs to Residential Community	CESS
Case-1	1	0	√	√	√
Case-2	0	1	√	√	√
Case-3	1	0	√	–	√
Case-4	0	1	√	–	√
Case-5	1	0	–	–	√
Case-6	0	1	–	–	√
Case-7	1	0	√	√	–
Case-8	0	1	√	√	–
Case-9	1	0	–	√	–
Case-10	0	1	–	√	–

Results from the simulations realized are presented in Table 6. It should be stated that the costs incurred in Case-1, Case-3, Case-5, Case-7 and Case-9, which are carried out for cost minimization purposes, are increasing gradually from Case-1 to Case-9. The minus expression here means that the community is making a profit. In Case-7 and Case-9, a fee is paid for electrical energy taken from the power grid. If it is compared to Case-1 and Case-3, the gain is reduced by USD 179.46 without the support of FCEVs to the residential community mode. It should be stated that the cost is highest in the absence of CESS and PVs, but CESS has a great effect on this result. In this respect, CESS is a more effective tool than the FCEV V2G mode and PV system. In each case study, where the cost is minimized, 16,206 kWh interruptions occur in interruptable loads. In the curtailment minimization problem, it was concluded that there is an amount of interruption (4271 kWh) in the resiliency-sensitive loads when CESS is not included. Considering the low number of dwellings and the fact that the flexible loads are in operation or not during the hours of a power grid outage, this can be expressed as the reason for the low amount of interruption. In Case-2, Case-4, Case-6 and Case-8, where curtailment minimization is aimed, the community does not make a profit and makes a payment. By considering both cost and curtailment minimization, it should be underlined that the best results are obtained in the proposed structure. However, if no interruption in the load is desired, the community must make a payment in addition to making a profit.

Table 6. Results for case studies.

Cases	Cost [TL]	Curtailment [kWh]
Case-1	−1263.81	16,206
Case-2	406.99	—
Case-3	−1084.35	16,206
Case-4	456.32	—
Case-5	−1022.63	16,206
Case-6	499.18	—
Case-7	201.41	16,206
Case-8	354.27	4271
Case-9	261.96	16,206
Case-10	409.32	5959

The total power consumption of 40 dwellings in Case-1 and Case-2 are presented in Figure 7. In Case-1, where the cost is minimized, the total amount of power purchased from the grid is reduced during the hours of a power outage, as resiliency-sensitive loads are cut in order to reduce the total cost. Here, power is supplied to the residential community from FCEVs. In Case-2, where the total amount of curtailment is minimized, the community is operated without any interruptions. Here, the power usage is the same as the expected power consumption. It should be stated that energy is provided by CESS and FCEVs in case there is no energy provided in the power grid. As a result, it is seen that resiliency-sensitive loads are de-energized in economic operation, while there is no curtailment in Case-2. It should be stated that approximately 16.21 kWh (all of the loads) of power outage occurred in Case-1. In order to save from such a value of the resiliency-sensitive load, as can be seen in Table 5, the community enters a position to pay while making a profit.

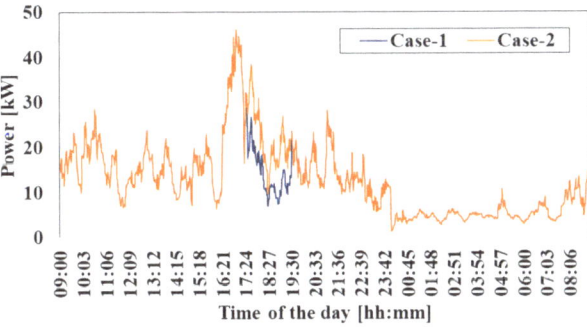

Figure 7. Total power consumption of dwellings in Case-1 and Case-2.

The power balance of dwelling25 and dwelling27 is given in Figure 8. Here, the PV productions are primarily evaluated at dwellings. Then, excess power is sold to the residential community. Due to the relatively small size of the PV capacity, approximately 1.51 kWh of energy is produced throughout the day. Dwelling25 and dwelling27 have an energy consumption of 5.05 and 4.04 kWh, respectively. During the day, 0.61 kWh and 0.56 kWh of energy are sold to the community, respectively, while 4.15 kWh and 3.08 kWh of energy are purchased from the community.

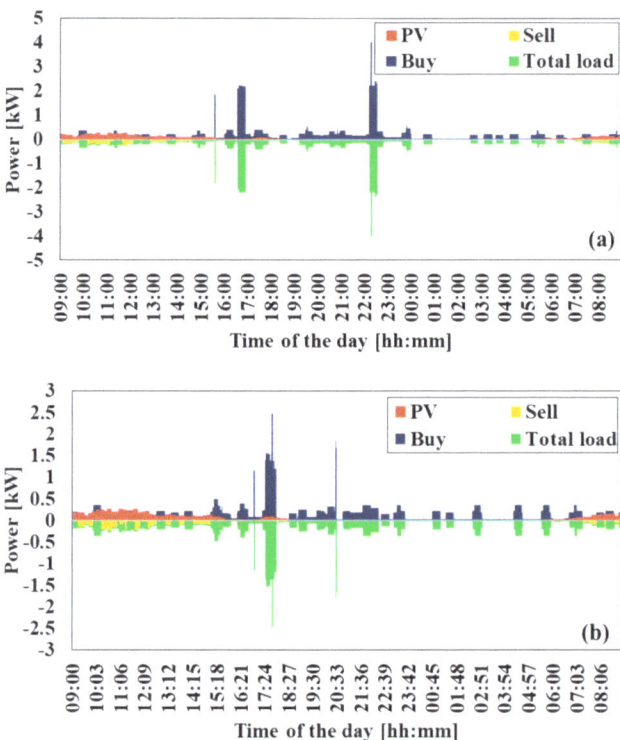

Figure 8. Power balance for dwelling25 (**a**) and dwelling27 (**b**).

Figure 9 presents the energy change in the common CESS owned by the community throughout the day. All case studies involving CESS were reviewed here. While comparing Case-1 and Case-2, which have different objective functions, CESS is less evaluated in Case-2, where the aim is only to reduce curtailment. In Case-1, where economic operation is provided, CESS is used more effectively, reducing costs. Case-3 and Case-5, where there is no energy support from FCEVs and no PVs, are slightly different from Case-1. However, in Case-6, where the curtailment is minimized, and there is no PV system and FCEV support, the energy change in CESS takes place very little during the day. An interesting result here is that CESS is not evaluated much during the hours of a mains power outage.

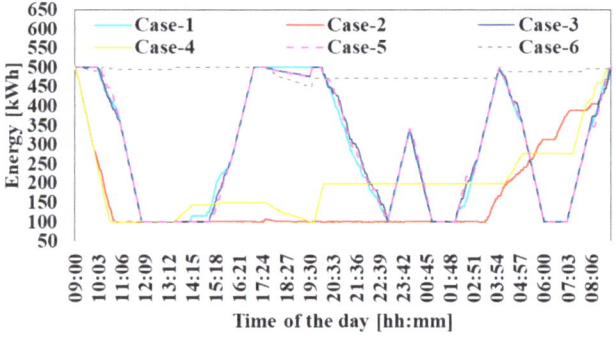

Figure 9. State of energy variation in CESS.

The variation in the hydrogen molar amount in the common hydrogen storage system is discussed in Figure 10. When Case-1 and Case-2 are compared, as in CESS, it can be said that hydrogen exchange is less in Case-2. Hydrogen refueling and releasing events occur more frequently in Case-1, where economical operation is performed. It should be noted here that FCEVs offer greater energy support to the residential community. In both case studies, it is concluded that the hydrogen molar amount decreases significantly during power outage times.

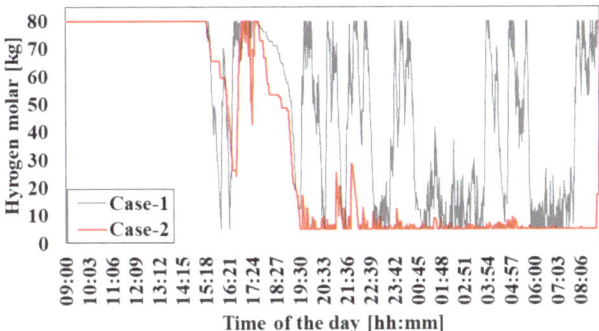

Figure 10. Hydrogen molar change in common hydrogen storage in Case-1, Case-2, Case-5 and Case-6.

The molar change in the hydrogen tank of FCEV18 in Case-1 and Case-2 is given in Figure 11. It should be stated that there is more refueling and releasing event in order to increase the gain in Case-1, as in the CESS and common hydrogen storage system. While the vehicle has 0.54 kg of hydrogen in its tank when it comes to the community, it has 5.9 kg of hydrogen, which is the desired hydrogen level when the vehicle leaves the community. In Case-2, it is seen that the amount of hydrogen in the vehicle's tank decreases when there is no energy support from the power grid.

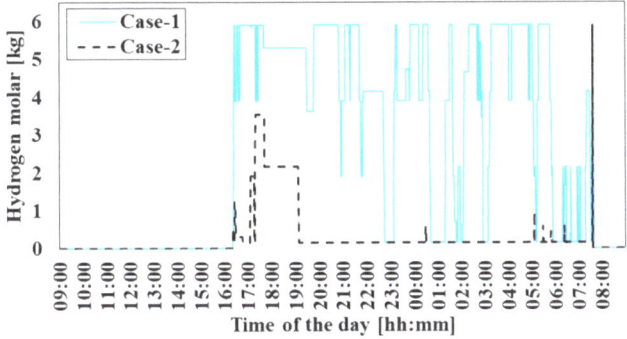

Figure 11. Variation in hydrogen molar at hydrogen tank of FCEV18 in Case-1 and Case-2.

In Case-2, hydrogen molar changes in hydrogen tanks of FCEV4 and FCEV16 are presented in Figure 12. Here FCEVs are considered to provide energy support to the residential community during power outages. FCEV16 provides more support, according to FCEV4. It should be stated that the hydrogen exchange in the tank is low according to Case-1 because the aim is to reduce the amount of interruption only during the power outage.

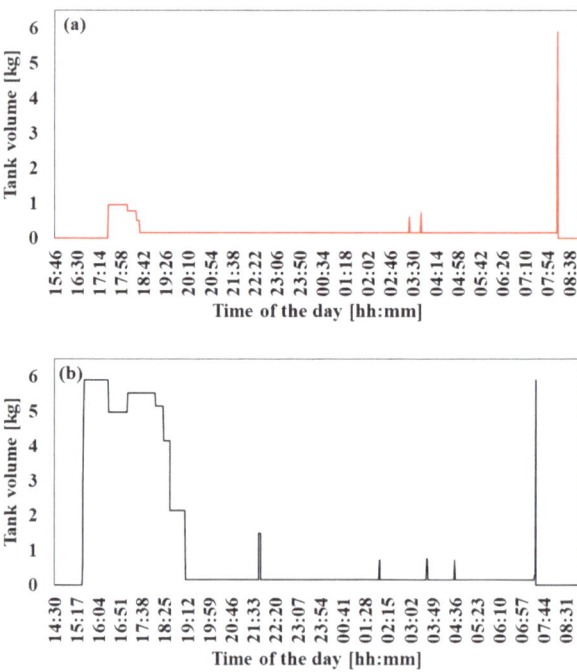

Figure 12. Variation in hydrogen molar at hydrogen tanks of FCEV4 (**a**) and FCEV16 (**b**) in Case-2.

Data on the amount of power curtailment in the absence of the power system for Case-1, Case-8 and Case-10 are given in Figure 13. It should be said that all of the resiliency-sensitive loads are interrupted in Case-1, which is economically operated. On the other hand, in Case-8 without CESS and Case-10 without CESS and PV (the aim in both case studies is to minimize curtailment), the curtailment is realized as 4.27 kWh and 5.96 kWh, respectively. CESS has a greater impact than PV systems and FCEV energy support. In addition, in Case-2, which is the recommended structure and curtailment is minimized, the total amount of interruption is 0.

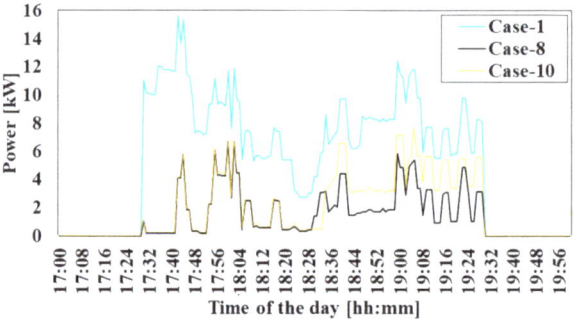

Figure 13. Power curtailment in Case-1, Case-8 and Case-10.

The total power consumption of the electrolyzer during the day in Case-1 and Case-2 is given in Figure 14. As can be seen from the behavior of the common hydrogen tank and FCEVs in Case-1 and Case-2, the electrolyzer is being further evaluated in order to minimize the total cost. In Case-2, on the other hand, there is only the aim of curtailment minimization. The electrolyzer consumes 651.86 kWh in Case-1 and 84.11 kWh in Case-2.

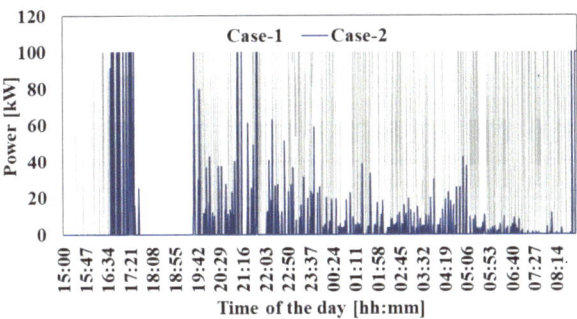

Figure 14. Power consumption of electrolyzer in Case-1 and Case-2.

Figure 15 provides data on the total power consumption from the power grid throughout the day for all components of the residential community. It should be stated that no power is drawn from the grid during the power outage. Approximately 1.932 kWh of energy is purchased for Case-1 and 619 kWh for Case-2. It should be noted that CESS and electrolyzer-induced energy purchases increased in Case-1.

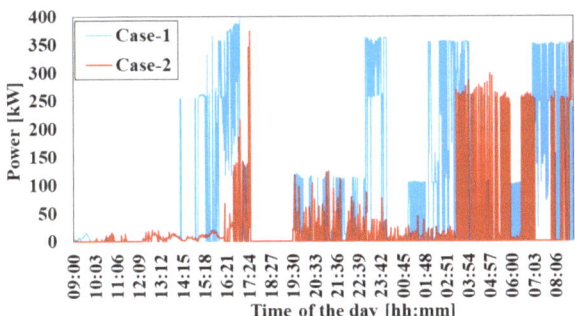

Figure 15. Power consumption of residential community in Case-1 and Case-2.

Moreover, for Case-1, data on the residential community and upstream grid power exchange are presented in Figure 16. While 1.932 kWh of energy is taken from the upstream grid throughout the day, 1.626 kWh of energy is sold to the upstream grid.

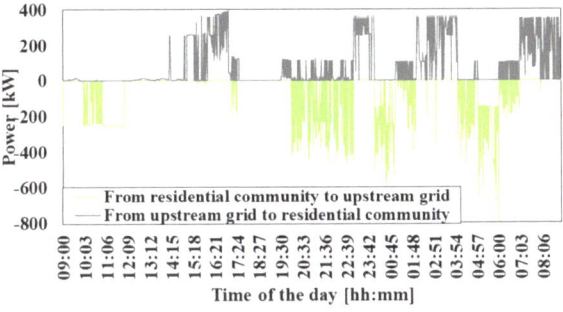

Figure 16. Power exchange data of the residential community with the upstream grid.

The total buying and selling power amounts of dwellings in Case-1 are given in Figure 17. It should be stated that some of the power produced in the PV system is sold, while a total of 20.2 kWh of energy is sold to the power grid. Dwellings have a net energy consumption of 227.89 kWh in total.

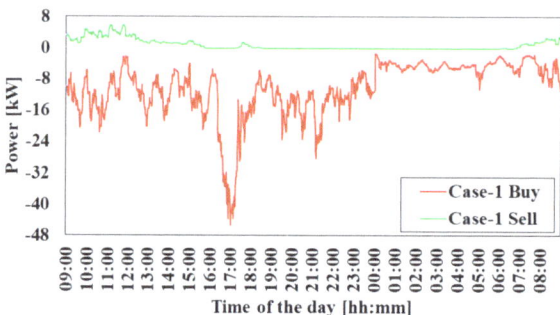

Figure 17. Total power exchange in dwellings with residential community.

4. Conclusions

In this study, a MILP model of the resiliency-sensitive decision-making mechanism of a residential community with common PV and rooftop PVs, common hydrogen storage and an electrolyzer was presented. In the study, two different objective functions were determined, namely, cost minimization and curtailment minimization.

According to the results obtained from the study, while the proposed structure works with the least cost in the economic working condition, it can minimize the amount of curtailment in the flexible loads in the curtailment minimization condition. However, in this case, the resulting cost increases as no interruption in flexible loads is ensured. Even in the case of economic operation, while the community earns a profit, it has to make a payment if no curtailment is requested. When the results were examined, it was seen that the share of CESS on the results has the highest value for both objective functions. Moreover, the worst results occur in the condition where PVs, CESS and FCEV support are absent. With the proposed structure, it is concluded that even in the absence of energy from the grid, the community can be operated uninterruptedly, and even profit can be obtained with a very small amount of interruption.

The consideration of the uncertainties regarding the PV-based renewable generation, FCEV-related parameters (arrival and departure times, arrival hydrogen levels, etc.), together with the combination of different residential communities under a single decision-making entity together with energy market participation possibilities, can be given as a future study.

Author Contributions: Conceptualization, F.G.E.; methodology, F.G.E.; software, F.G.E. and A.Ç.; validation, F.G.E., A.Ç. and O.E.; investigation, F.G.E., A.Ç. and O.E.; resources, F.G.E. and A.Ç.; data curation, F.G.E. and A.Ç.; writing—original draft preparation, F.G.E., A.Ç. and O.E.; writing—review and editing, F.G.E., A.Ç. and O.E.; visualization, F.G.E. and A.Ç.; supervision, O.E.. All authors have read and agreed to the published version of the manuscript.

Funding: This research received no external funding.

Conflicts of Interest: The authors declare no conflict of interest.

Nomenclature

Sets

t	Set of time periods.
h	Set of households in the residential community.
k	Set of FCEVs.
n	Set of resiliency-sensitive loads.

Parameters

A, B	Binary parameters to decide the structure of the objective function.
a_{H2-P}	Hydrogen amount to electric power conversion constant [kg/kW].
CE	Charging efficiency.
DE	Discharging efficiency.
$m_{H2-des,k}$	Desired hydrogen amount in FCEV k during departure time [kg].
$m_{H2-init,k}$	Initial hydrogen amount in FCEV k during arrival time [kg].
$m_{H2-max,k}$	Maximum allowable hydrogen amount in FCEV k [kg].
$m_{H2-min,k}$	Minimum allowable hydrogen amount in FCEV k [kg].
$m_{H2-CS-init}$	Initial hydrogen amount in common hydrogen storage unit [kg].
$m_{H2-CS-max}$	Maximum allowable hydrogen amount in common hydrogen storage unit [kg].
$m_{H2-CS-min}$	Minimum allowable hydrogen amount in common hydrogen storage unit [kg].
N	Sufficiently large positive constant.
$P_{CPVU,t}$	Power production of common PV unit in period t [kW].
$P_{inflexload,h,t}$	Inflexible load–demand of household h in period t [kW].
$P_{PV,h,t}$	PV power production of household h in period t [kW].
$P_{rs-load-profile,h,n,t}$	The expected load profile of resiliency-sensitive load n of household h in period t [kW].
$SoE_{CESS-init}$	Initial state-of-energy of common energy storage unit [kWh].
$SoE_{CESS-max}$	Maximum allowable state-of-energy of common energy storage unit [kWh].
$SoE_{CESS-min}$	Minimum allowable state-of-energy of common energy storage unit [kWh].
$T_{a,k}$	Arrival time of FCEV k.
$T_{d,k}$	Departure time of FCEV k.
$u_{grid,t}$	Grid availability binary parameter in period t.
$\tau_{buy,t}$	Buying price of energy from the upstream grid in period t [€/kW].
$\tau_{sell,t}$	Selling price of energy to the upstream grid in period t [€/kW].
ΔT	Time granularity [h].

Variables

$k_{h,n,t}$	Binary variable regarding the decision to curtail the resiliency-sensitive load n of household h in period t.
$m_{FC-inj,k,t}$	Amount of hydrogen injected into the hydrogen tank of FCEV k from the common hydrogen storage unit in period t [kg].
$m_{H2,k,t}$	Hydrogen amount in the hydrogen tank of FCEV k in period t [kg].
$m_{H2-cons,k,t}$	Hydrogen consumption of FCEV k during community support mode in period t [kg].
$m_{H2-CS,t}$	Hydrogen amount in common hydrogen storage unit in period t [kg].
$m_{H2-CS-prod,t}$	Hydrogen amount produced by the electrolyzer unit in period t [kg].
$P_{buy,h,t}$	Power procured by household h in period t [kW].
$P_{CESS-ch,t}$	Charging power of CESS unit in period t [kW].
$P_{CESS-disc,t}$	Discharging power of CESS unit in period t [kW].
$P_{elec,t}$	Electrolyzer power in period t [kW].
$P_{FCEV-disch,k,t}$	Discharging power of FCEV k in period t [kW].
$P_{FCEV-disch-tot,t}$	Total discharging power of FCEV in period t [kW].
$P_{H2RC-tot,t}$	Total power injected back to the residential community by the households in period t [kW].
$P_{RC2H-tot,t}$	Total power drawn from the residential community by the households in period t [kW].
$P_{RC2UG,t}$	Power injected back to the upstream grid by the residential community in period t [kW].
$P_{rs-load,h,n,t}$	The actual power demand of resiliency-sensitive load n of household h in period t [kW].
$P_{sell,h,t}$	Reverse power injection by household h in period t [kW].
$P_{totalload,h,t}$	The total load of household h in period t [kW].
$P_{UG2RC,t}$	Power drawn from the upstream grid by the residential community in period t [kW].
$SoE_{CESS,t}$	State-of-energy of common energy storage unit in period t [kWh].
$u_{1,t}, u_{2,t}, u_{3,t}, u_{4,t}$	Binary variables to prevent simultaneous occurrence of different power exchange conditions.

References

1. Wang, C.; Ju, P.; Wu, F.; Pan, X.; Wang, Z. A systematic review on power system resilience from the perspective of generation, network, and load. *Renew. Sustain. Energy Rev.* **2022**, *167*, 112567. [CrossRef]
2. Norouzi, F.; Hoppe, T.; Elizondo, L.R.; Bauer, P. A review of socio-technical barriers to Smart Microgrid development. *Renew. Sustain. Energy Rev.* **2022**, *167*, 112674. [CrossRef]
3. Yu, H.; Niu, S.; Shang, Y.; Shao, Z.; Jia, Y.; Jian, L. Electric vehicles integration and vehicle-to-grid operation in active distribution grids: A comprehensive review on power architectures, grid connection standards and typical applications. *Renew. Sustain. Energy Rev.* **2022**, *168*, 112812. [CrossRef]
4. Forero-Quintero, J.F.; Villafáfila-Robles, R.; Barja-Martinez, S.; Munné-Collado, I.; Olivella-Rosell, P.; Montesinos-Miracle, D. Profitability analysis on demand-side flexibility: A review. *Renew. Sustain. Energy Rev.* **2022**, *169*, 112906. [CrossRef]
5. Tan, K.M.; Babu, T.S.; Ramachandaramurthy, V.K.; Kasinathan, P.; Solanki, S.G.; Raveendran, S.K. Empowering smart grid: A comprehensive review of energy storage technology and application with renewable energy integration. *J. Energy Storage* **2021**, *39*, 102591. [CrossRef]
6. Jeddi, B.; Mishra, Y.; Ledwich, G. Distributed load scheduling in residential neighborhoods for coordinated operation of multiple home energy management systems. *Appl. Energy* **2021**, *300*, 117353. [CrossRef]
7. Gholinejad, H.R.; Adabi, J.; Marzband, M. An energy management system structure for Neighborhood Networks. *J. Build. Eng.* **2021**, *41*, 102376. [CrossRef]
8. Fatemeh, T.; Fard, H.H.; Collins, E.R.; Jin, S.; Ramezani, B. Optimization and energy management of distributed energy resources for a hybrid residential microgrid. *J. Energy Storage* **2020**, *30*, 101556. [CrossRef]
9. Paterakis, N.G.; Erdinç, O.; Pappi, I.N.; Bakirtzis, A.G.; Catalão, J.P.S. Coordinated Operation of a Neighborhood of Smart Households Comprising Electric Vehicles, Energy Storage and Distributed Generation. *IEEE Trans. Smart Grid* **2016**, *7*, 2736–2747. [CrossRef]
10. Gholinejada, H.R.; Lonib, A.; Adabia, J.; Marzband, M. A hierarchical energy management system for multiple home energy hubs in neighborhood grids. *J. Build. Eng.* **2020**, *28*, 101028. [CrossRef]
11. Akter, M.N.; Mahmud, M.A.; Amanullah, M.E.H.; Oo, M.T. An optimal distributed energy management scheme for solving transactive energy sharing problems in residential microgrids. *Appl. Energy* **2020**, *270*, 115133. [CrossRef]
12. Taşcıkaraoğlu, A.; Paterakis, N.G.; Erdinç, O.; Catalão, J.P.S. Combining the Flexibility From Shared Energy Storage Systems and DLC-Based Demand Response of HVAC Units for Distribution System Operation Enhancement. *IEEE Trans. Sustain. Energy* **2019**, *10*, 137–148. [CrossRef]
13. Rafique, S.; Hossain, M.J.; Nizami, M.S.H.; Irshad, U.B.; Mukhopadhyay, S.C. Energy Management Systems for Residential Buildings With Electric Vehicles and Distributed Energy Resources. *IEEE Access* **2021**, *9*, 46997–47007. [CrossRef]
14. Ancona, M.A.; Catena, F.; Ferrari, F. Optimal design and management for hydrogen and renewables based hybrid storage micro-grids. *Int. J. Hydrogen Energy* **2022**, in press. [CrossRef]
15. Liu, X.; Ji, Z.; Sun, W.; He, Q. Robust game-theoretic optimization for energy management in community-based energy system. *Electr. Power Syst. Res.* **2023**, *214*, 108939. [CrossRef]
16. Hu, M.; Xiao, F.; Wang, S. Neighborhood-level coordination and negotiation techniques for managing demand-side flexibility in residential microgrids. *Renew. Sustain. Energy Rev.* **2021**, *135*, 110248. [CrossRef]
17. Thirunavukkarasu, G.S.; Seyedmahmoudian, M.; Jamei, E.; Horan, B.; Mekhilef, S.; Stojcevski, A. Role of optimization techniques in microgrid energy management systems—A review. *Energy Strategy Rev.* **2022**, *43*, 100899. [CrossRef]
18. Denga, Y.; Mua, Y.; Wang, X.; Jin, S.; He, K.; Jia, H.; Li, S.; Zhang, J. Two-stage residential community energy management utilizing EVs and household load flexibility under grid outage event. *Energy Rep.* **2023**, *9*, 337–344. [CrossRef]
19. Beyazıt, M.A.; Taşcıkaraoğlu, A.; Catalão, J.P.S. Cost optimization of a microgrid considering vehicle-to-grid technology and demand response. *Sustain. Energy Grids Netw.* **2022**, *32*, 100924. [CrossRef]
20. Mobarakeh, A.I.; Sadeghi, R.; Esfahani, H.S.; Delshad, M. Optimal planning and operation of energy hub by considering demand response algorithms and uncertainties based on problem-solving approach in discrete and continuous space. *Electr. Power Syst. Res.* **2023**, *214*, 108859. [CrossRef]
21. Hou, L.; Dong, J.; Herrera, O.E.; Merida, W. Energy management for solar-hydrogen microgrids with vehicle-to-grid and power-to-gas transactions. *Int. J. Hydrogen Energy* **2022**, in press. [CrossRef]
22. Energy Exchange Istanbul (EXIST-EPIAS) Tranparency Platform. Available online: https://seffaflik.epias.com.tr/transparency/index.xhtml (accessed on 3 November 2022).

Article

Risk Assessment of Industrial Energy Hubs and Peer-to-Peer Heat and Power Transaction in the Presence of Electric Vehicles

Esmaeil Valipour [1], Ramin Nourollahi [2], Kamran Taghizad-Tavana [2], Sayyad Nojavan [1,*] and As'ad Alizadeh [3]

[1] Department of Electrical Engineering, University of Bonab, Bonab 55517-61167, Iran
[2] Faculty of Electrical and Computer Engineering, University of Tabriz, Tabriz 51666-16471, Iran
[3] Department of Civil Engineering, College of Engineering, Cihan University-Erbil, Erbil 44001, Iraq
* Correspondence: sayyad.nojavan@ubonab.ac.ir

Abstract: The peer-to-peer (P2P) strategy as a new trading scheme has recently gained attention in local electricity markets. This is a practical framework to enhance the flexibility and reliability of energy hubs, specifically for industrial prosumers dealing with high energy costs. In this paper, a Norwegian industrial site with multi-energy hubs (MEHs) is considered, in which they are equipped with various energy sources, namely wind turbines (WT), photovoltaic (PV) systems, combined heat and power (CHP) units (convex and non-convex types), plug-in electric vehicles (EVs), and load-shifting flexibility. The objective is to evaluate the importance of P2P energy transaction with on-site flexibility resources for the industrial site. Regarding the substantial peak power charge in the case of grid power usage, this study analyzes the effects of P2P energy transaction under uncertain parameters. The uncertainties of electricity price, heat and power demands, and renewable generations (WT and PV) are challenges for industrial MEHs. Thus, a stochastically based optimization approach called downside risk constraint (DRC) is applied for risk assessment under the risk-averse and risk-neutral modes. According to the results, applying the DRC approach increased by 35% the operation cost (risk-averse mode) to achieve a zero-based risk level. However, the conservative behavior of the decision maker secures the system from financial losses despite a growth in the operation cost.

Keywords: peer-to-peer energy transaction; distributed energy resources; downside risk constraint; risk-averse; risk-neutral

1. Introduction

In the power market, the role of local energy systems including energy storage units, wind farms, distributed energy resources, and solar photovoltaic (PV) systems has become significant [1]. The development of smart grid facilities along with the energy and prosumers communities has further accelerated the trend [2]. Hence, consumer-centric types of energy systems with modern market designs are required for future power systems to adapt to local energy systems and buildings for the management of DERs. An emerging method is to develop smaller units and collect them in energy hubs [3]. Therefore, due to the differences in the trading price of energy as well as losses, sharing DERs on a local scale can be quite effective [4]. The peer-to-peer (P2P) strategy is an emerging alternative that encourages neighbors in a community to share excess energy to manage peak power demands [5], through which both consumers and prosumers can be supplied. In this regard, the effectiveness of the DERs, self-consumption, local energy balance, and grid operation flexibility can be strengthened [6]. The P2P concept can promote not only the flexibility of a system (storage unit, demand response program) but also energy transaction based on local prices [7]. Hence, the grid utility tariff is a market-based feature that potentially influences P2P trading. For instance, the peak demand charge is a promising solution in Norway by which consumers (commercial and industrial) are incentivized to reduce their power demand, as they are already subject to the peak demand charge [8]. Due to the high level of energy, large consumers can supply a major part of their energy consumption using a distribution network [9].

1.1. Research Review

As P2P energy transaction is a new concept, there is no agreement on a pricing scheme or market design aiming to support the development of local markets. In this field, recent research has mainly focused on several perspectives. Firstly, the role of storage units in coordinating local resources is investigated in [10] to create more balance in the system functioning. Secondly, a bidding mechanism for local energy trading is developed in [11]. Then, while the digitalization of the system components is carried out in [12], the requirement of computational properties as well as the coordination algorithm is taken into account for a P2P structure in [13]. Many of the research works related to P2P power transaction are rarely considered for real-life projects, including the Enerchain [14], Brooklyn Microgrid [15], and Sonnen Community [16]. A further step is taken in [17] by considering possible market frameworks for the consumers and prosumers of a community participating in the electricity market. The development of smart grid technologies will facilitate the establishment of local P2P energy trading with consumer-based electricity markets having access to the wholesale electricity market [18]. In this regard, blockchain technology can be taken into account to create an affordable and secure platform for energy transaction [19].

Several pieces of research have also been conducted on the management of multi-energy hubs. In [20], a two-stage stochastic programming approach is used to manage several energy hubs, in which the reliability aspect of the system is investigated. In [21], the authors introduced a new method for the optimal design and operation of several energy hubs concerning the cost of cables and operation costs using a two-stage stochastic optimization method. In [22], researchers increased the operational flexibility of several microgrids by using P2P power transaction. In addition to the incrementing of renewable penetration, the emission cost of the system is intended to be minimized. Random EV charging as an uncertainty is investigated in [23] for the energy management of several EHs to minimize the operation cost along with the power losses and greenhouse gas emissions in a multi-objective study. In [24], the authors propose a software-defined grid system to facilitate energy sharing in an MEH using a transactive energy framework. The result of this implementation is the reduction of the overall cost of the EHs. In [25], the chance constraint method is applied for the optimal day-ahead planning of an MEH, considering environmental constraints.

In a P2P strategy, thermal energy generated by CHP units can improve not only the energy flexibility but the proficiency of the combined heat and power unit. In [26], the transaction of power and heat among smart energy hubs is investigated in a two-stage process. While energy trading is considered in the first stage, financial issues related to power and heat energy are evaluated in the second stage to ensure the privacy of the EHs. In [27], a genetic algorithm is used to analyze P2P multi-energy sharing to increase the local energy balance and minimize the energy cost of the proposed system. In [28], a new P2P trading model is taken into account among FC-CHP systems to enhance the resiliency and self-sufficiency of local energy systems by the alternating direction method of multipliers (ADMM) algorithm. In [29], the effectiveness of a novel P2P energy transaction of multiple energy hubs is validated by a cooperative game to establish a proper payoff allocation.

In the existing literature, the importance of P2P power trading among residential prosumers has been mainly considered to evaluate the operational and economic aspects of systems. However, a few of them focused on the importance of P2P heat and power trading among industrial energy hubs and the functions of system components under uncertainties.

1.2. Novelties and Contributions

In this paper, a framework of a Norwegian industrial multi-energy hub (MEH) is developed, in which the hubs are equipped with various energy resources to supply their power and heat demands and share their excess energy with other peers and the electricity network. The CHP unit, solar photovoltaic (PV) system, wind turbine (WT), and plug-in electric vehicles (PEVs) are the generation facilities being used in the industrial hubs. Also, load shifting is a flexibility asset that is considered for two hubs, upon which the decision-maker can shift a part of the energy demand from peak demand to valley demand. However, the performance of stochastic renewable generations as well as the thermal and electrical demands with market electricity price in the energy hubs (EHs) have potential effects on the optimal function of the system. In this concern, a scenario-based stochastic optimization procedure called the downside risk constraint (DRC) approach is applied to investigate the functions of the system components under uncertainties. The risk-averse ($\lambda = 1$) and risk-neutral ($\lambda = 0$) modes are used to forecast the impact of uncertain parameters, concisely. The contribution of this paper can be summarized below:

- ❖ Techno-economic analysis of an industrial MEH with P2P heat and power transaction.
- ❖ Risk analysis of an industrial MEH with the downside risk constraint method (DRC) as a stochastic optimization procedure.
- ❖ Load-shifting flexibility asset and distributed energy resources, namely WTs, PVs, convex and non-convex CHP units, and plug-in electric vehicles (PEVs) to support energy demands.

In order to achieve a better perception, the overall structure of the proposed system is represented in Figure 1.

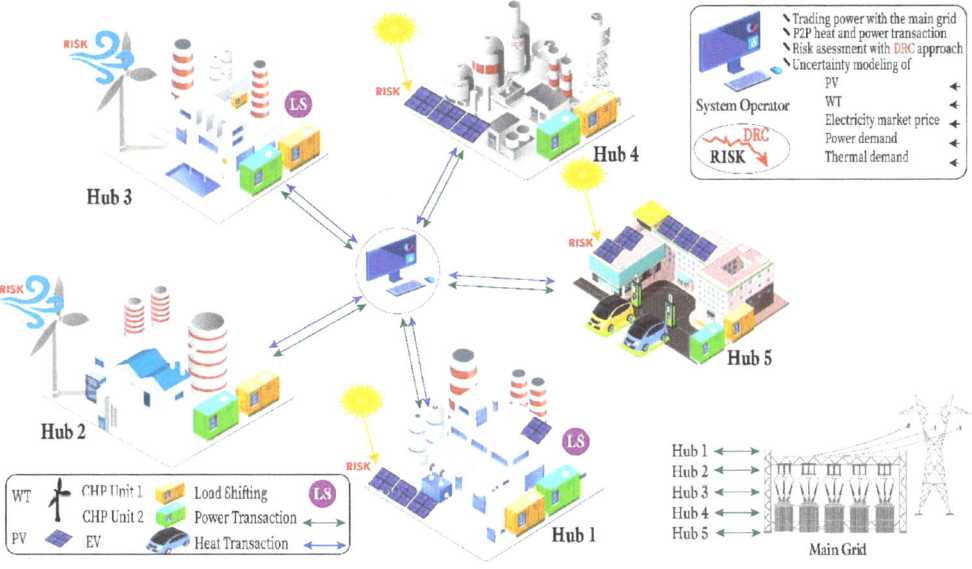

Figure 1. The structure of the proposed system.

1.3. Paper Structure

This paper is organized as follows: Section 2 expresses the objective function as well as the mathematical modeling of the MEHs. The constraints related to the DRC method are presented in Section 3. The study case along with the input data and the simulation results of the MEH are shown in Section 4. Finally, the conclusion of this paper is discussed in Section 5.

2. Mathematical Formulation

The mathematical model of the MEH is a linear optimization model representing the P2P interaction of the EHs and the operational decisions related to the DERs and flexibility assets. The objective function is to minimize the operational cost and analyze the risk level in the industrial EHs that are equipped with the CHP units and other distribution resources.

2.1. Objective Function

Based on the market features, the industrial EHs have the opportunity to trade their excess power through local P2P transaction. However, the operational cost arises in the cases of grid power consumption, load-shifting practice, and importing electricity from other peers. As there is a capability of trading power with the network and other peers, the units can obtain an income from grid feed-in and exporting electricity to a peer, which affects the optimal operation of the individual hubs. As shown in Equation (1), the cost of trading power with the power network is given in the first three terms. While the fourth term indicates the operational cost of the CHP units, the load-shifting cost is shown in the fifth term. Finally, the power and thermal energy transaction among the industrial peers is demonstrated in the last four terms.

$$\min_{\substack{\forall t \in T \\ \forall b \in B}} C = \sum_b \left\{ \begin{array}{l} \sum_t^T \left[\left(C_{g,eng} + C_{(t)}^{g,SP}\right) \cdot P_{(t,h)}^{g,buy}\right] + \sum_m^M \left[C_{g,fix} + C_{g,peak} \cdot P_{(t,h)}^{g,peak}\right] - \\ \sum_t^T \left[C_{(t)}^{feed-in} \cdot P_{(t,h)}^{g,sell}\right] + \sum_t^T \sum_h^H \sum_c^C \left[a\_chp \times P_{t,h,c}^{chp} + b\_chp + c\_chp \times T_{t,h,c}^{chp}\right] + \\ \sum_t^T \left[C_{(h)}^{LS} \cdot P_{(t,h)}^{ls,sh}\right] + \sum_t^T \left[C_{(t,h)}^{P,P2P} \cdot P_{(t,h)}^{imp} \cdot \frac{1}{\psi_{P,P2P}}\right] - \sum_t^T \left[\sum_{p \neq h}^H C_{(t,p)}^{P,P2P} \cdot P_{(t,h \to p)}^{exp,p}\right] + \\ \sum_t^T \left[C_{(t,b)}^{Th,P2P} \cdot T_{(t,h)}^{imp} \cdot \frac{1}{\psi_{Th,P2P}}\right] - \sum_t^T \left[\sum_{p \neq h}^H C_{(t,p)}^{Th,P2P} \cdot T_{(t,h \to p)}^{exp,p}\right] \end{array} \right\} \quad (1)$$

The overall cost of power network electricity for the Norwegian industrial MEHs is presented in Equation (2). As the local network company is responsible for determining the utility tariff, the utility tariff system in Norway differs from the flat rate tariffs towards time-of-use pricing, which has a peak-demand cost [30]. As shown in the equation, the first and second parenthetical terms represent buying electrical energy at a spot price and the price of energy during the peak hours, respectively. Moreover, in the second term, feed-in energy cost is calculated and considered for selling power to the power grid.

$$C_{(h)}^{g.tot} = \sum_{t \in T} \left(C_{(t)}^{g,SP} \cdot P_{(t,h)}^{g.buy} + C_{g,eng} \cdot P_{(t,h)}^{g.buy} \right) + \left(C_{g,fix} + C_{g,peak} \cdot P_{(h)}^{g.peak} \right) - \sum_{t \in T} \left(C_{(t)}^{feed-in} P_{(t,h)}^{g.sell} \right) \quad (2)$$

2.2. Energy Balance

P2P power transaction between the industrial hubs will affect the balance of the system. In this regard, the constraint of power balance is shown in Equation (3), in which the total amount of electrical demand is equal to the overall level of power generated by the DERs. Also, thermal energy transaction is considered to increase the energy flexibility, though the related constraint is given in Equation (4). As mentioned in the equation, the overall amount of heat energy produced by CHP units along with the imported power from other peers, bought power from the main grid, and discharged energy of PEVs must be equal to the overall level of electrical demand. Also, the thermal energy produced by the CHP unit and imported from other peers must be equivalent to the demand and the exported heat energy. Finally, the limitations of trading power with a power network are shown in Equations (5) and (6).

$$\sum_{c=1}^{C} P_{(t,h,c)}^{chp} + P_{(t,h)}^{dem} + P_{(t,h)}^{g.sell} + P_{(t,h)}^{exp} + P_{(t,h)}^{ev,ch} + P_{(t,h)}^{ls,dem} + P_{(t,h)}^{curtail} = P_{(t,h)}^{DER} + P_{(t,h)}^{g.buy} + P_{(t,h)}^{imp} + P_{(t,h)}^{ev.dch} + P_{(t,h)}^{ls.sh} \quad (3)$$

$$\sum_{c}^{C} \left(T_{t,h,c}^{chp} \right) + T_{t,h}^{imp} = T_{t,h}^{dem} + T_{t,h}^{exp} \quad (4)$$

$$0 \leq P^{g.buy}_{(t,h)} \leq P^{g.peak}_{(h)} \tag{5}$$

$$0 \leq P^{g.sell}_{(t,h)} \leq P^{max}_{feed-in} \tag{6}$$

2.3. Constraints of P2P Energy Transaction

As the interconnected industrial hubs have the opportunity to trade energy among each other, specific market mechanisms defined by local markets can secure the trading process [31]. As the proposed P2P energy transaction methodology is a general model, the implementation of the system on different decentralized platforms like blockchains can be carried out according to [32,33]. The total sums of exported and imported power by hub h are shown in Equations (7) and (8), respectively. As indicated in Equation (9), the imported electricity from a peer must be equal to the electricity exported from the peer to the EH, which involves power losses (ψ_{P2P}). Also, the total level of traded power among the EHs is given by Equation (10). The mentioned constraints are also indicated for thermal energy transaction between Equations (11) and (14)

$$P^{exp}_{(t,h)} = \sum_{p \neq h} P^{exp,p}_{(t,h \rightarrow p)} \tag{7}$$

$$P^{imp}_{(t,h)} = \sum_{p \neq b} P^{imp,p}_{(t,h \leftarrow p)} \tag{8}$$

$$P^{imp,p}_{(t,h \leftarrow p)} = P^{exp,p}_{(t,h \rightarrow p)} \times \psi_{P,P2P}, \forall p \neq h \tag{9}$$

$$\sum_{h}^{H} P^{exp}_{(t,h)} \times \psi_{P,P2P} = \sum_{h}^{H} P^{imp}_{(t,h)} \tag{10}$$

$$T^{exp}_{(t,h)} = \sum_{p \neq h} T^{exp,p}_{(t,h \rightarrow p)} \tag{11}$$

$$T^{imp}_{(t,h)} = \sum_{p \neq b} T^{imp,p}_{(t,h \leftarrow p)} \tag{12}$$

$$T^{imp,p}_{(t,h \leftarrow p)} = T^{exp,p}_{(t,h \rightarrow p)} \times \psi_{Th,P2P}, \forall p \neq h \tag{13}$$

$$\sum_{h}^{H} T^{exp}_{(t,h)} \times \psi_{Th,P2P} = \sum_{h}^{H} T^{imp}_{(t,h)} \tag{14}$$

2.4. Load-Shifting Constraints

Load shifting as a crucial strategy can help industrial hubs to reduce their operational cost by running a production process in the low–peak interval instead of the peak demand period. However, this strategy imposes productivity losses and labor rescheduling costs, both of which are considered penalty costs in the objective function. The mathematical modeling of load shifting is regarded in the form of a storage unit with 10% capacity to ease the computational burden [34], which is defined in the form of hourly rescheduled load in Equation (15). In this regard, while the storage balance for each EH is shown in Equation (16), their energy level with the maximum power shift is indicated in Equation (17).

$$0 \leq P^{ls,sh}_{(t,h)}, P^{ls,dem}_{(t,h)} \leq 0.1 \times P^{g,peak}_{(h)} \tag{15}$$

$$E^{ls}_{(t,h)} = E^{ls}_{(t-1,h)} + P^{ls,sh}_{(t,h)} - P^{ls,dem}_{(t,h)} \tag{16}$$

$$0 \leq E^{ls}_{(t,h)} \leq 0.4 \times P^{g,peak}_{(h)} \tag{17}$$

2.5. Electric Vehicle (EV) Constraints

Vehicle-to-grid (V2G) is an on-site flexibility option providing a bi-directional use of electricity for EHs as a fast-responding storage unit that can be used for spinning reserve and peak shaving. As the industrial units hold a large number of employees, considering V2G technology for the parking lots can be an alternative flexibility asset. As given in Equation (18), the EV parking lot is a storage unit balancing the energy consumption of EHs. Equation (19) shows the limitation of the charging/discharging process by the nominal capacity of the EV charger. Finally, the start and end of a workday are limited by Equation (20) and Equation (21), respectively [35].

$$E^{ev}_{(t,h)} = E^{ev}_{(t-1,h)} + \eta_{ev,ch} \cdot P^{ev,ch}_{(t,h)} - \frac{1}{\eta_{ev,ch}} \cdot P^{ev,dch}_{(t,h)} \tag{18}$$

$$0 \leq P^{ev,ch}_{(t,h)}, P^{ev,dch}_{(t,h)} \leq P^{nom}_{ev,charger} \cdot EV_{num} \tag{19}$$

$$E^{ev}_{(d_{start}(t),h)} = E^{nom}_{ev} \cdot EV_{num} \cdot E_{start} \quad d_{start}(t) \in T \tag{20}$$

$$E^{ev}_{(d_{end}(t),h)} \geq E^{nom}_{ev} \cdot EV_{num} \cdot E_{end} \quad d_{end}(t) \in T \tag{21}$$

2.6. Constraints of CHP Units

Due to the high proficiency of CHP units, they can be used to supply both power and thermal energy in the EHs. The concept of the feasible operation region (FOR), which is shown in Figure 2, is used to model the constraints of the CHP units. The installed CHP units are convex and non-convex; hence, the start-up and shutdown limitations are given in Equations (22) and (23), respectively. While the convex unit is modeled by Equations (24)–(28), the model of the non-convex unit is formulated by Equations (29)–(34).

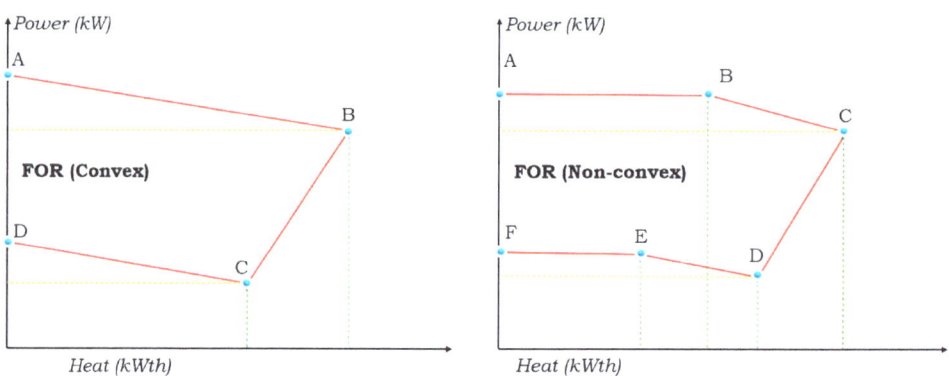

Figure 2. The feasible operation regions of the CHP units.

$$SU^{chp}_{t,h,s,c} \geq C^{SU}_{h,c} \left[V^{chp}_{t,h,s,c} - V^{chp}_{t-1,h,s,c} \right] \tag{22}$$

$$SD^{chp}_{t,h,s,c} \geq C^{SHD}_{h,c} \left[V^{chp}_{t-1,h,s,c} - V^{chp}_{t,h,s,c} \right] \tag{23}$$

$$P^{chp}_{t,h,s,c} - P^{chp,A}_{h,c} - \frac{P^{chp,A}_{h,c} - P^{chp,B}_{h,c}}{T^{chp,A}_{h,c} - T^{chp,B}_{h,c}} \left[T^{chp}_{t,h,s,c} - T^{chp,A}_{h,c} \right] \leq 0 \tag{24}$$

$$P^{chp}_{t,h,c} - P^{chp,B}_{h,c} - \frac{P^{chp,B}_{h,c} - P^{chp,C}_{h,c}}{T^{chp,B}_{h,c} - T^{chp,C}_{h,c}} \left[T^{chp}_{t,h,c} - T^{chp,B}_{h,c} \right] \geq -M \tag{25}$$

$$P_{t,s,c}^{chp} - P_{h,c}^{chp,C} - \frac{P_{h,c}^{chp,C} - P_{h,c}^{chp,D}}{T_{h,c}^{chp,C} - T_{h,c}^{chp,D}}\left[T_{t,s,c}^{chp} - T_{h,c}^{chp,C}\right] \geq -M \tag{26}$$

$$0 \leq P_{t,h,c}^{chp} \leq P_{h,c}^{chp,A} \cdot V_{t,h,c}^{chp} \tag{27}$$

$$0 \leq T_{t,h,c}^{chp} \leq T_{h,c}^{chp,B} \cdot V_{t,h,c}^{chp} \tag{28}$$

$$P_{t,h,c}^{chp} - P_{h,c}^{chp,B} - \frac{P_{h,c}^{chp,B} - P_{h,c}^{chp,C}}{T_{h,c}^{chp,B} - T_{h,c}^{chp,C}}\left[T_{t,h,c}^{chp} - T_{h,c}^{chp,B}\right] \leq 0 \tag{29}$$

$$P_{t,h,c}^{chp} - P_{h,c}^{chp,C} - \frac{P_{h,c}^{chp,C} - P_{h,c}^{chp,D}}{T_{h,c}^{chp,C} - T_{h,c}^{chp,D}}\left[T_{t,h,c}^{chp} - T_{h,c}^{chp,C}\right] \geq 0 \tag{30}$$

$$P_{t,h,c}^{chp} - P_{h,c}^{chp,D} - \frac{P_{h,c}^{chp,D} - P_{h,c}^{chp,E}}{T_{h,c}^{chp,D} - T_{h,c}^{chp,E}}\left[T_{t,h,s,c}^{chp} - T_{h,c}^{chp,D}\right] \geq -[1 - X_t^a]M \tag{31}$$

$$P_{t,h,c}^{chp} - P_{h,c}^{chp,D} - \frac{P_{h,c}^{chp,D} - P_{h,c}^{chp,F}}{T_{h,c}^{chp,D} - T_{h,c}^{chp,F}}\left[T_{t,h,c}^{chp} - T_{h,c}^{chp,D}\right] \geq -\left[1 - X_t^b\right]M \tag{32}$$

$$0 \leq P_{t,h,c}^{chp} \leq P_{h,c}^{chp,A} \cdot V_{t,h,c}^{chp} \tag{33}$$

$$0 \leq T_{t,h,c}^{chp} \leq T_{h,c}^{chp,C} \cdot V_{t,h,c}^{chp} \tag{34}$$

2.7. Constraints of Renewable Energy Sources

Wind turbines and solar photovoltaic (PV) systems are renewable resources that are considered in this model. The generated available power of wind turbines and PV systems are modeled in Equation (35) and Equation (36), respectively. As shown in the equations, while wind speed potentially affects the power generation of WTs, solar radiation influences the PV system function.

$$P_{t,h}^{wt} = \begin{cases} 0 & V_t < V_{cut-in}, V_t \geq V_{cut-out} \\ P_{rated}^{wt} \times \left(\frac{V_t - V_{cut-in}}{V_{rated} - V_{cut-in}}\right)^3 & V_{cut-in} \leq V_t \leq V_{rated} \\ P_{rated}^{wt} & V_{rated} \leq V_t \leq V_{cut-out} \end{cases} \tag{35}$$

$$P_{t,h}^{pv} = \begin{cases} P_{rated}^{pv}\left(\frac{I_t^2}{I_{std}I_C}\right) & I_t \leq I_C \\ P_{rated}^{pv}\left(\frac{I_t^2}{I_{std}}\right) & I_C \leq I_t \end{cases} \tag{36}$$

3. Downside Risk Constraint (DRC) Method

The downside risk constraint (DRC) method is applied to control the risk of financial losses. Regarding decision variables, convexity is the most important feature of the DRC method. Compared to other risk measures, the DRC method has more robustness [36], with substantial advantages from the risk-management point of view. As the DRC has a convex function with a set of minimum points, it simplifies the optimization and control of the uncertainty through mathematical programming. As shown in Equation (37), when the cost of the scenario is less than the expected cost, the risk level is zero; otherwise, the level of risk can be obtained from the difference between the scenario cost and the expected cost. Also, the expected risk of the DRC method is given in Equation (38), in which the operator aims to achieve a small value for $DRC(C_0)$.

$$RISK(\omega) = \begin{cases} Cost(\omega) - C_0 & if \quad Cost(\omega) > C_0 \\ 0 & if \quad Cost(\omega) \leq C_0 \end{cases} \tag{37}$$

$$DRC(C_0) = E[RISK(\omega)] = \sum_{\omega=\Omega} \pi(\omega) \cdot RISK(\omega) \qquad (38)$$

Equation (39) indicates that the scenario costs have more value than the expected cost. In this equation, the term $P(\omega|\text{Cost}(\omega) \geq C_0)$ shows the probability of a cost that is higher than the target cost. If the operator is not satisfied with the obtained risk level, a risk constraint like Equation (40) can be added to the main formulation as below:

$$DRC(C_0) = C_0 - \frac{1}{P(\omega|\text{Cost}(\omega) \geq C_0)} \times \sum_{\omega=1}^{\Omega} \{\pi(\omega) \cdot \max[(\text{Cost}(\omega) - C_0), 0]\} \qquad (39)$$

$$DRC(C_0) \leq DRC_0 \qquad (40)$$

where the term DRC_0 indicates the tolerance of the downside risk constraint. The flowchart of the proposed method with the strategy of the system operator is demonstrated in Figure 3.

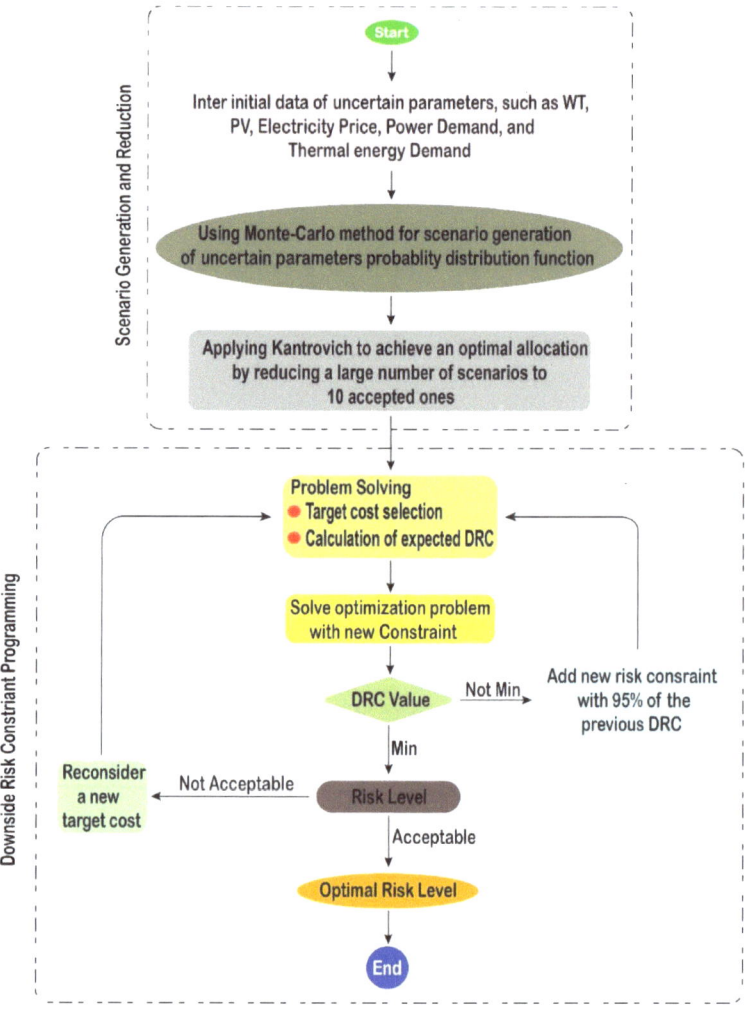

Figure 3. The framework of the DRC method.

As shown, the initial data related to the uncertain parameters are provided based on the historic data, upon which the scenario generation of uncertainties is performed by applying the Monte Carlo method for the wind turbine (WT), PV, electricity price, power, and thermal energy demands. In order to ease the computational burden, the Kantorovich method is used to achieve an optimal allocation of scenarios being used in the DRC programming.

4. Study Case

In this paper, a Norwegian industrial area with five energy hubs is considered, the prosumers of which are connected to each other to become more flexible by applying a P2P power and heat transaction strategy. These industrial units are related to food-processing industries, mechanical workshops, and manufacturing factories. In this regard, supporting their energy demand due to their high power consumption is a big challenge. The EHs are equipped with several energy sources, including PV systems and wind turbines as renewables, as well as EVs and CHP units. From a flexibility perspective, the load-shifting strategy increases the productivity of the energy hubs significantly. Intuitively speaking, load shifting means that the industrial building is willing to move demand from the peak demand period to valley demand, making a production process run at a later time. However, this strategy has rescheduling costs, namely overtime pay for laborers, rescheduling, and productivity losses. Moreover, the interconnected EHs have a connection with the main power grid, as the installed sources may not be able to meet the power demand of prosumers on their own. The industrial hubs are located in different places, and they are completely different in terms of load consumption and size [37]. As a final remark, the simulation of the proposed system is carried out through the GAMS software as a mixed-integer linear programming model with the CPLEX solver.

4.1. Input Data

This subsection presents the input data related to the generated scenarios for the uncertain parameters. The market electricity price, power demand, thermal energy demand, and renewable generations are taken into account as uncertainties. In this regard, Table 1 is given to show the price of electricity over the period of a 24 h scheduling horizon in 10 scenarios [37]. For each energy EH, 10 scenarios of power and heat demand are generated, which are shown in Figures 4 and 5, respectively. In fact, the scenario generation is carried out based on the Monte Carlo method, in which a probability distribution is used to generate 100 scenarios for each uncertainty [38]. Whereas a large number of scenarios results in a computational burden, the Kantorovich procedure is applied to select 10 scenarios with high probability [39]. Based on the figures, hub5 has more energy demand compared to the other hubs. Meanwhile, in Figure 6 the PV function is demonstrated in hub1, hub4, and hub5, Figure 7 shows the function of WTs in hub2 and hub3. Also, Table 2 gives related data about the convex and non-convex CHP units [40].

Table 1. The electricity price generated in 10 scenarios (Nok).

	SC = 1	SC = 2	SC = 3	SC = 4	SC = 5	SC = 6	SC = 7	SC = 8	SC = 9	SC = 10
t = 1	0.2125	0.2762	0.4314	0.3241	0.3631	0.3883	0.2539	0.3572	0.3039	0.2153
t = 2	0.3486	0.2504	0.2775	0.2111	0.2695	0.3523	0.4421	0.2276	0.3448	0.3589
t = 3	0.3888	0.4054	0.2231	0.3130	0.4031	0.3503	0.4162	0.4021	0.4889	0.1063
t = 4	0.2782	0.2776	0.1954	0.2638	0.4112	0.3078	0.3762	0.4273	0.2813	0.2723
t = 5	0.2078	0.2883	0.2545	0.2928	0.2923	0.2534	0.2211	0.2805	0.3109	0.2404
t = 6	0.3793	0.3360	0.3259	0.4281	0.5496	0.4225	0.4046	0.3490	0.3599	0.3993
t = 7	0.3250	0.2364	0.4177	0.3110	0.4144	0.2988	0.2590	0.2448	0.4037	0.5615
t = 8	0.403	0.4308	0.3975	0.4058	0.6106	0.3087	0.3684	0.3191	0.4522	0.4082
t = 9	0.359	0.4635	0.2927	0.4026	0.4105	0.5448	0.4436	0.3942	0.2596	0.2471
t = 10	0.4481	0.3221	0.3669	0.3056	0.4082	0.4127	0.5607	0.5679	0.4941	0.4963
t = 11	0.2697	0.5027	0.2914	0.2287	0.6829	0.4115	0.3180	0.3819	0.3877	0.1839
t = 12	0.3095	0.4268	0.3803	0.2202	0.3306	0.3628	0.3846	0.2187	0.4010	0.5211
t = 13	0.2244	0.4139	0.4983	0.4084	0.4279	0.3287	0.3557	0.2662	0.3616	0.3214
t = 14	0.3112	0.5073	0.3616	0.4311	0.4074	0.2780	0.1965	0.2389	0.3982	0.3444
t = 15	0.4501	0.3423	0.2532	0.3879	0.4824	0.4033	0.3606	0.3384	0.4982	0.4647
t = 16	0.3398	0.2525	0.2937	0.4124	0.2767	0.3438	0.3690	0.3179	0.3212	0.3591
t = 17	0.3301	0.2543	0.4533	0.2948	0.4819	0.3499	0.3534	0.5068	0.4078	0.3443
t = 18	0.3051	0.3345	0.4230	0.4660	0.3892	0.3372	0.4168	0.4491	0.3440	0.5233
t = 19	0.3493	0.4706	0.4773	0.3897	0.3761	0.4688	0.2110	0.3875	0.3909	0.4194
t = 20	0.2996	0.2491	0.1309	0.4372	0.4885	0.3420	0.6137	0.3232	0.2586	0.370
t = 21	0.3960	0.4517	0.3976	0.3122	0.2389	0.3292	0.2543	0.4556	0.3758	0.2071
t = 22	0.4198	0.2960	0.5145	0.3897	0.4067	0.3089	0.4350	0.3909	0.4055	0.3539
t = 23	0.4451	0.1496	0.3993	0.4839	0.3651	0.4388	0.3263	0.4000	0.3274	0.4932
t = 24	0.4254	0.5116	0.3767	0.3750	0.4105	0.5185	0.1532	0.2211	0.5013	0.5609

Table 2. Parameters related to the operation region of the CHP unit.

CHP Units	a ($/kW^2)	b ($/kW^2)	c ($)	d ($/kWth^2)	e ($/kWth)	f ($/kW.kWth)	Feasible Region Coordinates
CHP 1	0.0345	44.5	26.5	0.03	4.2	0.031	[1.258 0], [1.258 0.324], [1.102 1.356], [0.4 0.75], [0.44 0.159], [0.44 0]
CHP 2	0.0435	56	12.5	0.027	0.6	0.011	[2.47 0], [2.15 1.8], [0.81 1.048], [0.988 0]

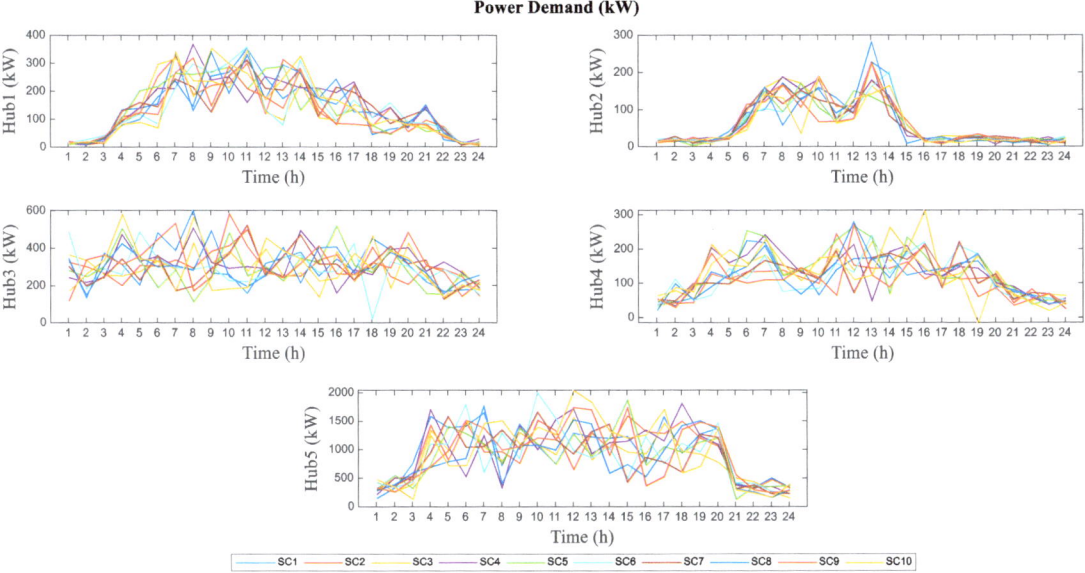

Figure 4. Power demand at each energy hub.

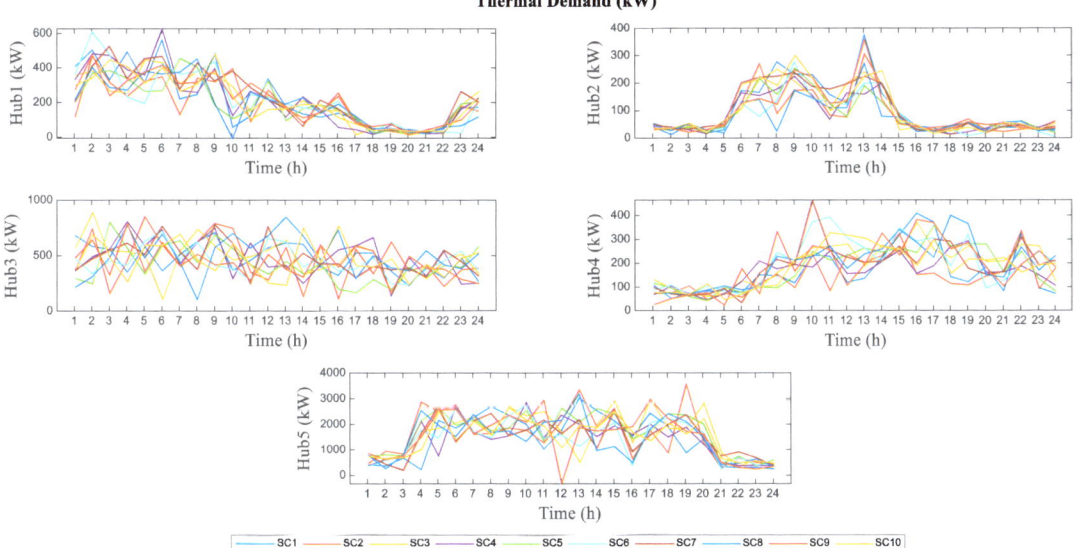

Figure 5. Thermal demand at each energy hub.

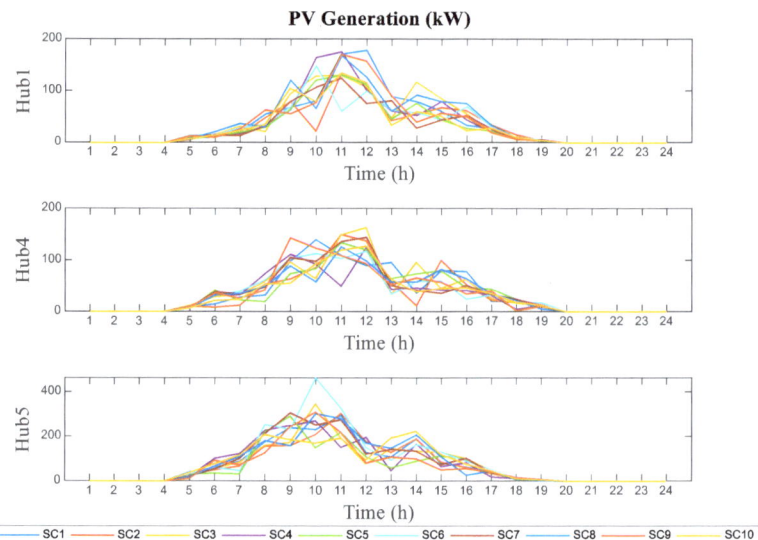

Figure 6. PV generation at three energy hubs.

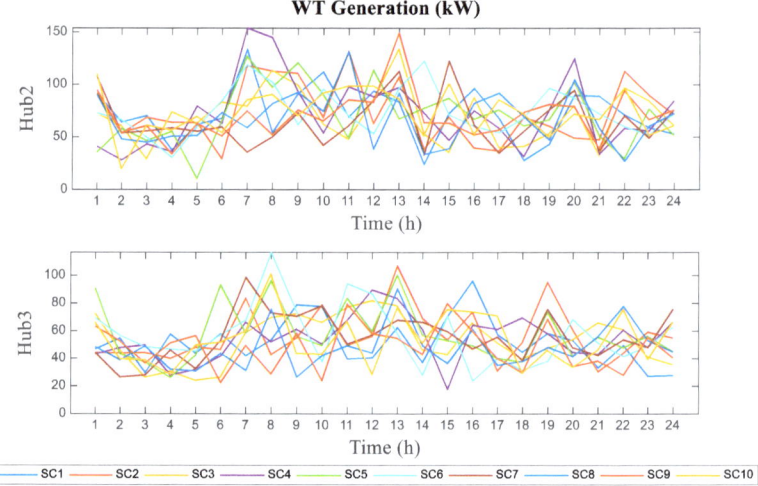

Figure 7. Wind generation at three energy hubs.

4.2. Numerical Results

In this section, the levels of operational cost and risk are shown for different iterations. In Figure 8, the simulation result is obtained in 11 iterations, in which the lambda value changes from zero ($\lambda = 0$) to one ($\lambda = 1$). In this case, $\lambda = 0$ and $\lambda = 1$ imply the risk-neutral and risk-averse modes, respectively. Despite an increment in operational cost, the existing risk level decreases as we approach a high value of λ. In order to ease the understanding of this concept, the amount of risk in each iteration is shown in Figure 9. As shown, In the risk-neutral mode, the amount of risk is at its maximum value, but in the risk-averse mode, the operational cost is at its maximum amount (4.5 NOK), while the amount of risk for the system operator is flattened. The numerical results of the operational cost and risk level are demonstrated in Tables 3 and 4, respectively. According to Table 3, the operational cost of the system for different lambdas is shown. When $\lambda = 0$ (risk-neutral), a low level of cost is obtained for all scenarios. However, in the second scenario, which is the worst-case scenario, a high level of cost (663866 NOK) is obtained because there is a probability of scenarios with maximum financial losses.

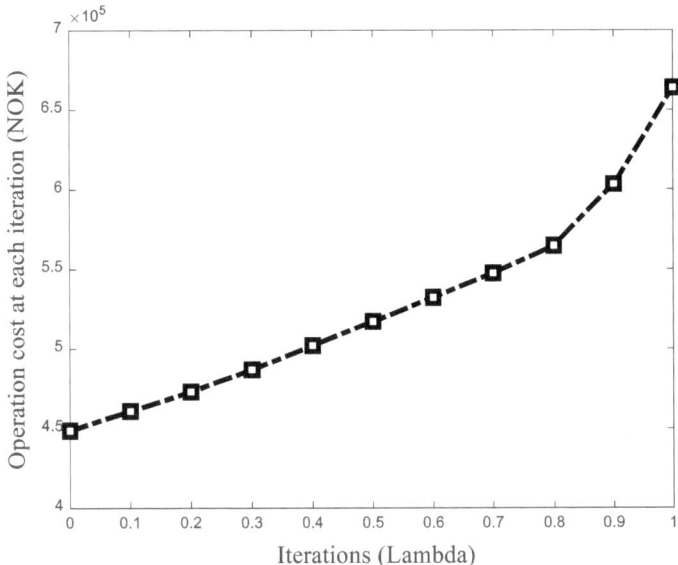

Figure 8. The operational cost performance.

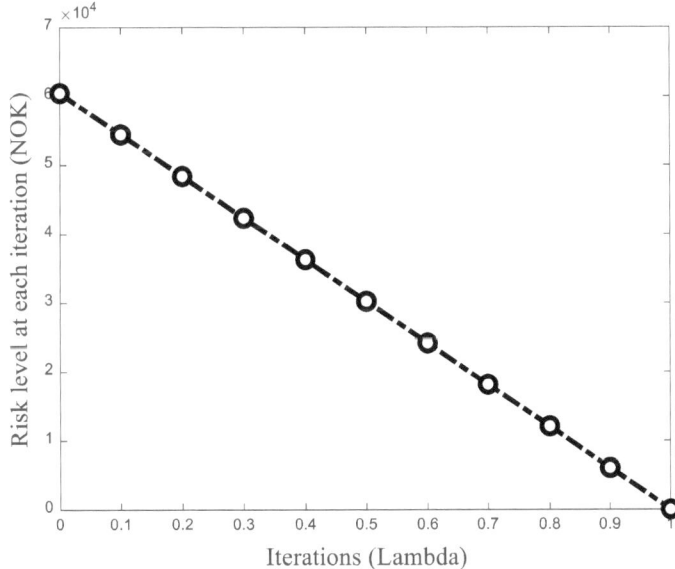

Figure 9. Risk level at each iteration.

Table 3. Operational cost for each scenario and iteration (NOK).

n1	SC = 1	SC = 2	SC = 3	SC = 4	SC = 5	SC = 6	SC = 7	SC = 8	SC = 9	SC = 10
λ = 0	223,296	663,866	478,104	419,689	252,088	586,506	377,805	366,694	560,148	560,861
λ = 0.1	239,965	663,866	478,104	441,956	279,744	586,506	411,143	387,756	560,148	560,861
λ = 0.2	295,246	663,867	478,105	454,059	301,644	586,506	416,584	414,024	560,148	560,861
λ = 0.3	282,311	663,867	486,978	478,361	339,428	586,506	448,238	463,084	560,148	560,861
λ = 0.4	283,285	663,867	502,102	492,357	406,741	586,506	470,921	494,234	560,148	560,861
λ = 0.5	322,415	663,867	517,226	477,362	484,446	586,506	482,204	517,226	560,148	560,861
λ = 0.6	381,447	663,867	532,350	515,415	532,350	586,506	485,118	505,438	560,148	560,861
λ = 0.7	547,474	663,867	547,474	547,474	464,808	586,506	489,965	506,162	560,148	560,861
λ = 0.8	514,006	663,867	564,691	564,691	564,691	586,506	494,385	564,691	564,691	564,691
λ = 0.9	603,371	663,867	603,371	603,371	598,991	603,371	603,371	547,256	603,371	603,371
λ = 1	663,867	663,867	663,867	663,867	663,867	663,867	663,867	663,867	663,867	663,867

Table 4. Risk level for each scenario and iteration (NOK).

C	SC = 1	SC = 2	SC = 3	SC = 4	SC = 5	SC = 6	SC = 7	SC = 8	SC = 9	SC = 10
λ = 0	0	214,960	29,198	0	0	137,600	0	0	111,242	111,955
λ = 0.1	0	202,861	17,099	0	0	125,501	0	0	99,143	99,856
λ = 0.2	0	190,762	5000	0	0	113,402	0	0	87,044	87,757
λ = 0.3	0	176,888	0	0	0	99,528	0	0	73,170	73,883
λ = 0.4	0	161,764	0	0	0	84,404	0	0	58,046	58,759
λ = 0.5	0	146,641	0	0	0	69,280	0	0	42,922	43,635
λ = 0.6	0	131,517	0	0	0	54,156	0	0	27,798	28,511
λ = 0.7	0	116,393	0	0	0	39,032	0	0	12,674	13,387
λ = 0.8	0	99,176	0	0	0	21,815	0	0	0	0
λ = 0.9	0	60,495	0	0	0	0	0	0	0	0
λ = 1	0	0	0	0	0	0	0	0	0	0

Figures 10 and 11 show the amounts of electrical and thermal energy exchanged among the EHs, respectively. According to Figure 10, all energy hubs have exported a large proportion of their generated power to meet the power demand of hub5. Because the load demand in hub5 is higher than in the other hubs (with a maximum value of 3000 kW), each of them has exported an amount of power between 1500 kW and 3000 kW to hub5. Also, the same trend is achieved for thermal energy transaction, as shown in Figure 11. As the evaluation of the energy transaction is carried out in both the risk-averse and risk-neutral modes, the level of transacted energy in the risk-averse mode is less than that in the risk-neutral mode due to the conservative behavior of the system operator.

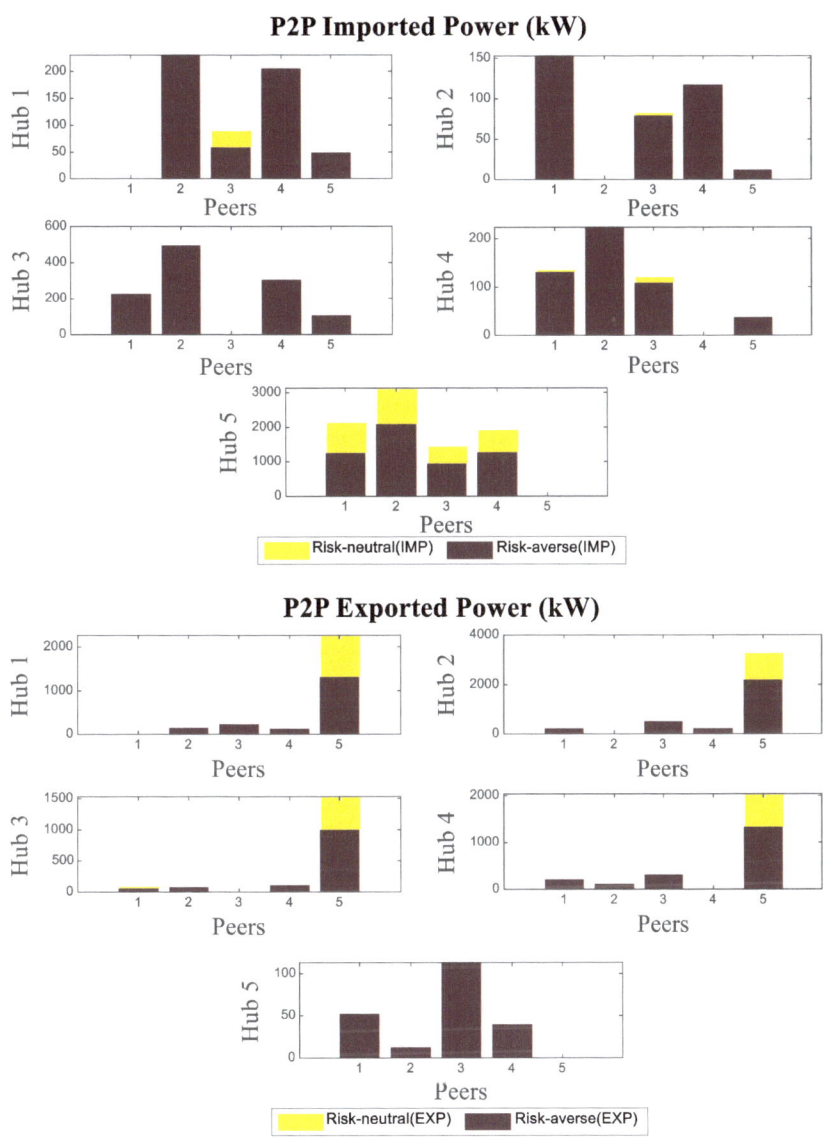

Figure 10. P2P power transaction between energy hubs.

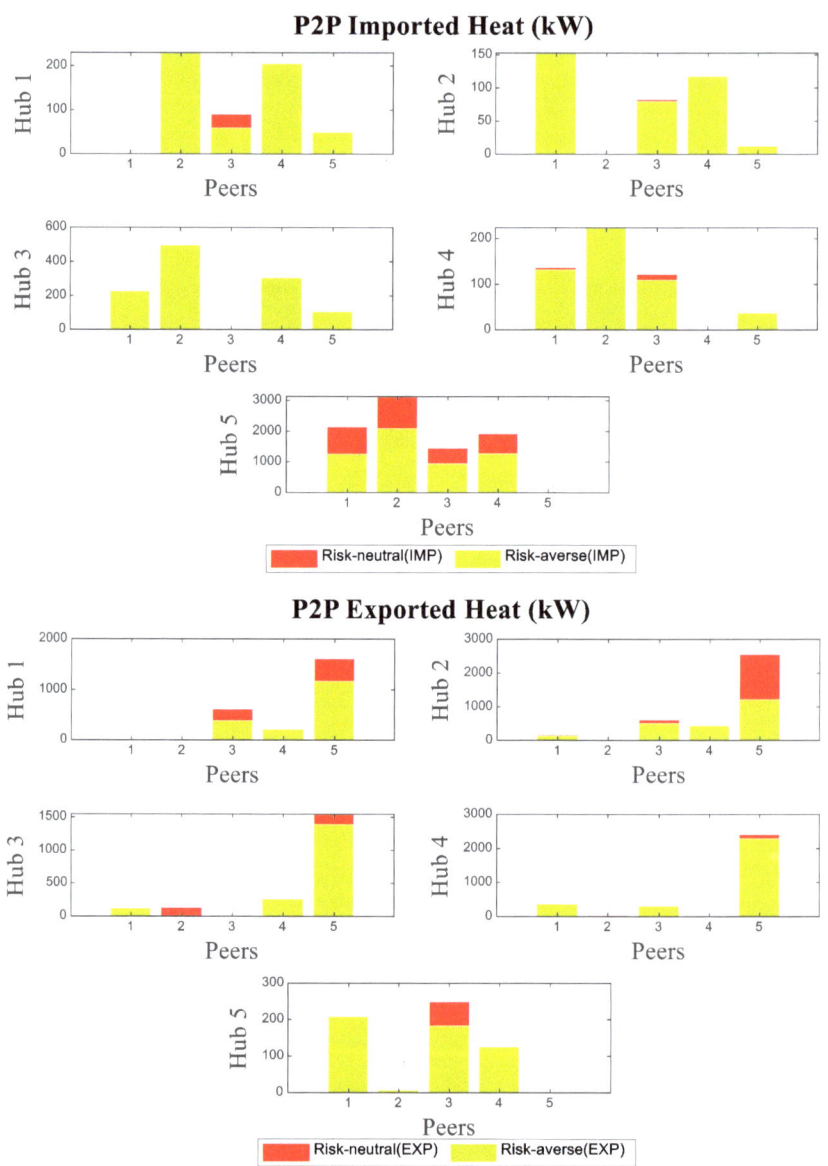

Figure 11. P2P heat transaction between energy hubs.

Figure 12 shows the amount of electrical energy exchanged between the network and the EHs that are connected to the power network. Although the first, third, and fifth hubs bought electricity from the network in the risk-neutral mode, all hubs sold a portion of their energy to the grid which was between 400 kW and 800 kW. As shown in the figure, the EHs sold 70% more electrical energy in the risk-neutral mode compared to the risk-averse mdoe.

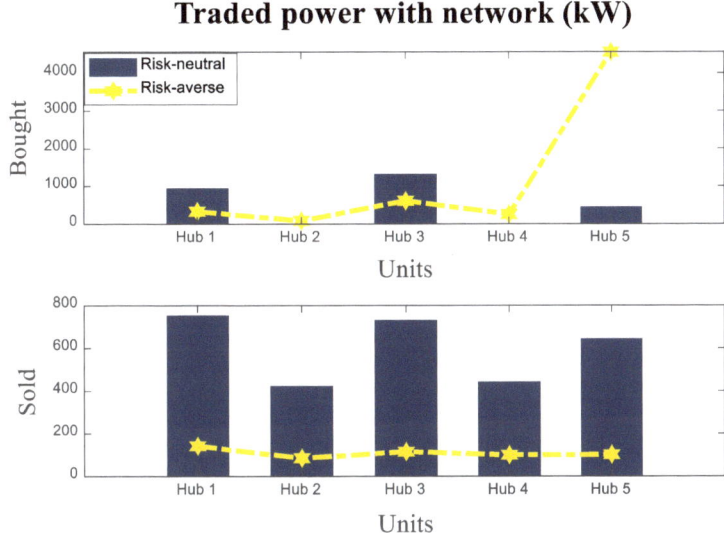

Figure 12. Power traded between the network and the energy hubs.

Figures 13 and 14 demonstrate the power and heat energy produced by the CHP units in the EHs, respectively. By comparing the figures, it can be deduced that the second CHP unit (non-convex) generates more power, but the first CHP unit (convex) produces more thermal energy. The functioning of the CHP units is considerable in hub5 (CHP1 (1600 kW in risk-neutral mode and 1250 kW in risk-averse mode) and CHP2 (3400 kW in risk-neutral mode and 2600 kW in risk-averse mode)).

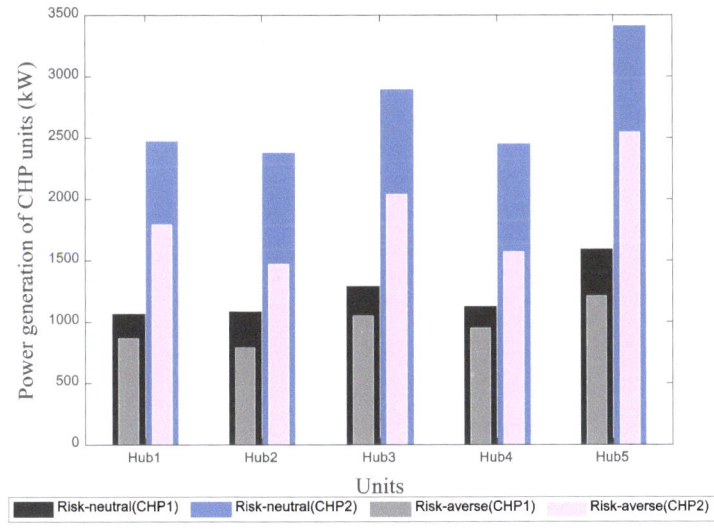

Figure 13. Power generation of CHP units.

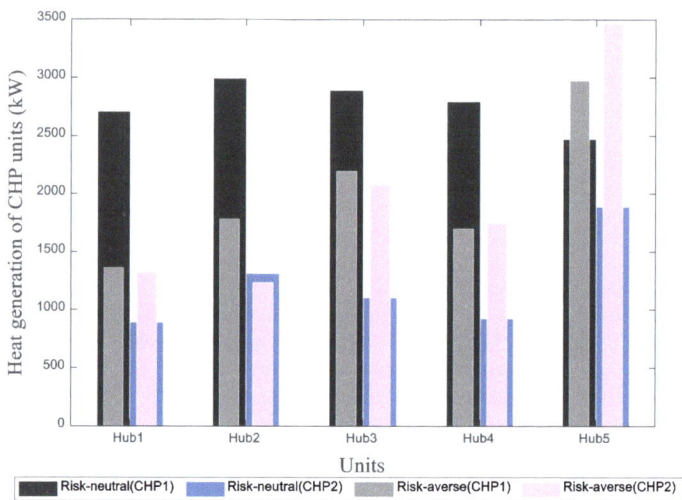

Figure 14. Heat generation of CHP units.

Figure 15 shows the charging and discharging statements of the EVs. Because we have considered the presence of EVs for 8 h in hub5, the functioning is obtained from 8 to 16. Due to the high peak demand, a significant amount of power has been delivered to hub5—about 20 kW. Finally, the function of the load-shifting strategy is demonstrated in Figure 16 for hub1 and hub3. This strategy is quite helpful in terms of energy cost reduction by moving the demand from peak demand to other time intervals. Based on this strategy, 20 kW and 15 kW of power loads in hub1 and hub3, respectively, are moved out of peak demand to reduce the energy cost of the hubs.

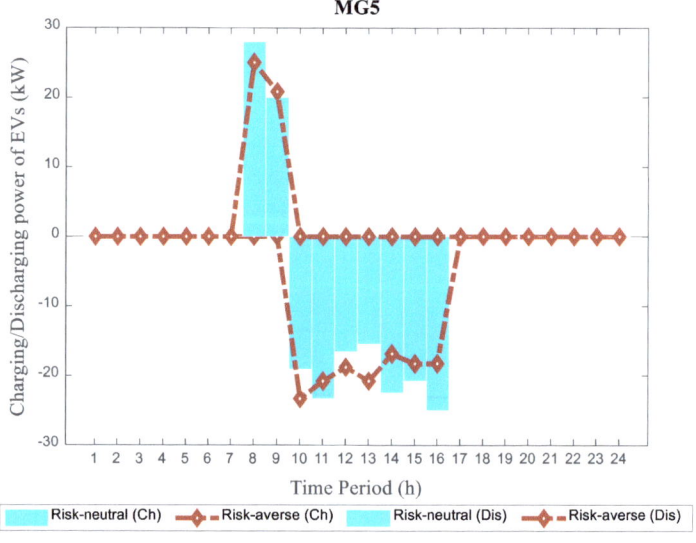

Figure 15. Charging and discharging power of EVs.

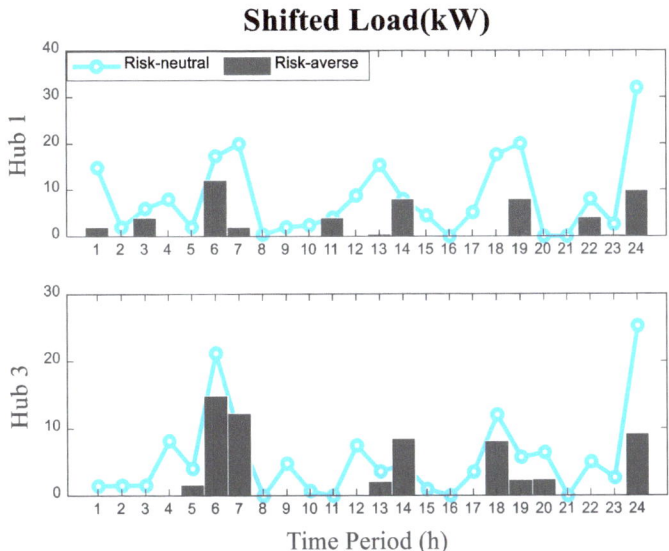

Figure 16. Load-shifting function in hub1 and hub3.

5. Results Validation

In this section, the generation of scenarios is carried out for different data representing a risky condition to validate the obtained results. Under such a condition, the standard deviation, as well as the fluctuation level, are a somewhat larger compared to the previous model. However, there is a slight difference when there is a comparison between uncertain parameters. In this regard, Figures 17–21 indicate the generated scenarios for PV systems and WTs, market price, power demand, and thermal energy demand, respectively. As shown in Table 5, the scenario costs are obtained for different lambda values, for which the operating costs were led to a certain number (663852 NOK) in the last iteration. In the previous section, however, the obtained value when $\lambda = 1$ is 663867 NOK, representing the fact that in all uncertain situations, the scenario cost is quite close to our results.

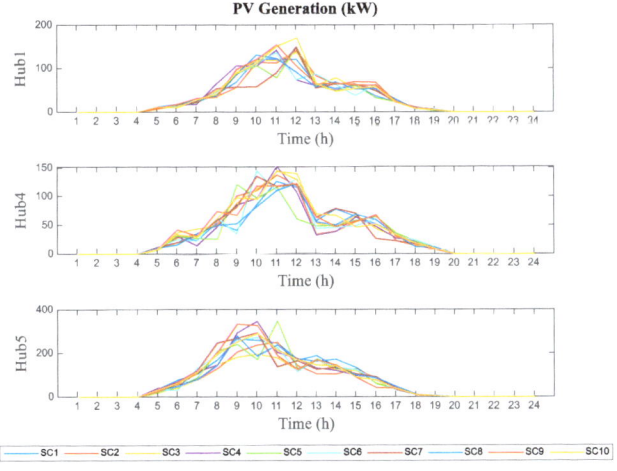

Figure 17. Generated scenario for PV system function.

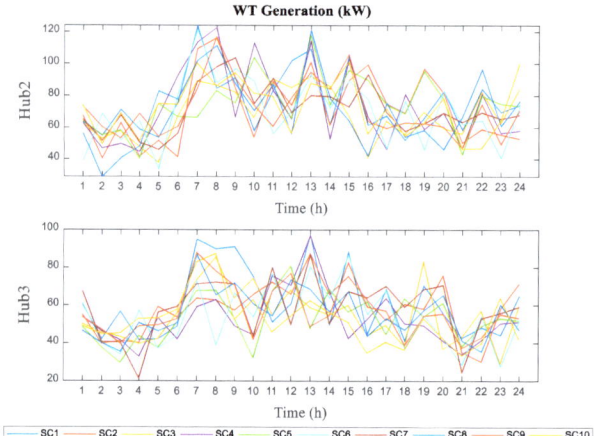

Figure 18. Generated scenario for WT system function.

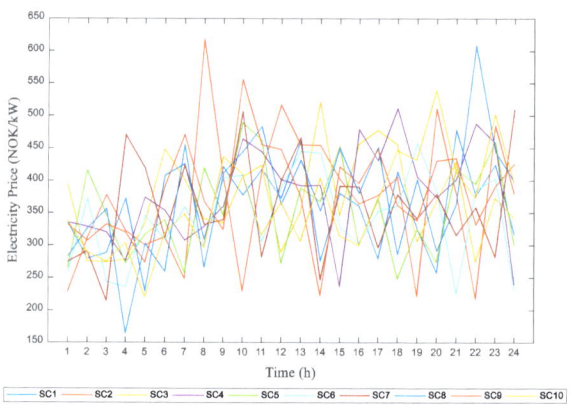

Figure 19. Generated scenario for electricity market price.

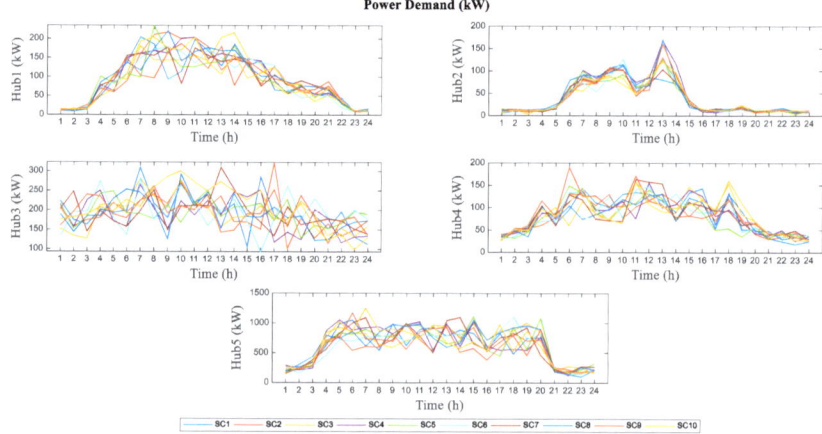

Figure 20. Generated scenario for power demand.

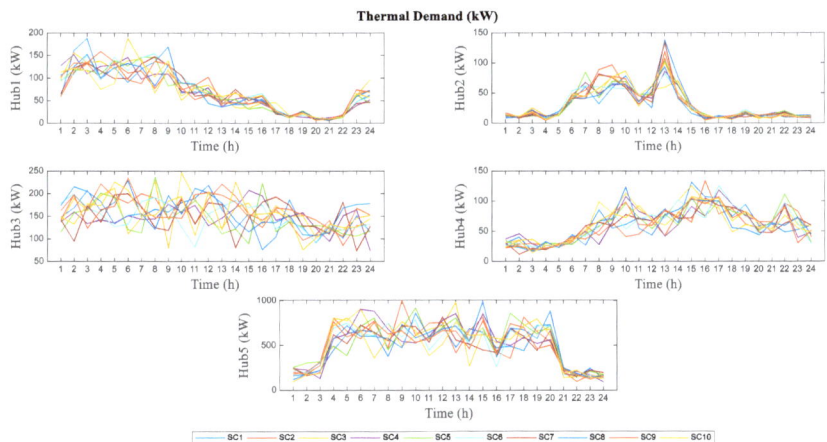

Figure 21. Generated scenario for thermal energy demand.

Table 5. Obtained scenario costs versus different levels of lambda.

	SC = 1	SC = 2	SC = 3	SC = 4	SC = 5	SC = 6	SC = 7	SC = 8	SC = 9	SC = 10
$\lambda = 0$	223,256	550,148	477,104	418,689	251,083	576,426	374,801	365,674	633,806	560,661
$\lambda = 0.1$	249,265	564,361	478,604	448,456	279,244	586,402	411,243	387,550	663,840	560,661
$\lambda = 0.2$	295,416	564,361	478,605	453,049	301,534	586,402	416,514	413,064	663,852	560,661
$\lambda = 0.3$	282,231	564,361	486,278	476,351	338,328	586,402	448,248	463,124	663,852	560,661
$\lambda = 0.4$	283,755	564,361	502,001	492,357	405,541	586,402	470,451	494,144	663,852	560,661
$\lambda = 0.5$	322,345	564,361	517,026	476,342	484,356	586,402	482,164	517,247	663,852	560,661
$\lambda = 0.6$	381,627	564,361	532,140	518,412	532,530	586,402	485,248	505,531	663,852	560,661
$\lambda = 0.7$	547,284	564,361	547,974	547,671	464,758	586,402	489,855	505,212	663,852	560,661
$\lambda = 0.8$	514,126	564,361	564,792	564,664	564,521	586,402	494,245	564,951	663,852	565,691
$\lambda = 0.9$	603,321	603,451	603,470	603,571	598,691	603,371	603,191	547,426	663,852	602,371
$\lambda = 1$	663,852	663,852	663,852	663,852	663,852	663,852	663,852	663,852	663,852	663,852

6. Conclusions

In this paper, the value of P2P heat and power trading in combination with different resources of on-site flexibility is investigated for a Norwegian industrial site, where industrial units are considered as EHs and equipped with energy sources including renewables (WTs and PV systems), CHP units (convex and non-convex), and EVs. Due to the presence of uncertain parameters that greatly reduce the flexibility of the system, the downside risk constraint (DRC) method is applied to evaluate the flexibility of the system under risk-averse and risk-neutral modes. In comparison with the risk-neutral mode, the operator acts more conservatively in the risk-averse mode. For instance, as hub5 has a significant level of electrical demand, more power is imported from the network (4500 kW), and a small amount of value is sold to the grid (100 kW) based on the conservative behavior of the decision maker. Also, by increasing the system's operational cost, the amount of risk is set at zero. In this concern, the operational cost to achieve a zero-risk condition is increased by nearly 36%. Because the CHP units produce power and heat simultaneously, thermal energy is exchanged among the EHs to meet their thermal loads. Also, the consideration of the load-shifting strategy in the first and third hubs resulted in a significant electrical load reduction in the risk-neutral mode (20 kW in hub1 and 15 kW in hub3). By and large, although the operational cost rises in the risk-averse mode, the decision maker becomes capable enough to face uncertain parameters.

In future research studies, the game-theoretic modeling of energy hubs could be examined based on free competition. Also, applying the ADMM algorithm to analyze the P2P heat and power transaction in a decentralized mode could be another research direction for studying the power flow among the energy hubs.

Author Contributions: Conceptualization, E.V. and S.N.; Methodology, R.N.; Software, K.T.-T. and A.A.; Formal analysis, E.V. and A.A.; Investigation, R.N.; Resources, K.T.-T.; Data curation, R.N.; Writing—original draft, E.V.; Writing—review & editing, S.N. and A.A.; Visualization, K.T.-T.; Supervision, S.N.; Project administration, S.N. All authors have read and agreed to the published version of the manuscript.

Funding: This research received no external funding.

Data Availability Statement: Data available on request due to restrictions e.g., privacy or ethical. The data presented in this study are available on request from the corresponding author. The data are not publicly available due to [personal reasons].

Conflicts of Interest: The authors declare no conflict of interest.

Nomenclature

Sets

t	Index of time interval
h	Index of energy hubs
s	Index of scenarios
c	Index of CHP units

Parameters

$C_{g,eng}$	Energy cost (NOK/kWh)
N_h^{wt}	The number of wind turbines in energy hubs
V_t	Wind speed
I_t	Solar radiation
P_{rated}^{wt}	The nominal capacity of wind turbine
V_{rated}	Rated wind speed
V_{cut-in}	Cut-in wind speed
$V_{cut-out}$	Cut-out wind speed
I_{std}	Solar radiation in a typical day
I_C	Radiation point
P_{rated}^{pv}	The nominal capacity of solar panels
$C_{g,fix}$	Utility tariff cost
ψ_{P2P}	The power loss of the distribution network and P2P transaction
Δt	Time duration of step t
$P_{feed-in}^{max}$	Maximum power to meet prosumers' needs (kWh)
$\eta_{ev,ch}$	The efficiency of EV charging unit
$\eta_{ev,dch}$	The efficiency of EV discharging unit
E_{ev}^{nom}	The nominal capacity of the storage unit
$P_{ev,charger}^{num}$	The nominal capacity of the EV charger (kWh)
C_h^{LS}	Load-shifting penalty for hub h
$C_{t,h}^{p,P2P}$	Price of P2P power transaction (kWh/NOK)
$C_{t,h}^{T,P2P}$	Price of P2P heat transaction
$C_t^{g,SP}$	Spot price of wholesale (kWh/ NOK)
a_chp	
b_chp	Cost coefficients of CHP units
c_chp	

Variables

$P_{t,h}^{g,buy}$	Power consumption from the grid
$P_h^{g,peak}$	The maximum power demand of hub h
$P_{t,h}^{g,sell}$	Power feed-in to the grid
$P_{(t,h)}^{imp}$	P2P electricity imported by hub h
$P_{(t,h\leftarrow p)}^{imp,p}$	P2P electricity imported by hub h from peer p
$P_{(t,h)}^{exp}$	P2P electricity exported by hub h
$P_{(t,h\rightarrow p)}^{exp,p}$	P2P electricity exported by hub h to the peer p
$P_{t,h,c}^{chp}$	Generated power by CHP units
$T_{t,h,c}^{chp}$	Produced heat by CHP units
$T_{t,h}^{imp}$	P2P heat energy imported by hub h
$T_{t,h\leftarrow p}^{imp,p}$	P2P heat energy imported by hub h from peer p
$T_{t,h}^{exp}$	P2P heat energy exported by hub h
$T_{t,h\leftarrow p}^{exp,p}$	P2P heat energy exported by hub h to the peer p
E_t^{ev}	The overall level of EV storage unit
$p_{t,h}^{ev,ch}$	Charged power to the EV storage unit
$p_{t,h}^{ev,dch}$	Discharged power from the EV storage unit
$p_{t,h}^{ls,sh}$	Shifted load by hub h
$p_{t,h}^{ls,dem}$	Rescheduled load by hub h
$E_{t,h}^{ls}$	The amount of shifted power

E_{start}	The energy level in EVs when they arrive at work	**Binary Variable**	
E_{end}	The energy level in EVs when they leave work	$u_{t,h}^{buy}$	Binary variable to buy power from the network
EV_{num}	The number of parked EVs during work time	$u_{t,h}^{sell}$	Binary variable to sell power to the network
$C_{g,peak}$	Peak power price of utility tariff (NOK/Month)	$SD_{t,h,c}^{chp}$	Strat-up status of CHP unit
$P_{t,h}^{dem}$	Power demand of energy hubs (kW)	$SU_{t,h,c}^{chp}$	Shut-down status of the CHP unit
$T_{t,h}^{dem}$	Heat demand of energy hubs (kW)	$v_{t,h,c}^{chp}$	Commitment status of the CHP unit

References

1. Lingcheng, K.; Zhenning, Z.; Jiaping, X.; Jing, L.; Yuping, C. Multilateral agreement contract optimization of renewable energy power grid-connecting under uncertain supply and market demand. *Comput. Ind. Eng.* **2019**, *135*, 689–701. [CrossRef]
2. Islam, M.M.; Zhong, X.; Sun, Z.; Xiong, H.; Hu, W. Real-time frequency regulation using aggregated electric vehicles in smart grid. *Comput. Ind. Eng.* **2019**, *134*, 11–26. [CrossRef]
3. Nikmehr, N. Distributed robust operational optimization of networked microgrids embedded interconnected energy hubs. *Energy* **2020**, *199*, 117440. [CrossRef]
4. Long, C.; Wu, J.; Zhang, C.; Thomas, L.; Cheng, M.; Jenkins, N. Peer-to-peer energy trading in a community microgrid. In Proceedings of the 2017 IEEE Power & Energy Society General Meeting, Chicago, IL, USA, 16–20 July 2017; pp. 1–5.
5. Haider, S.; Walewski, J.; Schegner, P. Investigating peer-to-peer power transactions for reducing EV induced network congestion. *Energy* **2022**, *254*, 124317. [CrossRef]
6. Mbungu, N.T.; Naidoo, R.M.; Bansal, R.C.; Siti, M.W.; Tungadio, D.H. An overview of renewable energy resources and grid integration for commercial building applications. *J. Energy Storage* **2020**, *29*, 101385. [CrossRef]
7. Nourollahi, R.; Zare, K.; Nojavan, S. Energy Management of Hybrid AC-DC Microgrid under Demand Response Programs: Real-Time Pricing Versus Time-of-Use Pricing. In *Demand Response Application in Smart Grids*; Springer: Berlin/Heidelberg, Germany, 2020; pp. 75–93.
8. Jian, P.; Guo, T.; Wang, D.; Valipour, E.; Nojavan, S. Risk-based energy management of industrial buildings in smart cities and peer-to-peer electricity trading using second-order stochastic dominance procedure. *Sustain. Cities Soc.* **2022**, *77*, 103550. [CrossRef]
9. Salyani, P.; Nourollahi, R.; Zare, K.; Razzaghi, R. A new MILP model of switch placement in distribution networks with consideration of substation overloading during load transfer. *Sustain. Energy Grids Netw.* **2022**, *32*, 100944. [CrossRef]
10. Dong, J.; Ye, C. Green scheduling of distributed two-stage reentrant hybrid flow shop considering distributed energy resources and energy storage system. *Comput. Ind. Eng.* **2022**, *169*, 108146. [CrossRef]
11. Luo, X.; Liu, Y. A multiple-coalition-based energy trading scheme of hierarchical integrated energy systems. *Sustain. Cities Soc.* **2021**, *64*, 102518. [CrossRef]
12. Liu, C.; Chai, K.K.; Zhang, X.; Chen, Y. Peer-to-peer electricity trading system: Smart contracts based proof-of-benefit consensus protocol. *Wirel. Netw.* **2021**, *27*, 4217–4228. [CrossRef]
13. Moret, F.; Baroche, T.; Sorin, E.; Pinson, P. Negotiation algorithms for peer-to-peer electricity markets: Computational properties. In Proceedings of the 2018 Power Systems Computation Conference (PSCC), Dublin, Ireland, 11–15 June 2018; pp. 1–7.
14. Burgwinkel, D. (Ed.) Blockchaintechnologie und deren Funktionsweise verstehen. In *Blockchain Technology: Einführung für Business- und IT Manager*; De Gruyter Oldenbourg: Berlin, Boston, Germany, 2016; pp. 3–50. [CrossRef]
15. Mengelkamp, E.; Gärttner, J.; Rock, K.; Kessler, S.; Orsini, L.; Weinhardt, C. Designing microgrid energy markets: A case study: The Brooklyn Microgrid. *Appl. Energy* **2018**, *210*, 870–880. [CrossRef]
16. Zhang, C.; Wu, J.; Zhou, Y.; Cheng, M.; Long, C. Peer-to-Peer energy trading in a Microgrid. *Appl. Energy* **2018**, *220*, 1–12. [CrossRef]
17. Xie, B.-C.; Lu, L.; Duan, N. Environmental efficiency assessment of China's integrated power system under the assumption of semi-disposability. *Comput. Ind. Eng.* **2022**, *167*, 108023. [CrossRef]
18. Nourollahi, R.; Gholizadeh-Roshanagh, R.; Feizi-Aghakandi, H.; Zare, K.; Mohammadi-Ivatloo, B. Power distribution expansion planning in the presence of wholesale multimarkets. *IEEE Syst. J.* 2022; *early access*. [CrossRef]
19. Noor, S.; Yang, W.; Guo, M.; van Dam, K.H.; Wang, X. Energy demand side management within micro-grid networks enhanced by blockchain. *Appl. Energy* **2018**, *228*, 1385–1398. [CrossRef]
20. Seyfi, M.; Mehdinejad, M.; Mohammadi-Ivatloo, B.; Shayanfar, H. Scenario-based robust energy management of CCHP-based virtual energy hub for participating in multiple energy and reserve markets. *Sustain. Cities Soc.* **2022**, *80*, 103711. [CrossRef]
21. Nasir, M.; Jordehi, A.R.; Matin, S.A.A.; Tabar, V.S.; Tostado-Véliz, M.; Mansouri, S.A. Optimal operation of energy hubs including parking lots for hydrogen vehicles and responsive demands. *J. Energy Storage* **2022**, *50*, 104630. [CrossRef]
22. Qu, Z.; Chen, J.; Peng, K.; Zhao, Y.; Rong, Z.; Zhang, M. Enhancing stochastic multi-microgrid operational flexibility with mobile energy storage system and power transaction. *Sustain. Cities Soc.* **2021**, *71*, 102962. [CrossRef]
23. Kandpal, B.; Pareek, P.; Verma, A. A robust day-ahead scheduling strategy for EV charging stations in unbalanced distribution grid. *Energy* **2022**, *249*, 123737. [CrossRef]
24. Zargar, R.H.M.; Yaghmaee, M.H. Energy exchange cooperative model in SDN-based interconnected multi-microgrids. *Sustain. Energy Grids Netw.* **2021**, *27*, 100491. [CrossRef]

25. Aghdam, F.H.; Kalantari, N.T.; Mohammadi-Ivatloo, B. A stochastic optimal scheduling of multi-microgrid systems considering emissions: A chance constrained model. *J. Clean. Prod.* **2020**, *275*, 122965. [CrossRef]
26. Sobhani, S.O.; Sheykhha, S.; Madlener, R. An integrated two-level demand-side management game applied to smart energy hubs with storage. *Energy* **2020**, *206*, 118017. [CrossRef]
27. Li, L.; Zhang, S. Peer-to-peer multi-energy sharing for home microgrids: An integration of data-driven and model-driven approaches. *Int. J. Electr. Power Energy Syst.* **2021**, *133*, 107243. [CrossRef]
28. Nguyen, D.H.; Ishihara, T. Distributed peer-to-peer energy trading for residential fuel cell combined heat and power systems. *Int. J. Electr. Power Energy Syst.* **2021**, *125*, 106533. [CrossRef]
29. Gan, W.; Yan, M.; Yao, W.; Wen, J. Peer to peer transactive energy for multiple energy hub with the penetration of high-level renewable energy. *Appl. Energy* **2021**, *295*, 117027. [CrossRef]
30. Nourollahi, R.; Salyani, P.; Zare, K.; Mohammadi-Ivatloo, B.; Abdul-Malek, Z. Peak-Load Management of Distribution Network Using Conservation Voltage Reduction and Dynamic Thermal Rating. *Sustainability* **2022**, *14*, 11569. [CrossRef]
31. Lüth, A.; Zepter, J.M.; del Granado, P.C.; Egging, R. Local electricity market designs for peer-to-peer trading: The role of battery flexibility. *Appl. Energy* **2018**, *229*, 1233–1243. [CrossRef]
32. Howell, A.; Saber, T.; Bendechache, M. Measuring node decentralisation in blockchain peer to peer networks. *Blockchain Res. Appl.* **2022**, 100109. [CrossRef]
33. Dong, J.; Song, C.; Liu, S.; Yin, H.; Zheng, H.; Li, Y. Decentralized peer-to-peer energy trading strategy in energy blockchain environment: A game-theoretic approach. *Appl. Energy* **2022**, *325*, 119852. [CrossRef]
34. McIlvenna, A.; Herron, A.; Hambrick, J.; Ollis, B.; Ostrowski, J. Reducing the computational burden of a microgrid energy management system. *Comput. Ind. Eng.* **2020**, *143*, 106384. [CrossRef]
35. Sperstad, I.B.; Helseth, A.; Korpås, M. Valuation of Stored Energy in Dynamic Optimal Power Flow of Distribution Systems with Energy Storage. In Proceedings of the 2016 International Conference on Probabilistic Methods Applied to Power Systems (PMAPS), Beijing, China, 16–20 October 2016; pp. 1–8.
36. Zhang, H.; Cai, J.; Fang, K.; Zhao, F.; Sutherland, J.W. Operational optimization of a grid-connected factory with onsite photovoltaic and battery storage systems. *Appl. Energy* **2017**, *205*, 1538–1547. [CrossRef]
37. Sæther, G.; Del Granado, P.C.; Zaferanlouei, S. Peer-to-peer electricity trading in an industrial site: Value of buildings flexibility on peak load reduction. *Energy Build.* **2021**, *236*, 110737. [CrossRef]
38. Jafarikia, S.; Feghhi, S. Built in importance estimation in forward Monte Carlo calculations. *Ann. Nucl. Energy* **2022**, *177*, 109298. [CrossRef]
39. Gökçer, T.Y.; Aslan, İ. Approximation by Kantorovich-type max-min operators and its applications. *Appl. Math. Comput.* **2022**, *423*, 127011. [CrossRef]
40. Rezaei, N.; Pezhmani, Y.; Khazali, A. Economic-environmental risk-averse optimal heat and power energy management of a grid-connected multi microgrid system considering demand response and bidding strategy. *Energy* **2022**, *240*, 122844. [CrossRef]

Article

Agnostic Battery Management System Capacity Estimation for Electric Vehicles

Lisa Calearo [1,2], Charalampos Ziras [1], Andreas Thingvad [3] and Mattia Marinelli [1,*]

1. Department of Wind and Energy Systems, Technical University of Denmark (DTU), Risø Campus, 2800 Roskilde, Denmark
2. Ramboll Danmark A/S, 2300 Copenhagen, Denmark
3. Hybrid Greentech ApS, 4000 Roskilde, Denmark
* Correspondence: matm@dtu.dk

Abstract: Battery degradation is a main concern for electric vehicle (EV) users, and a reliable capacity estimation is of major importance. Every EV battery management system (BMS) provides a variety of information, including measured current and voltage, and estimated capacity of the battery. However, these estimations are not transparent and are manufacturer-specific, although measurement accuracy is unknown. This article uses extensive measurements from six diverse EVs to compare and assess capacity estimation with three different methods: (1) reading capacity estimation from the BMS through the central area network (CAN)-bus, (2) using an empirical capacity estimation (ECE) method with external current measurements, and (3) using the same method with measurements coming from the BMS. We show that the use of BMS current measurements provides consistent capacity estimation (a difference of approximately 1%) and can circumvent the need for costly experimental equipment and DC chargers. This data can simplify the ECE method only by using an on-board diagnostics port (OBDII) reader and an AC charger, as the car measures the current directly at the battery terminals.

Keywords: battery capacity; electric vehicle; DC charger; on-board charger; BMS data

Citation: Calearo, L.; Ziras, C.; Thingvad, A.; Marinelli, M. Agnostic Battery Management System Capacity Estimation for Electric Vehicles. *Energies* **2022**, *15*, 9656. https://doi.org/10.3390/en15249656

Academic Editors: Cesar Diaz-Londono and Yang Li

Received: 24 November 2022
Accepted: 15 December 2022
Published: 19 December 2022

Publisher's Note: MDPI stays neutral with regard to jurisdictional claims in published maps and institutional affiliations.

Copyright: © 2022 by the authors. Licensee MDPI, Basel, Switzerland. This article is an open access article distributed under the terms and conditions of the Creative Commons Attribution (CC BY) license (https://creativecommons.org/licenses/by/4.0/).

1. Introduction

1.1. Motivation

Due to the rapid growth of electric vehicle (EV) adoption, it is becoming increasingly important to understand how batteries degrade over a vehicle's lifetime. Li-ion battery packs used in EV applications are always equipped with a battery management system (BMS) [1]. This measures, controls, and manages battery usage [2], while keeping the voltage, current, and temperature of the battery in a safe operating area [3]. In addition, a BMS estimates capacity, a metric used to evaluate battery capacity loss. However, capacity estimations are not standardized between car manufacturers, and internal BMS estimations can vary from car to car depending on the applied method, frequency of recalibration, etc. Additionally, a few commercially available solutions have been developed to estimate the capacity of EV batteries, by using charge or discharge processes and relying on the BMS data. However, we are left with the question, Are EV BMS capacity estimations always reliable and accurate?

1.2. Capacity Estimation Techniques

BMS estimation techniques are divided into two groups: adaptive and experimental [4]. In adaptive methods capacity is estimated from parameters that are sensitive to the degradation of the battery cell. Examples are neural networks [5] or Kalman filters [6], which can provide accurate results. However, high computational needs and costs limit their application in commercial systems [4]. In experimental methods the cycling data history of the battery is stored, and capacity is estimated as a comparison with previously gained knowledge. The computational effort of experimental methods is lower, simplifying their implementation to the disadvantage of lower accuracy. An example is given when BMS capacity estimations are performed onboard by correlating the ampere hours charged or discharged with the voltage difference [1]. Estimation errors accumulate when ampere-hour counting is performed over a long period of time, resulting in inaccuracies and the need for recalibration. Nevertheless, thanks to its simplicity, the combination of Coulomb counting and state-of-charge (SOC)–open circuit voltage relation is used in current BMSs.

Hybrid adaptive/experimental methods, which first characterize chemical reactions and aging mechanisms are also proposed. They are based on approaches such as incremental capacity analysis (ICA) and differential voltage analysis [7], which have been mainly used as reliable offline tools, and have been investigated for online BMS applications [7]. The ICA method relies on plotting the derivative of the capacity with respect to voltage as a function of voltage (incremental capacity (IC) signature) [8,9]. As the battery experiences degradation, the peaks of the IC signature change location. Peaks and valleys of an aged battery can then be compared to the ones of a new battery, and thereby derive the capacity of the aged one [10]. This method has been applied at the EV level in [10,11], showing comparable characteristic peaks and valleys of the IC signature between cells and pack. However, the authors of [12,13] claim that the pack signature may not be directly extrapolated from the already available cells, given that those are not always subject to similar conditions.

Given the wide range of commercially available BMSs and the lack of transparency, it is important to be able to estimate battery capacity with a methodology that is agnostic to BMS data processing and can be applied on any EV battery chemistry, size, and usage. A methodology with such potential is an empirical capacity estimation (ECE) method, used for the first time in our previous work [14], but only applied to a 24 kWh Nissan Leaf. The method consists of a full charge of the battery without disassembling it from the vehicle and violating the warranty. When charging with onboard (AC) chargers, battery voltage and current are not accessible for measurement due to the presence of the AC/DC converter. Therefore, an external DC charger is used, where charging voltage and current can be measured with external equipment at the DC charger terminals. Capacity is then determined as the energy flowing into the battery during the full charge. The disadvantage of this method is that it is time consuming and requires the use of external equipment (EE) that is expensive and not readily accessible.

1.3. What Data Is Available

In series-produced cars, valuable BMS data can be read from the central area network (CAN)-bus via the on-board diagnostics port (OBDII). Available data includes instantaneous measurements, like battery voltage and current, and BMS-derived battery capacity estimations. On the one hand, this allows the use of BMS voltage/current measurements in the ECE method instead of that from EE, after first evaluating their accuracy. On the other hand, BMS estimations can be evaluated and compared with values obtained through a BMS-agnostic method.

The three levels of data which are considered in this work are displayed in Figure 1. They are

- voltage and current measured with EE during a full charge (in green), used to estimate capacity with the ECE method;
- voltage and current BMS measurements read through the OBDII port (in blue), used to estimate capacity with the ECE method; and
- capacity readings from the CAN-bus (in red), which are internally estimated by the BMS, the exact estimation process of which is unknown to the authors.

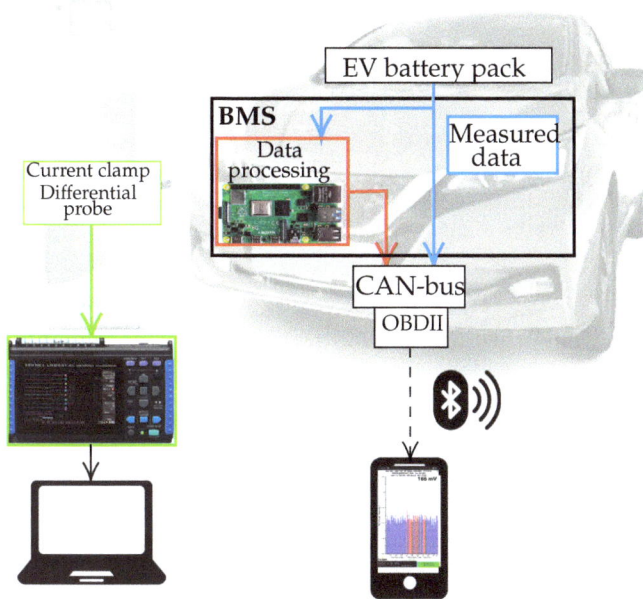

Figure 1. Data collection overview. On the left, current and voltage measurements are collected from the DC charger with current clamp, differential probe, and datalogger. On the right, raw measurements (in blue) are collected from the BMS and CAN-bus, together with estimations derived by the EV microcomputer.

1.4. Paper Contributions

In this work, we investigate and compare three capacity estimation approaches for six different EV batteries, without disassembling them from the vehicles. The main objective is to assess whether BMS readings can be used to circumvent the need for costly and invasive experimental measurements.

The main contributions can be summarized as follows.

- First, capacity readings from the CAN-bus are compared with estimations from the ECE method, while providing insight regarding the observed differences.
- Secondly, the validity of BMS instantaneous current and voltage measurements is assessed by comparing them with EE measurements.
- Thirdly, EE and BMS current/voltage datasets are used to estimate battery capacity with the ECE method, and a comparison between the two is provided.

1.5. Paper Organization

The rest of the paper is structured as follows. Section 2 presents the theoretical background for the capacity derivation. Section 3 presents the measurement methodology for the estimation of EV battery capacity. Section 4 overviews the case study, along with battery pack information and vehicle usage characteristics. Section 5 compares the battery capacity estimations with the different datasets. Section 6 concludes the manuscript with the main outcomes.

2. Theoretical Background

2.1. EV Battery Capacity

The total capacity of a battery (Q) is the amount of energy the pack can hold. This is a function of the initial energy capacity (Q_i), and it decreases over time due to irreversible degradation mechanisms, calendar, and cycle aging. Q_i represents the amount of energy that the battery can theoretically hold when it is new. The total battery capacity at time t is expressed as

$$Q(t) = Q_i(1 - (q_{cal}(t) + q_{cycle}(t))). \qquad (1)$$

q_{cal} and q_{cycle} are the accumulated calendar and cycle degradation, respectively, expressed as a percentage of Q_i. Calendar aging is a function of time, temperature, and SOC, and occurs even when the battery is not used. Cycle aging is a function of the active usage, in terms of full equivalent charge cycles at a certain temperature and current C-rate [14,15]. To maintain the battery lifespan of EVs, BMSs can restrict capacity usage by introducing energy reserves [16]. Thus, EV battery pack capacity can be distinguished between total and usable. Total capacity is the amount of energy the pack can hold without accounting for external restrictions. Usable capacity is the amount of energy that can be stored in the pack, limited by the BMS to protect the battery. If there is no reserve, then the usable capacity coincides with the total capacity. Moreover, it is important to point out that capacity depends on the test conditions and cannot be defined irrespective of them. Indeed, battery capacity changes with different temperatures and C-rate, and the test conditions are not standardized [17].

2.2. ECE Method

The usable energy capacity of a battery can be derived based on the ampere-hour exchange, or the energy exchange during a full charge or discharge cycle. Capacity in Ah is used for the vehicle internal capacity estimation, whereas capacity in Wh is usually provided as nameplate rating by EV manufacturers. Therefore, in this article, we consider both definitions.

Without disassembling the battery from the EV, the usable capacity can be measured during a full EV battery pack charging, and this corresponds to the total capacity if there are no reserves. In Ah (Q^{Ah}), capacity can be derived by integrating the current $I(k)$ during the full charge. With a time resolution of $\tau = 1$ s, $\Delta T = 3600$ s/h and N^s being the number of seconds on the full charge, Q^{Ah} is derived as

$$Q^{Ah} = \frac{1}{\Delta T} \sum_{k=1}^{N^s} I(k)\tau. \qquad (2)$$

Notice that time index t is dropped in Q^{Ah} and subsequent capacity values to simplify notation. These values will refer to the time when an experiment to estimate capacity was conducted.

When considering the battery capacity in Wh, the charging capacity accounts also for the heat dissipation in the battery internal resistance [14]. If charging is conducted with a low current C-rate, the heat dissipation should be limited and influence the results by a few percentage points. The capacity in Wh (Q^{Wh}) is derived by integrating the product between the pack voltage $V(k)$ and current $I(k)$ as

$$Q^{Wh} = \frac{1}{\Delta T} \sum_{k=1}^{N^s} I(k)V(k)\tau. \qquad (3)$$

3. Measurements Methodology

The battery needs to be fully discharged and then fully charged to measure its capacity. The measurable capacity, without disassembling the battery from the EV and violating the warranty, is the usable capacity, which coincides with the total if no reserve is present. This section presents the system—EV and charger—used for conducting the measurements in Section 3.1, the collected datasets in Section 3.2, and finally the methodology for performing the tests in Section 3.3.

3.1. System Layout

EVs can be charged via DC or AC chargers. When using an AC charger, power is first converted from the AC/DC onboard charger in the vehicle, and then flows into the Li-ion EV battery, see Figure 2. By using a DC charger, the power-dependent losses of the AC/DC on-board charger are avoided. The DC charger directly injects power into the 400 V bus, as shown in Figure 2. While charging, the motor side is off and no power is absorbed. Therefore, the power going to the 400 V bus is shared between the Li-ion EV battery and the 12 V bus supplying the auxiliary systems.

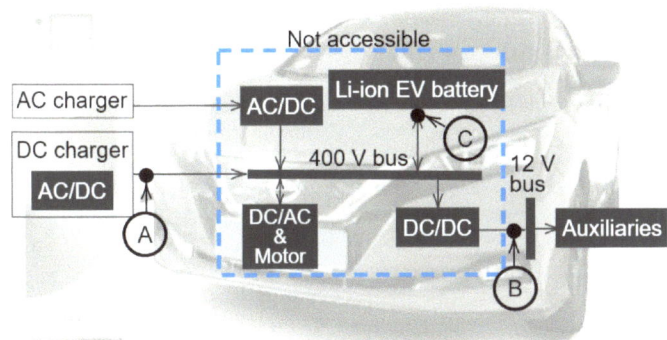

Figure 2. Overview of EV power flows. Modified from [14].

3.2. Data Collection

As shown in Figure 1, three types of data are available to determine EV battery capacity with three different estimation methods.

3.2.1. EE Data

This dataset consists of the voltage and current measured at the DC side of the charger (point A) and at the 12 V bus (point B) (see Figure 2). Current and voltage values are collected by using EE: current clamps for the former and voltage differential probes for the latter, with an overall measurement accuracy of 2.3% [14]. Measurements and estimated capacity are referred to as EE data.

3.2.2. BMS Data

This dataset consists of the voltage and current measured on point C in Figure 2 from the BMS of the vehicle. This data is collected with a maximum resolution of one value per second. It is read through the CAN-bus and OBDII port by using the Nissan Leaf Spy app [18], and becomes available to the user in a spreadsheet form. The accuracy of the EV internal measurement equipment is unknown to the authors and will be further investigated in this article. These measurements are referred to as BMS data.

3.2.3. CAN-Bus Data

The last dataset consists of battery capacity readings from the CAN-bus OBDII port. These values are internally estimated by the vehicle. The estimated capacity values are referred in the following as CAN-bus readings.

3.3. Measurement Process

EV battery pack capacity is measured with the ECE method, which is explained and extended in this section. The method is applicable for all car brands that can be charged with DC power via an external charger [14].

The ECE method involves a full charge of the battery pack from the minimum to the maximum SOC. The charging process consists of two phases. The first is constant current, in which the current is kept constant until voltage reaches the maximum value. The second is constant voltage, where battery voltage is at its maximum value, and current decreases until the charger stops charging. During the measurements it was observed that DC chargers stop charging when the current drops to approx. 3 A. This limitation was experienced with four chargers of two different brands and all investigated vehicles. This behaviour is assumed to be a common feature of DC charging due to the unnecessarily long charging time with very low efficiency. After the DC charger stops, the battery can still be charged if connected to an AC charger (see Figure 3). The amount of energy depends on the minimum current reached by the DC charger. The higher the minimum current, the higher the energy that can be charged with the AC charger.

If the battery pack is small, this energy can be a significant share of the total capacity. Therefore, the methodology in [14] has been revised in this work as follows. After the DC charger stops, the charging process is complemented with the final tail obtained by connecting an AC charger (AC charging tail). Two such examples are provided in Figure 3. Differently from the DC charger, the power coming from the AC charger goes through the AC/DC inverter and then to the Li-ion EV battery (see Figure 2). Measuring the current at the AC side would account for the inverter losses; therefore, this is avoided by considering the current measured at the terminals of the battery (point C in Figure 2). Without disassembling the battery, these values can only be obtained by the BMS.

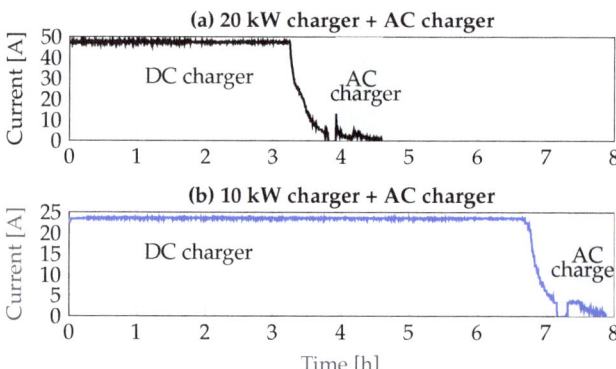

Figure 3. Current charging profiles of a 62 kWh battery pack with a 10 kW and a 20 kW DC charger, including the final tail with an AC charger.

Battery capacity can be derived by considering the current and voltage measured during the charging period with the two datasets (EE and BMS). Figure 4 shows the respective measurement locations. Voltage and current read from the BMS are internally measured at the battery terminals in point C, whereas EE data is measured at the DC charger side (point A) and at the 12 V bus (point B). To compare voltage and current, EE data are processed to derive the current at the battery terminals. This is calculated as the difference between the current in points A and B*. To derive the current in B* ($I^{B*}(k)$), the current in B is scaled to the 400 V bus by considering the voltage measured in A (400 V bus, $V^A(k)$) and B (12 V bus, $V^B(k)$) as in (4):

$$I^{B*}(k) = I^B(k) \cdot \frac{V^B(k)}{V^A(k)}. \tag{4}$$

DC/DC converter losses between the 400 V and 12 V buses are assumed to be negligible.

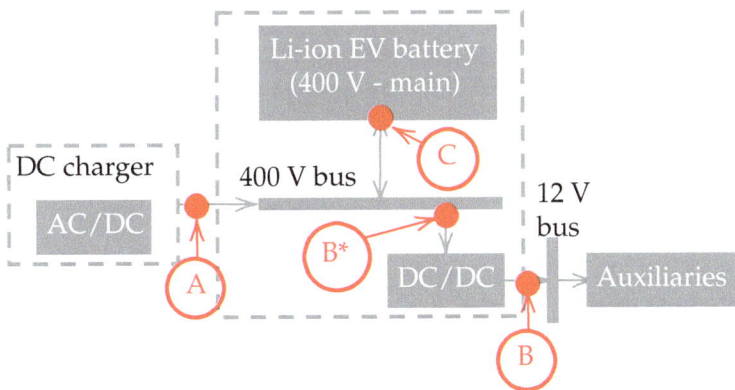

Figure 4. Measurement location overview. In A and B, voltage and current are measured with the external equipment. The BMS voltage and current data are measured from point C. B* is the derived current measurement with the external equipment.

By using (2), capacity is derived by (5) considering the external measurements in points A and B and the additional AC tail, and by (6) considering the BMS current measurements in point C:

$$Q_{EE}^{Ah} = \frac{1}{\Delta T}\left(\sum_{k=1}^{N_{DC}^s} \left(I^A(k) - I^{B^*}(k)\right)\tau + \sum_{k=N_{DC}^s+1}^{N_{DC}^s+N_{AC}^s} I^C(k)\tau \right), \quad (5)$$

$$Q_{BMS}^{Ah} = \frac{1}{\Delta T} \sum_{k=1}^{N_{DC}^s+N_{AC}^s} I^C(k)\tau, \quad (6)$$

where N_{DC}^s and N_{AC}^s is the number of seconds while charging with the DC and AC charger, respectively. Similarly, battery capacity can be derived in Wh by adapting (3).

In the following sections, the normalized capacity q (ratio between the measured and the initial energy capacity) will be used. The superscript Ah or Wh will denote the convention used to express capacity, and the subscript will refer to the used dataset (EE, BMS or CAN).

4. Case Study

4.1. Battery Characteristics

Six EVs with different battery size, chemistry, and usage have been chosen, to demonstrate that results are applicable independently of these factors. Additionally, to account for the rapid technology development during the last decade, EVs introduced in 2014 and 2020 are considered. The EVs names and their characteristics are provided in Table 1.

The Ah nominal capacity (Q_i) of the EVs can be read through the BMS and Leaf Spy app [18]. Nominal voltage is derived as the average open-circuit voltage measured during the constant current phase of the full charge. To the authors' knowledge, the chemistry of the 30 kWh is still unknown in the literature; however, it is expected to be similar to previous and newer battery versions.

Table 1. Vehicles' battery pack and cell characteristics. The number next to E and L indicates the nominal capacity in kWh.

EVs	Env-200 24 kWh	Env-200 24 kWh	LEAF 30 kWh	LEAF 30 kWh	LEAF 40 kWh	LEAF 62 kWh
Name	E24-1	E24-2	L30-1	L30-2	L40	L62
Chemistry	LMO [19]		LMO + NMC(?)		NMC [19]	NMC [19]
Voltage [V]	369.6		360.0		350.4	350.4
Number of cells	192		192		192	288
Cells in series	96		96		96	96
Cells in parallel	2		2		2	3
Capacity [Ah]	65.4		79.5		115.4	176.4
Capacity [kWh]	24.2		28.6		40.4	61.8

4.2. Vehicle Daily Usage

All four E24 and L30 vehicles are driven during the day by the local municipality of the Danish island of Bornholm, and provide frequency regulation (FR) during the night since the end of 2016 [20]. Frequency control is provided for approximately 14 h during the weekdays, and during the weekends for the entire day. An external ± 10 kW vehicle-to-grid (V2G) charger with CHAdeMO connector is used to provide FR. L40 is parked in the laboratory of the Technical University of Denmark, and is only used for measurements a few times per year [15]. L62 is privately owned and driven daily in Denmark [21]. Usage characteristics are summarized in Table 2. EV battery production date is not provided to the owners; therefore, it is here considered to be two months prior to the registration date. Only for L40 is the battery production set eight months prior because the vehicle was previously used for exhibition purposes [15]. The energy throughput for the distance driven per day is derived considering an average consumption of 6 km/kWh [22]. The energy throughput for the FR service of all four E24 and L30 vehicles is considered as in [14], because the service is based on the same frequency and control strategy.

Table 2. Average vehicles usage, distance driven, and FR throughput. * The vehicle provided FR only during the first year.

EV	E24-1	E24-2	L30-1	L30-2	L40	L62
Registration date	7 July 2016	23 June 2017	21 September 2017	6 December 2016	1 August 2018	30 November 2020
Distance per day [km/day]	10	21	20	21	0	35
Throughput drive [kWh/day]	3.3	7	6.6	7	0	11.7
FR	Yes	No *	Yes	Yes	No	No
Throughput FR [kWh/day]	45	45	45	45	0	0
Tot. throughput [kWh/day]	48.3	52	51.6	52	0	11.7
Active cooling	Yes	Yes	No	No	No	No

4.3. Charging C-Rate

External 10 kW DC chargers with CHAdeMO connector are used for charging the EV batteries. During the constant current phase, the current is approximately 24 A. For the considered vehicles, this corresponds to a C-rate (defined as the current divided by the nominal capacity in Ah) between 0.37 for the smallest battery and 0.14 for the largest one, see Table 3. In both cases, C-rate should be sufficiently low to keep the battery heat dissipation limited to a few percentage units and estimate the battery capacity [14].

Table 3. Current and C-rate during the constant current phase of the charging process.

EV	E24-1	E24-2	L30-1	L30-2	L40	L62
Capacity [Ah]	65.4		79.5		115.4	176.4
Current [A]	24		24		24	24
C-rate [-]	0.37		0.30		0.21	0.14

5. Results

Results are presented in three main steps, as shows in Table 4. During the first step (Section 5.1), capacity estimations over five years derived via testing with ECE method and EE data, and readings from the CAN-bus are compared. This step shows the uncertainties arising from the nontransparent BMS estimations. Thus, in the second step (Section 5.2), the instantaneous current and voltage values provided by the BMS are compared with those from EE. These values are then used in the last step (Section 5.3) to compare the capacity estimated with the two datasets, i.e., EE and BMS. Finally, Section 5.4 concludes the section with field test insights on capacity estimation methods and data collection.

Table 4. Steps for results comparison.

STEP 1:	EE capacity estimate (q_{EE})	VS	CAN-bus capacity estimate (q_{CAN})
STEP 2:	EE current and voltage data	VS	BMS current and voltage data
STEP 3:	EE capacity estimate (q_{EE})	VS	BMS capacity estimate (q_{BMS})

5.1. Step 1: EE and CAN-Bus Readings Capacity Comparison

5.1.1. Capacity EE Estimation

Figure 5 compares the normalized capacity of the different vehicles versus their age, both in Ah and Wh. As discussed in Section 3.3, DC chargers stop charging at low current values and more energy can still flow to the battery via AC charging. Despite the fact that this energy is limited for newer and larger batteries, it is important to consider this effect in the overall capacity estimation. A more detailed explanation is provided in Appendix A ?

The used initial battery capacity values in Ah and Wh are provided in Table 1 and used in this section for normalizing the measured capacity values. By comparing the normalized capacity (q_{EE}^{Ah} and q_{EE}^{Wh}) versus age, the different battery chemistry and size do not seem to have a large impact on the degradation trend. L40 ages more slowly, which can be explained by the sole existence of calendar aging, constant battery temperature of 22 °C, and SOC of 50% [15]. Another interesting finding is that kWh capacity values are 3–4% higher than the Ah values. The difference is caused by the battery joule losses that depend on the C-rate during charging. For example, taking a vehicle with battery resistance in p.u. as 6%, if charged with 1 C-rate losses are 6%, whereas if charged with 0.2 C-rate, losses are approximately 1.2%. Therefore, the smaller the battery, the larger the C-rate and the joule losses, and the difference between Ah and Wh normalized values.

Based on Figure 5 the measured capacity does not present a smooth, or even monotonous, decrease. This can be due to different factors. First, battery temperature varied during testing. Despite the fact that measurements were conducted in spring and autumn with similar ambient temperatures, it is not possible to keep the temperature of the batteries constant. Battery temperature variations during the charging phase are kept below 8 °C for most of the cases, with temperatures ranging between 15 °C and 25 °C (see Appendix A.1). Only twice were the battery temperatures of E24-1 and L40 approximately 35 °C, due to usage before the measurements.

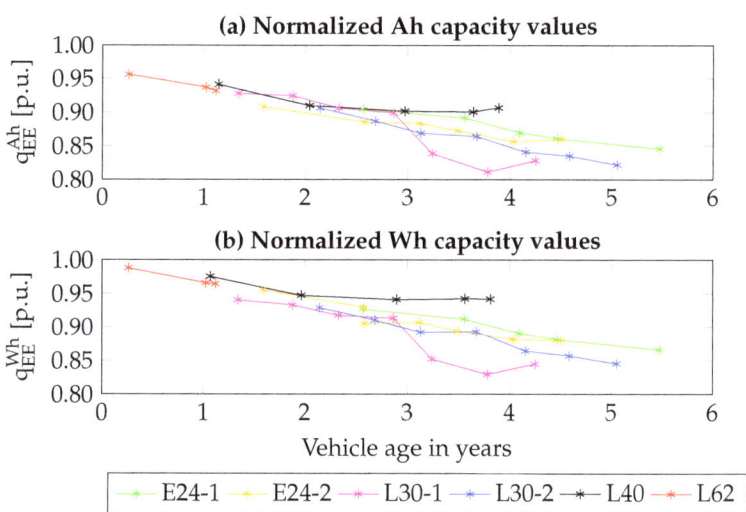

Figure 5. Normalized capacity measurements versus age of the vehicles in years.

Secondly, the ±1% accuracy of current and ±0.1% accuracy of voltage measurements are propagated in the final capacity with an accuracy of 2.3% [14]. Thirdly, for what concerns the discharging process, the reached minimum voltage is not always the same, and it does not always correspond to the same minimum SOC (see Table A1). This is because during the discharging phase the BMS stops the discharging process when the lowest cell voltage reaches a level between 2.8 and 3.1 V. Thus, the minimum voltage can differ from test to test, due to a different cell imbalance each time. Nevertheless, because voltage increases quickly in the beginning of the tests (due to the initial steep relationship between SOC and open circuit voltage of Li-ion batteries), the difference of the minimum pack voltage has a limited effect on the measured battery capacity [14].

5.1.2. Comparison of EE and CAN-Bus Readings

Figure 6 compares the normalized measured capacity via the ECE method and EE data (q_{EE}^{Ah}) with those collected via the CAN-bus readings (q_{CAN}^{Ah}). Because vehicle internal estimations are usually based on Ah values, this section is only focused on those.

CAN-bus capacity readings are always higher than the measured ones, with the exception of L30^2. Moreover, CAN-bus readings above the initial capacity value (larger than 1 p.u.) are observed for E24-1 and E24-2, whereas L30-1 dropped from 0.78 p.u. to 0.69 p.u. in less than 6 months. For older vehicles, E24 and L30, larger differences between the EE measurements and CAN-bus readings are observed. The CAN-bus capacity readings cannot be fully explained by the authors, because they depend on internal vehicle estimation.

Furthermore, the computing power, available memory and accuracy of the current measurements can impact capacity estimations [1]. Although it is not possible to assess the first two (and the method used by the EV microcomputer), current and voltage measurements at the battery terminals can be collected from the BMS through the OBDII. Therefore, in the next subsection the accuracy of voltage and current measurements is investigated by comparing them with EE values.

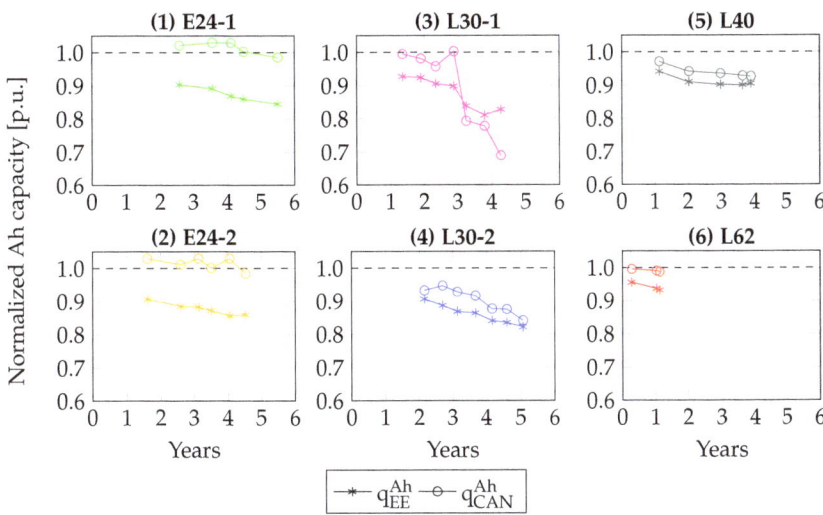

Figure 6. Normalized Ah capacity comparison between the measured EE (represented by asterisks) and CAN-bus readings (represented by circles).

5.2. Step 2: EE and BMS Current and Voltage Comparison

In this subsection, the accuracy of the battery voltage and current BMS measurements is investigated. This is done by comparing the values measured from the BMS with those measured by the EE dataset. As shown in Figure 4, the EE current at the battery terminal (point B*) is derived from the EE measurements in A and B considering (4). Measurements are compared in terms of the instantaneous percentage difference in Figure 7 during the first charging hour of the measurements. Table 5 provides the standard deviation (SD) and mean values of the percentage difference of the current and voltage values during the constant current phase of the charging process.

Subplot (a) shows that the voltage difference is always limited to ±0.5%, whereas in (b) the current difference varies between ±8% for E24 and L30, and ±2% for L40 and L62. In addition, there seems to be a bias in the current difference of E24-2 and L30-2 of approximately 1 A. The offset of EE current at the beginning of each measurement is always reset to zero, whereas this cannot be done with the BMS because there is no control over the measurement equipment. Perhaps the aforementioned biases can be attributed to such calibration issues. By comparing the SD in Table 5, it is visible that current differences are much more volatile than the voltage ones.

Table 5. Standard deviation and mean values of the percentage difference during constant current phase between EE and BMS datasets.

	E24-1	E24-2	L30-1	L30-2	L40	L62
Voltage difference SD [%]	0.05	0.04	0.04	0.03	0.03	0.01
Voltage difference mean [%]	0.11	0.21	0.22	0.27	0.1	0.03
Current difference SD [%]	3.46	2.82	3.43	1.35	0.35	0.28
Current difference mean [%]	1.92	3.79	1.26	4.96	0.39	1.39

Additionally, we should be reminded that the current measured with the EE also accounts for the DC/DC inverter losses present between the 400 V and 12 V buses, which can also be an explanation of the current differences. Furthermore, the unknown performance of the measurement equipment inside the EV is also expected to affect the accuracy of current values. However, given that the differences of voltage and current with the BMS and the EE are limited for most of the cars, in the next subsection capacity is estimated with the two datasets to determine the impact of the different measurements on capacity estimation.

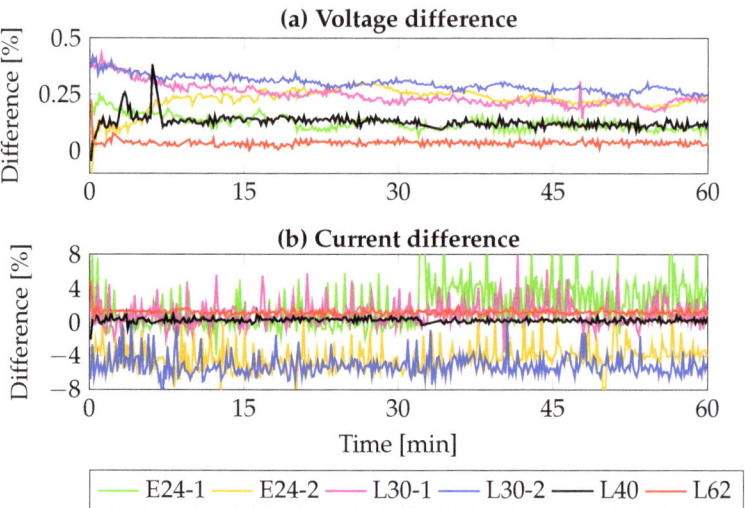

Figure 7. Comparison of voltage and current difference measured between EE and BMSs datasets. Subplot (**a**) shows the percentage voltage difference whereas (**b**) the percentage current difference.

5.3. Step 3: EE and BMS Capacity Estimation Comparison

The capacity estimated with the EE and BMS voltage and current datasets is reported in Table 6, in Ah and kWh. The difference between EE and BMS is limited to 3.8% for E24-2 and L30-2, and less than 1.5% for the remaining ones. This is in accordance with the findings from Figure 7b and Table 5, wherein currents for E24-2 and L30-2 prove to have an initial offset. Thus, the larger currents lead to a higher capacity estimation.

Thanks to the limited difference between the capacity estimated with the two datasets, it can be concluded that the BMS current and voltage values are accurate enough for estimating capacity with the ECE method. Because the BMS current and voltage data are directly collected at the battery terminals, this also means that the limitation of using DC chargers in the ECE method is lifted, and both chargers (AC and DC) can be used.

Table 6. Comparison of capacity derived with the ECE method with EE and BMS datasets in Ah and in kWh.

	Capacity in Ah		Capacity in kWh	
Data	EE	BMS	EE	BMS
E24-1	55.3	55.5	20.9	20.9
E24-2	56.3	58.2	21.3	22.0
L30-1	65.9	65.1	24.2	23.8
L30-2	65.3	67.9	24.2	24.9
L40	104.4	104.0	38.1	37.9
L62	165.3	163.0	59.7	58.9
	164.4	162.5	59.6	58.9

5.4. Discussion

This section compares the findings, highlighting advantages and disadvantages of each estimation approach. The main results are summarized in Table 7.

Table 7. Data collection comparison.

Characteristic/Data	EE	BMS	CAN-bus
Measurement accuracy	High	Medium/high, still unknown	Medium/high, still unknown
Measurement location	DC charger and 12 V bus	Battery terminals	Battery terminals
Equipment	Expensive	Limited (app to read data)	Limited (app to read data)
Electrical knowledge	Advanced	Limited	Limited
Data processing info	Full knowledge	Full knowledge	Limited knowledge

The CAN-bus capacity readings cannot be fully interpreted by the authors, due to restricted knowledge on methodology and internal vehicle calculations. Therefore, this subsection mainly focuses on capacity measurements using EE and BMS current data, and insights regarding the quality and accessibility of the datasets is provided.

First, with the ECE method and EE measurements both a DC and an AC charger, current clamps, voltage differential probes, and a data logger are necessary. Such equipment with a reasonable accuracy is expensive and not readily available. In contrast, battery current is continuously measured by the BMS, but the accuracy of the measurements is unknown to the authors. The instantaneous difference between the current measured with EE and the BMS was found to be higher for older vehicles, and limited to 2% for the newest Nissan Leafs. In this comparison, it should be taken into account that EE measurements also include the DC/DC converter losses between the 400 V and the 12 V buses, which are instead bypassed with the BMS current measurement.

Secondly, DC chargers are used to bypass the AC/DC converter located between the AC charger and the 400 V bus. In addition to being more expensive, DC chargers have higher charging currents that result in higher joule losses during charging. By considering BMS data, current is directly measured at the battery terminals, meaning that both DC and AC chargers can be used. Given the observed limited difference between current readings from BMS and EE, it can be concluded that capacity estimation can be performed by using only onboard chargers and the BMS, without the need for expensive experimental setups.

Thirdly, to connect the external equipment to the 400 V side of the DC charger, it has to be possible to open the charger door and have access to the correct terminals, which also means that electrical component knowledge is required. On the other hand, BMS data is collected from the OBDII-CAN bus of the vehicle. For Nissan Leaf vehicles, information from the OBDII is made accessible by the Leaf Spy app, but similar applications could be developed for other cars.

The accessibility to BMS current measurements could greatly simplify ECE applicability by only using an OBDII, a mobile phone, and an AC charger. Thus, costs are decreased and collection time is limited, e.g., by charging the vehicle during night. Nevertheless, this comes with a need for BMS data reading and translation availability, which is at the moment accessible only for a few vehicles. A few commercial solutions are already using charging/discharging events to estimate battery capacity. These rely on data from the BMS, e.g., current, voltage, etc., to estimate the battery capacity. Because the full methodology is unknown to the user, our future research will compare the capacity estimations from these solutions with the methodology presented in this work.

6. Conclusions

The present paper investigated and compared capacity estimation approaches for six different EV batteries without disassembling them from the vehicles. The main objective of this work was to assess whether BMS readings can be used to circumvent the need for costly experimental measurements.

By connecting an OBDII to the vehicle CAN-bus, it is possible to read capacity estimates derived from the BMS. These were compared with the estimates from an empirical capacity method based on experimental measurements, showing large differences for older and smaller vehicles but acceptable deviations in newer and larger EVs. However, CAN-bus estimates are not transparent, they depend on the car manufacturer, and the underlying method may change over time, so no certain conclusions can be drawn regarding their use.

The empirical capacity estimation method, which consists of a full charge of the EV battery with a DC charger, was also used to estimate battery capacity. A DC charger is needed to bypass the AC/DC converter in the EV, and external measurement equipment is used to obtain reference capacity estimations independently of the ones reported by the BMS. However, with the OBDII connection it is also possible to collect current and voltage data directly measured at the battery terminals from the OBDII. This gives the possibility to estimate capacity with the empirical method by using the BMS current and voltage.

The instantaneous current and voltage measured from the BMS and the EE were compared, showing differences limited to $\pm 2\%$ for the newest vehicles and resulting in a capacity estimation difference of 1.5 percentage points. This confirms that BMS current values can be used to derive capacity, and that EV battery capacity tests can be greatly simplified by using an AC charger and an OBDII, without any electrical equipment know-how.

Future work should focus on the development of translation tools/apps to access and download BMS data. The tools/apps should be simple to understand and to apply, and should be compatible with as many EV brands and versions as possible. Finally, it is important to observe that the approaches presented in this paper are expected to be applicable to all car brands. However, complications in understanding the results can occur in the event that, for certain car models, the car releases battery capacity over the vehicle lifetime. This aspect will be further investigated in our future work.

Author Contributions: Conceptualization, L.C., C.Z., A.T., and M.M.; methodology, L.C., C.Z., and M.M.; validation, L.C.; formal analysis, L.C.; investigation, L.C.; data curation, L.C. and M.M; writing—original draft, L.C.; writing—review & editing, L.C., C.Z., A.T., and M.M.; visualization, L.C.; supervision, C.Z. and M.M.; funding acquisition, M.M.. All authors have read and agreed to the published version of the manuscript.

Funding: The work in this paper is supported by the research projects CAR (EU-Interreg grant nr: STHB.03.01.00-SE-S112/17) and ACDC (EUDP grant nr: 64019-0541).

Institutional Review Board Statement: Not applicable.

Informed Consent Statement: Not applicable.

Data Availability Statement: Data is not publicly available.

Acknowledgments: The icons used in Figure 1 were designed by Freepik from Flaticon.

Conflicts of Interest: The authors declare no conflict of interest. The funders had no role in the design of the study; in the collection, analyses, or interpretation of data; in the writing of the manuscript; or in the decision to publish the results.

Abbreviations

The following abbreviations are used in this manuscript:

MDPI	Multidisciplinary Digital Publishing Institute
FR	frequency regulation
SOC	state-of-charge
EV	electric vehicle
BMS	battery management system
SOH	state of health
V2G	vehicle-to-grid
ICA	incremental capacity analysis
NMC	nickel manganese cobalt
LMO	lithium manganese oxide

OBDII   on-board diagnostics port
IC   incremental capacity
CAN   central area network
OCV   open circuit voltage
ECE   empirical capacity estimation
SD   standard deviation
EE   external equipment

Appendix A

Appendix A.1. BMS Data

Table A1. Leaf Spy data (N/A stands for not available).

	Vehicle Years	Distance [km]	V_{in} [V]	SOC_{in} [%]	SOC_{end} [%]	T_{in} [°C]	T_{end} [°C]	T_{out} [°C]
E24-1	2.6	9073	277	4.6	92.5	20	16	N/A
	3.5	14,380	282	8.9	91.8	35	15	5
	4.1	16,374	291	4.7	91.0	19	16	17
	4.5	17,061	296	7.5	90.3	20	16	19
	5.5	18,422	286	6.0	91.6	27	17	N/A
E24-2	1.6	14064	275	5.6	94.2	16	21	N/A
	2.6	22,687	274	10.9	97.8	20	25	11
	3.1	24,724	308	4.9	94.0	19	24	14
	3.5	26,735	303	3.8	94.0	22	28	17
	4.0	30,999	307	9.4	94.1	13	19	10
	4.5	33,644	N/A	1.8	93.4	21	27	16
L30-1	1.3	8147	266	3.2	97.7	26	25	N/A
	1.9	13,152	265	2.1	95.9	16	20	N/A
	2.3	17,058	258	2.4	97.6	22	24	12
	2.9	20,248	272	2.4	91.8	16	19	14
	3.2	22,999	N/A	0.7	97.7	21	22	18
	3.8	26,657	289	0.2	96.8	9	16	10
	4.2	30,719	286	0.0	97.8	25	28	16
L30-2	2.1	17506	277	2.2	97.7	26	25	N/A
	2.7	21,676	285	3.4	96.5	19	21	N/A
	3.1	25,310	277	2.5	96.0	17	21	10
	3.7	28,202	264	3.2	97.7	14	20	13
	4.6	34,040	272	3.4	96.8	15	20	10
	5.1	38,524	297	0.6	97.8	23	27	16
L40	1.1	35	271	0.9	N/A	37	32	23
	2.0	38	304	1.5	93.8	23	30	22
	2.9	43	294	0.1	93.9	24	32	23
	3.6	43	294	1.1	98.0	23	31	23
	3.8	43	283	1.2	98.0	24	31	23

Table A1. *Cont.*

Vehicle Years	Distance [km]	V_{in} [V]	SOC_{in} [%]	SOC_{end} [%]	T_{in} [°C]	T_{end} [°C]	T_{out} [°C]
	0.3	961	302	1.8	96.5	18	26	22
L62	1.0	12,631	290	1.8	96.9	26	30	22
	1.1	14,343	292	0.4	96.4	18	25	20

Appendix A.2. Effect of AC Charger Tail

DC chargers stop charging at approximately 3 A. If an AC charger is then connected, more energy can be stored in the battery. Consequently, a more accurate measurement of the actual capacity can be achieved by force-charging each vehicle in AC mode. Table A2 provides values for energy measured in A, B, and C, battery capacity, AC charging energy tail, and derived capacity, both with Ah and kWh.

The additional AC charged energy is limited, but not always negligible. For both E24 it represents 3–4% of the total capacity, with a minimum DC charging current of 3–4 A. This value was also observed for the E24 vehicles investigated in [14]. A lower value of 0.6% is measured for the L30-1, which is expected due to the low minimum current values reached with the DC charger of 1–2 A. L30-2 results are not provided because the DC charger stopped charging when the current was still constant at 24 A, due to equipment malfunction while conducting the experiment. For L40, the additional energy of 4% is caused by the considerable minimum DC charging current of 6 A. During the L62 measurements, the DC current reached 3 A, resulting in an additional energy of 0.7%. Given that the minimum current is typically 3 A, its influence on capacity is greater for smaller batteries, and for 62 kWh models it seems negligible. Results cannot be easily generalized though, because the minimum DC charging current also plays a role, and it seems to depend both on the DC charger and vehicle. Nevertheless, because it is still unclear why and when DC chargers stop charging, it is recommended to check the minimum DC current and consider the impact of the additional AC charging tail.

Table A2. Energy, battery capacity, and share of AC charging both considering Ah and kWh values.

	Energy in A	Energy in B	Energy AC Charge (in C)	Battery Capacity	Share AC Charge
			[Ah]		[%]
E24-1	55.1	2.0	2.2	55.3	4.0
E24-2	56.2	1.9	2.0	56.3	3.5
L30-1	66.6	1.1	0.4	65.9	0.6
L40	101.9	1.8	4.3	104.4	4.0
L62	167.1	2.9	1.1	165.3	0.7
	166.2	2.9	1.1	164.4	0.7
			[kWh]		[%]
E24-1	20.8	0.8	0.9	20.9	4.3
E24-2	21.2	0.7	0.8	21.2	3.7
L30-1	24.4	0.4	0.2	24.2	0.8
L40	37.0	0.6	1.7	38.1	4.5
L62	60.3	1.1	0.5	59.7	0.8
	60.2	1.0	0.4	59.6	0.7

References

1. Waag, W.; Fleischer, C.; Sauer, D.U. Critical review of the methods for monitoring of lithium-ion batteries in electric and hybrid vehicles. *J. Power Sources* **2014**, *258*, 321–339. [CrossRef]
2. Stroe, D.I.; Schaltz, E. SOH Estimation of LMO/NMC-based Electric Vehicle Lithium-Ion Batteries Using the Incremental Capacity Analysis Technique. In Proceedings of the 2018 IEEE Energy Conversion Congress and Exposition, ECCE 2018, Portland, OR, USA, 23–27 September 2018; pp. 2720–2725. [CrossRef]
3. Hu, G.; Huang, P.; Bai, Z.; Wang, Q.; Qi, K. Comprehensively analysis the failure evolution and safety evaluation of automotive lithium ion battery. *eTransportation* **2021**, *10*, 100140. [CrossRef]
4. Berecibar, M.; Gandiaga, I.; Villarreal, I.; Omar, N.; Van Mierlo, J.; Van Den Bossche, P. Critical review of state of health estimation methods of Li-ion batteries for real applications. *Renew. Sustain. Energy Rev.* **2016**, *56*, 572–587. [CrossRef]
5. Gou, B.; Xu, Y.; Feng, X. State-of-Health Estimation and Remaining-Useful-Life Prediction for Lithium-Ion Battery Using a Hybrid Data-Driven Method. *IEEE Trans. Veh. Technol.* **2020**, *69*, 10854–10867. [CrossRef]
6. Xiong, R.; Tian, J.; Shen, W.; Sun, F. A Novel Fractional Order Model for State of Charge Estimation in Lithium Ion Batteries. *IEEE Trans. Veh. Technol.* **2019**, *68*, 4130–4139. [CrossRef]
7. Farmann, A.; Waag, W.; Marongiu, A.; Sauer, D.U. Critical review of on-board capacity estimation techniques for lithium-ion batteries in electric and hybrid electric vehicles. *J. Power Sources* **2015**, *281*, 114–130. [CrossRef]
8. Gong, X. Modeling of Lithium-ion Battery Considering Temperature and Aging Uncertainties. Ph.D. Thesis, University of Michigan-Dearborn, Dearborn, MI, USA, 2016.
9. Stroe, D.; Schaltz, E. Lithium-Ion Battery State-of-Health Estimation Using the Incremental Capacity Analysis Technique. *IEEE Trans. Ind. Appl.* **2020**, *56*, 678–685. [CrossRef]
10. Schaltz, E.; Stroe, D.I.; Nørregaard, K.; Stenhøj Kofod, L.; Christensen, A. Incremental Capacity Analysis for Electric Vehicle Battery State-of-Health Estimation. In Proceedings of the 2019 Fourteenth International Conference on Ecological Vehicles and Renewable Energies (EVER), Monte-Carlo, Monaco, 8–10 May 2019.
11. Schaltz, E.; Stroe, D.I.; Nørregaard, K.; Ingvardsen, L.S.; Christensen, A. Incremental Capacity Analysis Applied on Electric Vehicles for Battery State-of-Health Estimation. *IEEE Trans. Ind. Appl.* **2021**, *57*, 1810–1817. [CrossRef]
12. Tanim, T.R.; Shirk, M.G.; Bewley, R.L.; Dufek, E.J.; Liaw, B.Y. Fast charge implications: Pack and cell analysis and comparison. *J. Power Sources* **2018**, *381*, 56–65. [CrossRef]
13. Dubarry, M; Beck, D.; Perspective on Mechanistic Modeling of Li-Ion Batteries. *Accounts Mater. Res.* **2022**, *3*, 843–853. https://10.1021/accountsmr.2c00082. [CrossRef]
14. Thingvad, A.; Calearo, L.; Andersen, P.B.; Marinelli, M. Empirical Capacity Measurements of Electric Vehicles Subject to Battery Degradation from V2G Services. *IEEE Trans. Veh. Technol.* **2021**, *70*, 7547–7557. [CrossRef]
15. Calearo, L.; Thingvad, A.; Ziras, C.; Marinelli, M. A methodology to model and validate electro-thermal-aging dynamics of electric vehicle battery packs. *J. Energy Storage* **2022**, *55*, 105538. [CrossRef]
16. Meng, J.; Cai, L.; Stroe, D.I.; Luo, G.; Sui, X.; Teodorescu, R. Lithium-ion battery state-of-health estimation in electric vehicle using optimized partial charging voltage profiles. *Energy* **2019**, *185*, 1054–1062. [CrossRef]
17. EN IEC 62660-1:2019(MAIN) Secondary Lithium-Ion Cells for the Propulsion of Electric Road Vehicles—Part 1: Performance Testing. Available online: https://standards.iteh.ai/catalog/standards/clc/8ea64757-cca6-4dff-b2f9-af5997976f3c/en-iec-62660-1-2019 (accessed on 1 June 2022).
18. Pollock, J. *LeafSpy Help Version 1.50, User Manual*; LeafSpy Pro: Cupertino, CA, USA, 2022.
19. Electric Vehicle Lithium-Ion Battery. Available online: https://www.nissan-global.com/EN/TECHNOLOGY/OVERVIEW/li_ion_ev.html (accessed on 20 October 2021).
20. Marinelli, M.; Thingvad, A.; Calearo, L. *Across Continents Electric Vehicles Services Project: Final Report*; Technical University of Denmark.: Lyngby, Denmark, 2020.
21. Marinelli, M.; Calearo, L.; Engelhardt, J.; Rohde, G. Electrical Thermal and Degradation Measurements of the LEAF e-plus 62-kWh Battery Pack. In Proceedings of the 2022 International Conference on Renewable Energies and Smart Technologies IEEE, Tirana, Albania, 28–29 July 2022.
22. Calearo, L.; Marinelli, M.; Ziras, C. A review of data sources for electric vehicle integration studies. *Renew. Sustain. Energy Rev.* **2021**, *151*, 111518. [CrossRef]

Article

Soft Switched Current Fed Dual Active Bridge Isolated Bidirectional Series Resonant DC-DC Converter for Energy Storage Applications

Kiran Bathala *, Dharavath Kishan and Nagendrappa Harischandrappa

Electrical and Electronics Engineering Department, National Institute of Technology Karnataka, Surathkal 575025, India
* Correspondence: 177ee005.kiranbathala@nitk.edu.in

Abstract: This paper proposes a high-frequency isolated current-fed dual active bridge bidirectional DC–DC series resonant converter with an inductive filter for energy storage applications, and a steady-state analysis of the converter is carried out. The performance of the proposed converter has been compared with a voltage-fed converter with a capacitive output filter. The proposed converter topology is operated in continuous conduction mode with zero circulation current (ZCC), less current stress and high efficiency. The conditions required for soft switching are determined, and it is found that the converter operates with soft switching of all switches for a wide variation in load and input voltage without loss of duty cycle. Current-fed converters are suitable for low-voltage renewable energy applications because of their inherent boosting capability. An inductive output filter is chosen to make the output current ideal for fast charging and high-power-density battery storage applications. Simple single-phase shift control is used to control the switches. The performance of the converter is studied using PSIM simulation software. These results are confirmed by an experiment on a 135 W converter on an OPAL-RT real-time simulator. The maximum efficiency obtained in simulation is 96.31%. Simulation and theoretical results are given in the comparison table for both forward and reverse modes of operation. A breakdown of the losses of this converter is also presented.

Keywords: dual active bridge; energy storage systems; current-fed; voltage fed; soft switching; photovoltaic; fuel cell

Citation: Bathala, K.; Kishan, D.; Harischandrappa, N. Soft Switched Current Fed Dual Active Bridge Isolated Bidirectional Series Resonant DC-DC Converter for Energy Storage Applications. *Energies* 2023, 16, 258. https://doi.org/10.3390/en16010258

Academic Editors: Cesar Diaz-Londono and Yang Li

Received: 29 October 2022
Revised: 25 November 2022
Accepted: 22 December 2022
Published: 26 December 2022

Copyright: © 2022 by the authors. Licensee MDPI, Basel, Switzerland. This article is an open access article distributed under the terms and conditions of the Creative Commons Attribution (CC BY) license (https://creativecommons.org/licenses/by/4.0/).

1. Introduction

With increasing pollution and an alarming climate situation, it has become imperative to switch from conventional energy sources to renewable energy sources. These renewable energy resources can produce cleaner energy than conventional energy resources. Renewable energy resources are abundant but intermittent in availability. Among the renewable energy resources, solar energy is the most abundant in nature. To utilize this energy, power electronics play a crucial role. In power electronics, DC-DC converters, in combination with photovoltaic systems, can convert solar energy into electrical energy. Due to the non-availability of solar energy at night and lack of sufficient technological advancements, this energy has been given less importance in previous decades. Due to the advances in photovoltaic technology, and pollution damaging the environment, it has become necessary for industry, policymakers and academia shift the focus to solar energy [1,2]. Policymakers are encouraging the use of renewable energy sources such as fuel cells, photovoltaics, wind, etc., for power generation. Currently, solar PVs are becoming more popular and are being installed on rooftops of individual houses. In this scenario, using DC-DC converters and recent advancements in energy storage systems, the drawback of the non-availability of solar energy during the night is no longer a severe problem and soon DC-DC converters will enable solar power to be stored in energy storage systems [3].

Isolated bidirectional dual active full-bridge converters (IBDC) have more advantages in terms of DC-DC converters. Galvanic isolation separates both bridges electrically. Dual active-bridge converters are symmetrical in structure and easy to analyze and control [4–6]. In isolated dual active-bridge converters, both voltage-fed non-resonant [7–10] and resonant [11–15] converters have been discussed. In voltage-fed converters [7,11], the output current has a negative component, which is circulating and incurs more losses. To avoid the circulating current, a discontinuous mode of operation has been chosen for these converters [8,9] in which no power is transferred to the load for a short time. A method has been discussed for a series-resonant DAB converter to reduce the current stress, increase the soft switching range and efficiency, and reduce the effect of dead-time on power transmission and soft switching [16]. A topology has been discussed without an isolation transformer to integrate the photovoltaic system and the grid. The absence of a transformer makes this converter compact but results in a lack of protection through isolation [17]. In [18], a hybrid bridge has been discussed to integrate the DC bus and energy storage system. This topology is a solution when the transformation ratio 'n' is away from unity and it has been operated with high efficiency and soft switching even when 'n' is not close to unity [18]. A controller has been discussed in [19] to regulate power transmission and reduce losses using a minimum-current-point-tracking technique. This controller is generic in design, as it does not depend on the circuit parameters and complicated circuit modeling [19].

In [20], a topology has been discussed to transfer more power than a conventional dual active-bridge (DAB) converter. This topology has twice the power transmission capacity of conventional DAB converters and has found application in electric-powered aircraft [20]. In a topology with a high step-down ratio, inverter bridges are stacked in series through a capacitor, and rectifier bridges are connected in parallel. Series-connected bridges lead to additions of voltage at the input. Parallel-connected bridges at the output lead to less voltage and more current. This converter uses GaN semiconductor switches and offers the highest efficiency of 99% and the least of 97.5% [21,22]. In [23], a scheme has been discussed to increase the dynamic response, especially for a wide range of load changes. Specifically, this topology is a good candidate for DES, solid-state transformers and energy storage applications [23]. A hybrid-switching scheme has been proposed to reduce the losses for a DAB converter with advanced switching devices like SiC; it has achieved 98.96% efficiency [24].

Voltage-current-fed isolated bidirectional DC-DC (IBDC) converters without a resonant network [25–27] have been discussed. The magnetizing current in these converters results in more current stress and losses. Voltage-current-fed IBDC converters with resonant networks [28–30] can offer a nearly sinusoidal current. This feature provides lower current stress and conduction losses. Voltage-current-fed converters are unsymmetrical structures due to extra inductors on one of the bidirectional operations. This structure leads to the uneven distribution of current on the switches. Current-fed isolated bidirectional DC-DC resonant converters (IBDC), in both the forward and reverse modes of operation, are symmetrical. These converters require less gate drive than voltage-fed converters, making them suitable for low-voltage applications due to their inherent boosting capability. A search of the literature shows that current-fed isolated bidirectional DC-DC series resonant converters with an inductive filter have not been studied in detail. In this work, a current-fed IBDC with an inductive filter has been studied. The important contributions of this paper are: (i). The converter is operated with an inductive output filter (ii). Continuous conduction mode without loss of duty cycle, (iii) The circulating current at the load is zero to avoid the current stress on the switches and corresponding losses. Due to their bidirectional power transfer capability and high gain, these converters have become an essential part of the conventional structure of plug-in hybrid electric vehicles [6,10]. Current-fed converters allow solar energy generation utilizing PV with a small voltage output, even for individual households. This article presents the operational performance of a high-frequency series-resonant IBDC with a current source for energy storage application. Section 2 describes the proposed circuit in forward and reverse modes of operation. Section 3 describes the steady-state analysis and soft switching boundary conditions for the current-fed IBDC converter. Simulation and experimental results are presented in Section 4; Section 5 gives a comparison of results, and Section 6 presents the conclusion.

2. Proposed Circuit: Current-Fed Isolated Dual Active Bridge Bidirectional DC-DC Series Resonant Converter

Renewable energy resources require power electronics to make use of the energy that they generate. The importance of power electronics in distributed energy systems is shown in Figure 1. Different types of low-voltage DC DES, which are in operation across the globe, are discussed along with their power and voltage ratings [31]. The voltage-fed DAB and proposed current-fed isolated DAB (CFIBDC) converters are shown in Figures 2a and 2b, respectively. These converters are very useful in charging the battery during daylight hours and discharge the same power from the battery whenever necessary [5]. The CFIBDC shown in Figure 2b consists of a boost stage with 'V_{pv}' as the input voltage, and inductor 'L_b', capacitor 'C_b', and inductor 'L_{dc}' are used for maintaining a constant current. The switches Q_1 to Q_4 are on the primary bridge, and Q_5 to Q_8 are on the second bridge. Diodes D_1 to D_8 are the anti-parallel body diodes of the switches Q_1 to Q_8, as shown in Figure 2b. The high-frequency transformer has a turns ratio of 1: n. The elements L_r and C_r are part of the series-resonant circuit. Circuit element C_r aids in avoiding transformer saturation [13–15,32]. An inductor L_o can reduce ripples in the load current [28].

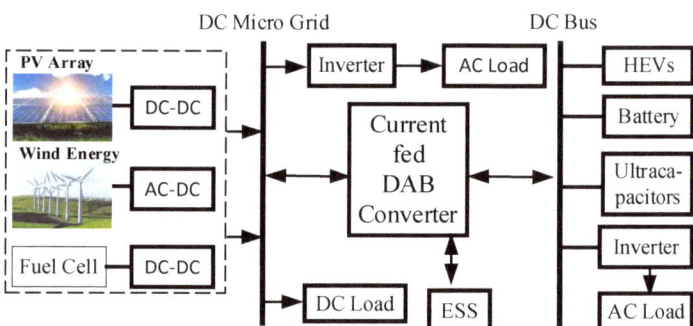

Figure 1. Block diagram of a distributed energy system.

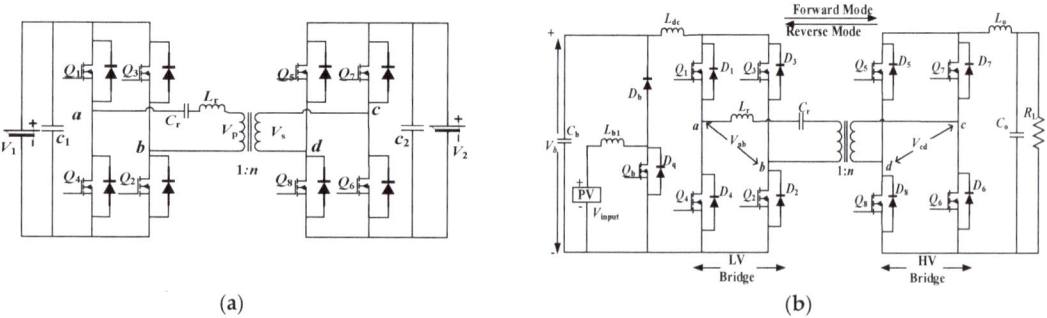

Figure 2. Isolated Bidirectional DC-DC Converter (**a**) Voltage-fed DAB converter (**b**) Current-fed DAB converter.

The model operating waveforms of the voltage-fed converter and current-fed converter for both forward and reverse operation modes are shown in Figure 3a,b and Figure 3c,d, respectively. These features can be avoided in the current-fed IBDC converter. Figure 3c,d shows a DC current without any negative component at the output of the second bridge in both modes of operation. This means no circulating current or stress on the switches due to the circulating current. This feature enhances the efficiency of the converter [9]. Due to technological advancements in power semiconductor switches, a high-frequency current-fed isolated bidirectional DC-DC converter (CFIBDC) can offer low input-current ripples, built-in short circuit protection, no duty cycle loss, higher gain, easy-to-control current and high-power density. The merits of these CFIBDC converters, which are given in [27], make them suitable for various applications, such as electric vehicles, battery storage power quality improvement and fuel cell EVs [27].

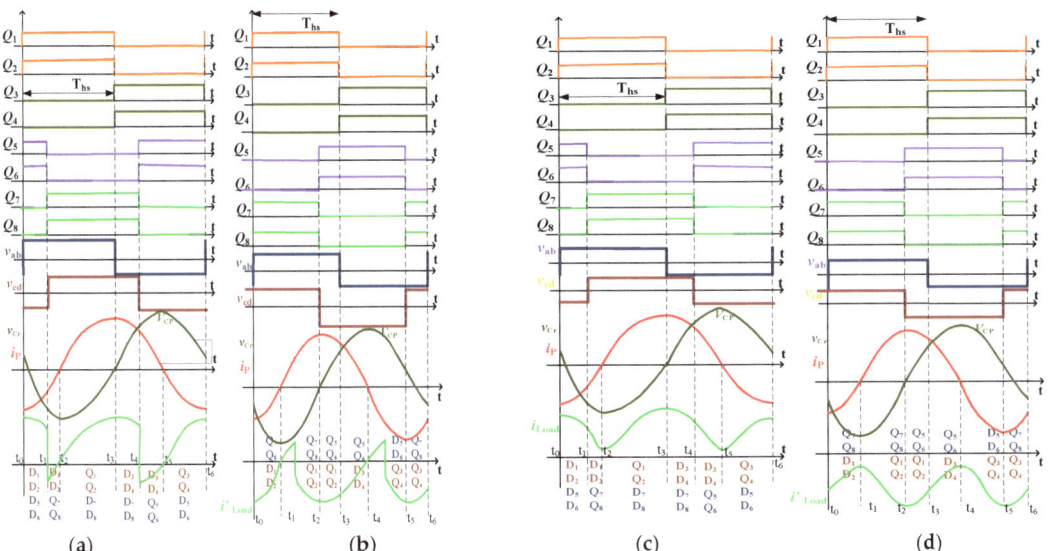

Figure 3. Model waveforms for voltage DAB converter (**a**) Forward mode (**b**) Reverse Mode and Current-fed converter (**c**) Forward mode (**d**) Reverse Mode.

For the CFIBDC converter, a single-phase shift control (SPS) control technique is applied. This technique is the simplest and most-preferred technique for isolated bidirectional dual active-bridge converters. In this technique, diagonal switch pairs in two bridges are turned ON to produce square waveforms having a 50% duty cycle for the respective switches. The square waveforms on either side of the isolation transformer have a phase difference; the leading voltage-side bridge delivers the power to the lagging voltage-bridge side of the transformer. Only a phase-shift ratio (or angle) 'φ' is chosen as the freedom of control; adjusting 'φ' allows power between the bridges to be controlled [33]. The various operation intervals in the converter's forward and reverse modes of operation are described in the following subsections. The equivalent circuits during the forward and reverse modes of operation are shown in Figure 4.

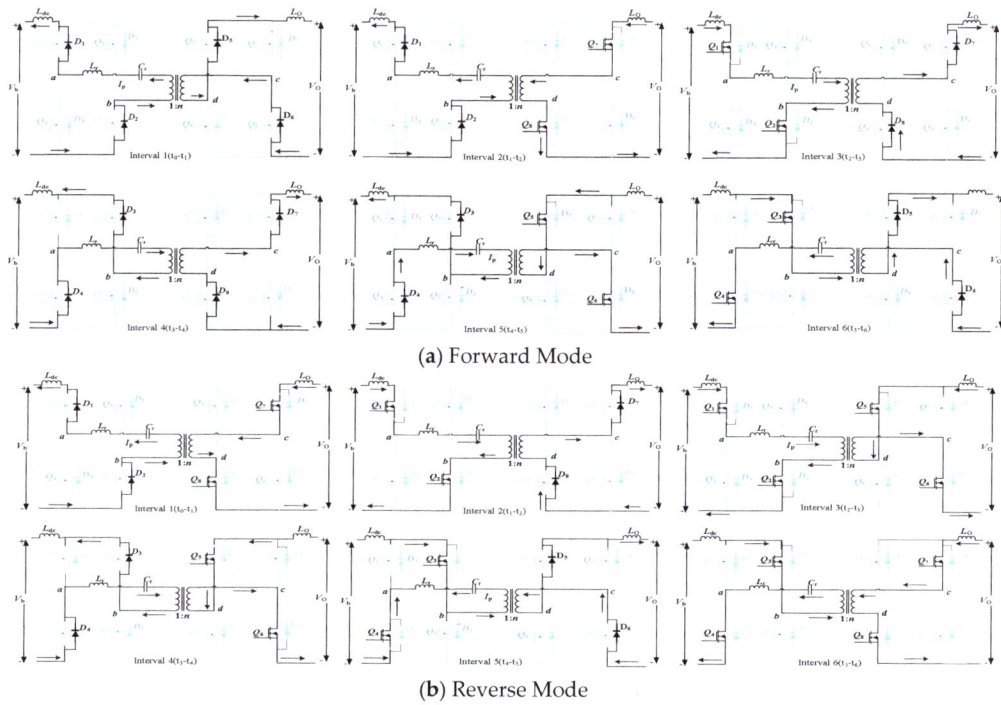

Figure 4. Modes of operation.

2.1. Forward Mode of Operation

Mode-1 (t_0–t_1): for $t \geq t_0$, Q_1 and Q_2 are given the pulses, and since the primary current i_p is negative, Q_1 and Q_2 remain off. Primary current flows through the anti-parallel diodes D_1 and D_2 on the primary side. Q_5 and Q_6 are triggered and are reverse-biased; negative current on the secondary side flows through diodes D_5 and D_6 through the load till $t = t_1$. $V_{ab} = V'_o$, $V_{cd} = -V_o$, $v_{cr}(t_0^+)$ is zero and $i_p(t_0) = (-nI_o)$.

Mode-2 (t_1–t_2): At $t \geq t_1$, primary current continues to flow through the anti-parallel body diodes D_1 and D_2 till $t = t_2$. On the secondary side, switches Q_7 and Q_8 are triggered and a secondary current starts flowing through Q_7 and Q_8; the current reaches zero at $t = t_2$. $V_{ab} = V'_o$, $V_{cd} = V_o$, $v_{cr}(t_2) = -V_{cp}$ and $i_p(t_2) = 0$.

$$i_p(t_2) = C_r \frac{dv_{cr}}{dt} \text{ and } L_r \frac{di_p}{dt} + v_{cr}(t_2) = \frac{-V_o}{n} \quad (1)$$

$$i_p(t) = I_{Lr} \cos(\omega_r t) + \frac{(v_{ab} - nv_{cr} - V_o)}{nZ_c} \sin(\omega_r t) \quad (2)$$

$$v_{cr}(t) = \frac{-V_o}{n} + I_{Lr} z_c \sin(\omega_r t) + \left(v_{cr} + \frac{V_o}{n} - V_{ab}\right) \cos(\omega_r t) + V_{ab} \quad (3)$$

Mode-3 (t_2–t_3): At t_2, $i_p \geq 0$, switches Q_1 and Q_2 conduct on the primary side, and the primary current increases till the end of this mode. Q_7 and Q_8 didn't conduct; increasing i_p makes Q_7 and Q_8 reverse-biased. D_7 and D_8 conduct on the secondary side. At t_2, the end of this mode, Q_1 and Q_2 turn off. At t_3, I_p reaches a maximum and then decreases for $t > t_3$. $V_{ab} = V'_o$, $V_{cd} = V_o$,

$$i_p(t) = i_p(t_2) \cos \omega_r(t - t_2) + \frac{n(v_{ab} - nv_{cr}) - V_o}{nZ_c} \sin \omega_r(t - t_2) \quad (4)$$

$$v_{cr}(t_2) = V_{ab} - \frac{V_o}{n} + i_p(t_2)z_c \sin \omega_r(t-t_2) + (v_{Cr}(t_2) + \frac{V_o}{n} - V_{ab})\cos \omega_r(t-t_2) \quad (5)$$

Mode-4 (t_3–t_4): Since Q_1 and Q_2 turn off at t_3, Switches Q_3 and Q_4 are triggered but reverse-biased by the I_P. I_P decreases linearly till t_4 and discharges through D_3 and D_4 on the primary side; its voltage remains zero during this interval. The anti-parallel diodes D_7 and D_8 still conduct on the secondary side, Q_7 and Q_8 are still reverse-biased by the I_p and the current starts decreasing from its peak value from t_3 onward.

$$V_{ab} = -V'_o, \; V_{cd} = +V_o, \; i_p(t_3) = I_{p(peak)} \quad (6)$$

$$i_p(t) = i_p(t_3)\cos \omega_r(t-t_3) + \frac{n(v_{ab} - nv_{cr}) - V_o}{nZ_c}\sin \omega_r(t-t_3) \quad (7)$$

Mode-5 (t_4–t_5): At t_4, Q_3 and Q_4 are still reverse-biased. I_P is positive and continues to freewheel through D_3 and D_4, linearly decreasing to zero at t_5. On the secondary side, switches Q_5 and Q_6 are triggered, and current flow through the load in the opposite direction, making the net current reach zero at $t = t_5$. During this interval $V_{ab} = -V'_o$, $V_{cd} = -V_o$, $i_p(t_5) = 0$.

Mode-6 (t_5–t_6): For $t \geq t_5$, the primary current is negative and starts increasing linearly; Q_3 and Q_4 are already triggered. These switches start to conduct and the $-I_p$ reaches a maximum at $t = t_6$. At $t \geq t_5$, Q_5 and Q_6 are reverse-biased, and the secondary current gets discharged through the anti-parallel diodes D_5 and D_6. Switches Q_3 and Q_4 are turned off at $t = t_6$. $V_{ab} = -V'_o$, $V_{cd} = -V_o$, $i_p(t)$ is negative.

$$i_p(t) = -i_p(t_4)\cos \omega_r(t-t_4) + \frac{n(-v_{ab} - v_{cr}(t_4) + V_{cd})}{nZ_c}\sin \omega_r(t-t_4) \quad (8)$$

$$v_{cr}(t) = \frac{V_{cd}}{n} - V_{ab} + i_p(t_4)z_c \sin \omega_r(t-t_4) + (v_{Cr}(t_4) - \frac{V_{cd}}{n} + V_{ab})\cos \omega_r(t-t_4) \quad (9)$$

2.2. Reverse Mode of Operation

Mode-1 (t_0–t_1): At t_0, Q_7 and Q_8 are turned on; since the secondary current i_s discharging and is negative, these switches (Q_7 and Q_8) are forward biased, and the voltage (v_{cd} & v_{ab}) across the secondary and primary is positive. Current on the primary flows through the anti-parallel body diodes D_1 and D_2 in this reverse mode of operation.

$$V_{ab} = V'_o, \; V_{cd} = V_{Battery}, \; v_{cr}(t_1) \text{ is } V_{cp} \text{ (peak) and } i_p(t_o) = (-nI_o).$$

$$i_p(t) = C_r \frac{dv_{cr}}{dt}, \text{ and } L_r \frac{di_p}{dt} + v_{cr}(t_1) = -nV_o \quad (10)$$

$$v_{cr}(t) = \frac{V_o}{n} + I_{Lr}z_c \sin(\omega_r t) + (v_{cr} + \frac{V_o}{n} - V_{ab})\cos(\omega_r t) - V_{ab} \quad (11)$$

Mode-2 (t_1–t_2): At t_1, $i_p \geq 0$, switches Q_7 and Q_8 are triggered, and the positive i_p makes these switches reverse-biased. Current conducts through the anti-parallel diodes D_7 and D_8 on the secondary side. On the primary, this change in the direction of the current makes Q_1 and Q_2 forward-biased and current flows through Q_2 and Q_1. The current increases linearly till the end of the following mode. $V_{ab} = V'_o$, $V_{cd} = V_{Battery}$, $v_{cr}(t_1)$ is $-V_{cp}$ (peak value); for $t = t_1^+$ voltage across the resonant capacitor starts decreasing towards zero, and $i_p(t_2) = 0$, $i_p(t)$ reaches a maximum by the end of this mode.

$$I_p(t) = i_p(t_1)\cos \omega_r(t-t_1) + \frac{n(v_{ab} - nv_{cr}) + V_o}{nZ_c}\sin \omega_r(t-t_1) \quad (12)$$

$$v_{cr}(t_2) = \frac{V_o}{n} - V_{ab} + i_p(t_1)z_c \sin \omega_r(t-t_1) + (v_{Cr}(t_1) - \frac{V_o}{n} + V_{ab})\cos \omega_r(t-t_1) \quad (13)$$

Mode-3 (t_2–t_3): For $t = t_2$, the Q_5 and Q_6 switches are turned on. For $t > t_2$, the secondary current i_s increases linearly and passes through Q_5 and Q_6 on the HV-bridge side. Switches Q_1 and Q_2 continue to conduct on the primary side of the converter. The current reaches its peak value at $t = t_3$, the end of this mode. Switches Q_1 and Q_2 on the primary side are turned off at the end of this mode. $V_{ab} = V'_o$, $V_{cd} = -V_{Battery}$, $v_{cr}(t_3)$ is zero; and $i_p(t_2) = I_p$ (peak). $i_p(t)$ starts decreasing from $t = t_2$.

Mode-4 (t_3–t_4): For $t > t_3$, the Q_5 and Q_6 switches continue to conduct on the secondary side. The current i_s starts decreasing linearly from its peak value. The current starts flowing through the anti-parallel body diodes D_3 and D_4 on the primary side. The current reaches zero at $t = t_4$, the end of this mode. $V_{ab} = -V'_o$, $V_{cd} = -V_{Battery}$, v_{cr} increases with a positive slope and reaches a maximum V_{cp} at $t = t_4$; $i_p(t)$ starts decreasing from $t = t_3$ and reaches zero at $t = t_4$, $i_p(t_4) = 0$.

$$V_{ab} = -V'_o,\ V_{cd} = +V_o,\ i_p(t_3) = I_{p(peak)} \qquad (14)$$

$$i_p(t) = i_p(t_3)\cos\omega_r(t-t_2) + \frac{n(v_{ab}-nv_{cr})-V_o}{nZ_c}\sin\omega_r(t-t_2) \qquad (15)$$

Mode-5 (t_4–t_5): For $t \geq t_4$, the current changes its direction, and switches Q_5 and Q_6 are reverse-biased. Now, the current i_s starts flowing through the anti-parallel diodes D_5 and D_6 during this interval on the secondary side. This change in the direction of the current makes the switches Q_3 and Q_4 forward-biased on the primary side. So, the current starts linearly increasing for $t \geq t_4$. $V_{ab} = -V'_o$, $V_{cd} = -V_{Battery}$, v_{cr} decreases with a negative slope and reaches zero $t = t_5^+$; and. $i_p(t)$ starts increasing negatively from $t = t_4$ and reaches $-I_p$ (peak).

$$I_p(t) = -i_p(t_4)\cos\omega_r(t-t_4) + \frac{n(-v_{ab}-v_{cr}(t_4)+V_{cd})}{nZ_c}\sin\omega_r(t-t_4) \qquad (16)$$

$$v_{cr}(t) = \frac{V_{cd}}{n} - V_{ab} + i_p(t_4)z_c\sin\omega_r(t-t_4) + \left(v_{Cr}(t_4)-\frac{V_{cd}}{n}+V_{ab}\right)\cos\omega_r(t-t_4) \qquad (17)$$

Mode-6 (t_5–t_6): For $t > t_5$, switches Q_7 and Q_8 are forward-biased and start conducting on the secondary side. Switches Q_3 and Q_4 continue to conduct on the primary side till the end of this mode. i_s reaches its peak value at $t = t_5$. $V_{ab} = -V'_o$, $V_{cd} = V_{Battery}$, v_{cr} decreases with a negative slope and reaches zero at $t = t_5^+$; and. $i_p(t)$ starts increasing negatively from $t = t_4$ and reaches $-I_p$ (peak).

To summarise, a current-fed isolated bidirectional DC-DC series resonant converter has been analyzed for different operating intervals. Next, a 135 W converter with the specification given in Table 1 is studied through PSIM simulation and experiment, and the results are presented in Section 4.

Table 1. Specifications of the converter.

Parameter	Specification
Output power (P_o)	135 W
Output voltage (V_o)	48 V
Switching frequency (f_s)	100 kHz
DC Input supply (V_{PV})	12–18 V

3. Steady-State Analysis of the Current-Source IBDC Converter

The current-fed isolated DAB bidirectional DC-DC converter transfers power by a phase shift between the gating signal between the primary and secondary sides on either side of the isolation transformer. To carry this analysis to the proposed topology using approximate analysis, the exceptions are given in [12,34–36]. The circuit shown in Figure 2b can be implemented in the forward mode of operation for PV and FC applications; the input is controlled by using the switch Q_{b1} to maintain a constant voltage, as shown in Figure 5a. Figure 5b shows input-voltage control by turning off the switch Q_{b1}. Voltage v_{cd} is taken as 'V_{rt}' in Figure 6a [11]. The phasor equivalent circuit of the proposed topology is shown in Figure 6b. The equivalent circuit of the proposed topology with fundamental voltages 'v_{ab}' and 'v_{cd}' is shown in Figure 6c. The expression for the fundamental components of the voltages shown in Figure 6c is given as

$$v_{ab1} = V_{ab} \sin(\omega t) \tag{18}$$

$$v_{cd1} = v_{cd} \cos(\omega t - \varphi) \tag{19}$$

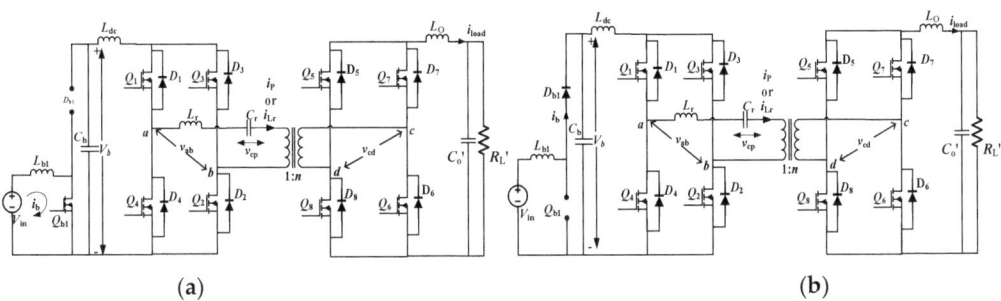

Figure 5. (a) CFIBDC Converter with switch Q_{b1} conducting in FM (b) CFIBDC Converter with switch Q_{b1} off.

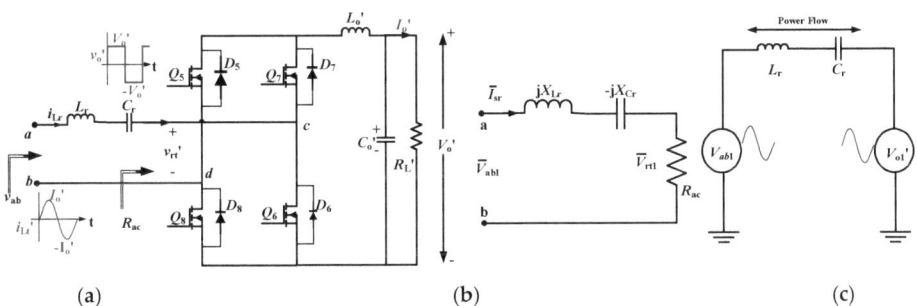

Figure 6. (a) Equivalent circuit of the converter with secondary referred to the primary side, (b). Phasor equivalent circuit at the output terminals of the inverter, (c). Equivalent circuit with the fundamental components of voltages.

The current transferred from one source to another, as shown in Figure 6b, is given as

$$\overline{I_{Lr}} = \frac{v_{ab1} - v_{cd1}}{j\left(\omega L_r - \frac{1}{\omega C_r}\right)} \tag{20}$$

The power transferred to load in forward mode $P_{cd}(t)$ is given as $p_{cd}(t) = v_{cd}(t)i(t)$, and the resultant power then becomes the average power over one full cycle as

$$P_{cd}(t) = \frac{v_{ab} \, v_{cd}}{\left(\omega L_r - \frac{1}{\omega C_r}\right)} \sin(\omega t) \cos(\omega t - \varphi) \tag{21}$$

3.1. R_{ac} for the Inductive Output Filter

The output current (i_o) referred to the primary side, as shown in Figure 6a, is given by the following equation

$$i'_o = \frac{2}{T_s}\int_0^{T_s/2} \frac{I_{Lr}}{n}\sin(\omega t)d(\omega t) \tag{22}$$

so the peak current of the i'_o is given as $I'_o = \frac{2I_{Lr}}{n\pi}$; The series-resonant tank current in terms of I'_o is given as $i_{Lr}(t) = \frac{n\pi}{2}I'_o\sin(\omega t)$. By using Fourier-series analysis, the peak and fundamental component of the voltage input to the rectifier bridge (at the 'c' and 'd' terminals) in terms of V'_o (output voltage referred to primary side) is

$$V_b = \frac{4V'_o}{n\pi}; \ v_b(t) = \frac{4}{n\pi}V'_o\sin(\omega t) \tag{23}$$

$$R_{ac} = \frac{v_b(t)}{i_{Lr}(t)} = \frac{8}{\pi^2}R'_L \tag{24}$$

The sinusoidal current flowing in the series-resonant components is expressed as

$$i_{Lr}(t) = I_{Lr}\sin(\omega t - \phi) \tag{25}$$

The peak current stress is the same for the resonant elements and switches, and it is expressed as

$$I_{Lr} = \frac{V_{ab1(peak)}}{|Z_{ab}|} \tag{26}$$

The impedance (Z_{ab}) offered by the circuit across the terminals 'a' and 'b' is given as;

$$Z_{ab} = R_{ac} + jX_{ab} \tag{27}$$

From Z_{ab}, the expression for ϕ is given as

$$\phi = \tan^{-1}(X_{ab}/R_{ac}) \tag{28}$$

The sign of the magnitude of the resonant current $i_{Lr}(t)$ at $\omega t = 0$ decides the kind of soft switching (either zero-voltage switching (ZVS)/zero-current switching (ZCS)). If $i_{Lr}(0)$ is negative then switches are turned-on with ZVS.

$$i_{Lr}(0) = I_{Lr}\sin(-\phi) \tag{29}$$

When $i_{Lr}(0)$ is positive, this indicates that the switches are turned off with ZCS. The ZVS/ZCS schemes minimize switching losses. The maximum voltage across C_r is V_{CP} is given as $V_{Cp} = I_{Csp} \cdot |X_{Cr}|$.

3.2. Soft Switching

The performance of the current-fed DAB converter can be analyzed by determining the soft-switching range and conditions of the converter. These boundary conditions of the converter are the critical factors that define the soft switching of the converter [33]. This converter is operated under the above resonance method to achieve ZVS for the switches; the corresponding switching pulses, voltage and current waveforms are shown in Figure 7. The two expressions given in (30) are derived by using simple trigonometry for the current waveshapes shown in Figure 7a. The corresponding equivalent circuit, which depicts the waveforms of Figure 7a during the turn-on transient time in the forward mode, is shown in Figure 8a,b, where $C_{stray,Qz}$ ($z = 1\sim8$) represents stray capacitance in parallel with the switches (Q_1 to Q_8).

$$\left. \begin{array}{l} I_{t2} = \frac{\pi n V_b - (\pi - 2\varnothing)n^2 V_o}{2\pi L_r f_s} \\ I_{t1} = \frac{\pi n^2 V_o - (\pi - 2\varnothing)n V_b}{4\pi L_r f_s} \end{array} \right\} \tag{30}$$

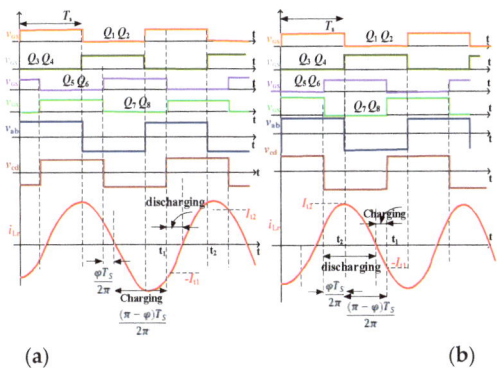

(a)   (b)

Figure 7. Triggering pulses for switches (Q_1 to Q_8), voltage (v_{ab} & v_{cd}) and current (i_p or i_{Lr}) of the converter (**a**) Forward mode (**b**) Reverse mode.

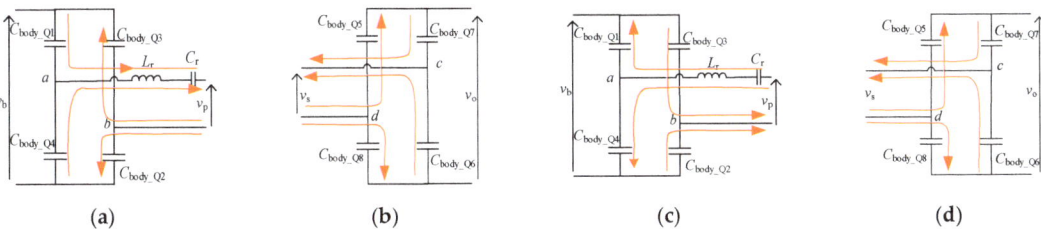

(a)   (b)   (c)   (d)

Figure 8. Equivalent circuits with the stray capacitance of the switches (**a**) input bridge and (**b**) output bridge in forward mode (**c**) output bridge (**d**) input bridge during the reverse mode of operation.

The bridges are symmetrical, so events belonging to Q_4 on the input and Q_7 on the output bridge are considered. The dynamic expressions (31) belong to the capacitor $C_{stray,Q4}$ and the series inductor are shown in Figure 8a,b.

$$\left. \begin{aligned} L_r \frac{di_p(t)}{dt} &= v_{Cstray,Q4}(t) - [V_b - v_{Cstary,Q3}(t)] - V'_o \\ &= 2v_{Cstray,Q4}(t) - V'_o - V_b \\ C_{stray,Q4} \frac{dv_{Cstray,Q4}(t)}{dt} &= -\tfrac{1}{2}i_p(t) \\ i_p(0) &= -nI_{t2} \\ v_{Cstray,Q4}(0) &= V_b \end{aligned} \right\} \quad (31)$$

The expression for the capacitor voltage can be given with a frequency ω_1

$$v_{Cstray,Q4}(t) = \frac{V_b - V'_o}{2}\cos(\omega_1 t) - \frac{nI_{t2}\omega_1 L_r}{2}\sin(\omega_1 t) + \frac{V_b + V'_o}{2} \quad (32)$$

where $\omega_1 = \frac{1}{\sqrt{L_r C_{stray,Q4}}}$. By reorganizing (32), the expression for θ_1 is expressed in (33).

From (32), ZVS of Q_4 can occur when its capacitance-voltage reaches zero by resonance. The ZVS condition for the input bridge is given in (34).

$$\theta_1 = \sin^{-1}\left[\frac{nI_{t2}\omega_1 L_r/2}{\sqrt{\left(\frac{V_b - V'_o}{2}\right)^2 + \left(\frac{nI_{t2}\omega_1 L_r}{2}\right)^2}}\right] \quad (33)$$

$$I_{t2} \geq \frac{2}{n}\sqrt{\frac{V_b V'_o C_{stray,Q4}}{L_r}} \quad (34)$$

The dead-time is essential and must be controlled so that switch turns on when the capacitor voltage of the respective switch discharges to a minimum value, and this will reduce the turn-on loss. The dead-time for the input bridge in the forward mode is derived [33] by substituting ω_1 and (33) into (35) and is given as (36)

$$T_{\text{dead,ab}} = \begin{cases} \theta_1/\omega_1 \; ; (V_b < V_o') \\ (\pi - \theta_1)/\omega_1 \; ; (V_b \geq V_o') \end{cases} \quad (35)$$

$$T_{\text{dead,ab}} = \begin{cases} \sqrt{L_r C_{\text{stray,Q4}}} \sin^{-1}\left[\sqrt{\dfrac{I_{t2}^2 L_r}{n^2 C_{\text{stray,Q4}}(V_b - nV_o)^2 + I_{t2}^2 L_r}}\right]; \\ \qquad (V_b < V_o') \\ \sqrt{L_r C_{\text{stray,Q4}}}\left\{\pi - \sin^{-1}\left[\sqrt{\dfrac{I_{t2}^2 L_r}{n^2 C_{\text{stray,Q4}}(V_b - nV_o)^2 + I_{t2}^2 L_r}}\right]\right\}; \\ \qquad (V_b \geq V_o') \end{cases} \quad (36)$$

For the bridge on the secondary, its ZVS analysis is simple, after $t = t_1$, i_p has a relatively small di/dt. It is assumed constant as I_1 over a short duration. During the transient time of the switch Q_7, its capacitor $C_{\text{stray,Q7}}$ discharges by a current $I_{t1/2}$, instead of resonance between L_r and $C_{\text{stray,Q7}}$. The voltage across the switch Q_7 capacitor ($C_{\text{stray,Q7}}$) drops from V_o to zero and is supposed to satisfy the condition given in equation (37). This equation provides the condition for achieving ZVS for the output bridge given in (38)

$$\frac{I_{t1}}{2} T_{\text{dead,l,c}} \geq C_{\text{stray,Q7}} V_o \quad (37)$$

$$I_{t1} \geq \frac{2C_{\text{stray,Q7}} V_o}{T_{\text{dead,l,c}}} \quad (38)$$

The switches (Q_1 to Q_8) for the input and output bridges are shown in Figure 8c,d. The expression for θ_2, I_{t1} and, during this mode, the charging and discharging of the capacitor across the switches, are shown in Figure 7b. The corresponding equivalent circuits of the input and output full bridges are shown in Figure 8b, I_{t2} and I_{t1} are still aligned with the turn-off of Q_1 at t_2 and Q_8 at t_1. The soft-switching conditions are evaluated for the single-phase shift control scheme, and it is as follows.

$$\frac{nI_{t2}}{2} T_{\text{dead,h,d}} \geq C_{\text{stray,Q4}} V_b \quad (39)$$

$$I_{t2} \geq \frac{2nC_{\text{stray,Q4}} V_b}{T_{\text{dead,l,c}}} \quad (40)$$

Using regular trigonometry, the expression for I_{t2} and I_{t1} in the discharging or reverse modes are the same as the expressions (30) derived in forward mode. The voltage and current waveforms for the reverse mode are shown in Figure 7b. Equivalent circuits representing the capacitors of the switches (Q_1 to Q_8) for the input and output bridges are shown in Figure 8c,d. The expression for θ_2, I_{t1} and dead-time ($T_{\text{dead,cd}}$) for reverse operation are derived just as in the forward mode of operation. These expressions are given in (41) to (44)

$$\theta_2 = \sin^{-1}\left[\frac{n^2 I_{t1} \omega_2 L_r / 2}{\sqrt{\left(\dfrac{nV_o - V_b}{2n}\right)^2 + \left(\dfrac{n^2 I_{t1} \omega_2 L_r}{2}\right)^2}}\right] \quad (41)$$

$$I_{t1} \geq 2\sqrt{\frac{nV_b V_o C_{stray,Q1}}{L_r}} \qquad (42)$$

$$T_{dead,cd} = \begin{cases} (\pi - \theta_2)/\omega_2 \, ; (V_b < nV_o) \\ \theta_2/\omega_2 \, ; (V_b \geq nV_o) \end{cases} \qquad (43)$$

$$T_{dead,cd} = \begin{cases} \frac{\sqrt{L_r C_{stray,Q7}}}{n} \sin^{-1}\left[\sqrt{\frac{I_{t1}^2 L_r}{C_{stray,Q7}(nV_o - V_b)^2 + I_{t1}^2 L_r}}\right]; \\ \qquad (V_b \geq nV_o) \\ \frac{\sqrt{L_r C_{stray,Q7}}}{n}\left\{\pi - \sin^{-1}\left[\sqrt{\frac{I_{t1}^2 L_r}{C_{stray,Q7}(nV_o - V_b)^2 + I_{t1}^2 L_r}}\right]\right\}; \\ \qquad (V_b < nV_o) \end{cases} \qquad (44)$$

4. Simulation and Experimental Results

This section presents the simulation and experimental results of the proposed current-fed isolated DAB bidirectional DC-DC series resonant converter (CFIBDC) in Sections 4.1 and 4.2, respectively.

4.1. Simulation Results

The specifications and design values for the proposed converter are shown in Tables 1 and 2. Results are presented for three cases with an output power of 135 Watts. Case 1: In forward mode with an input voltage $V_{in(min)}$ = 12 Volts, V_o = 48 Volts at 100% and 10% of full load. Case 2: In forward mode with a $V_{in(max)}$ = 18 Volts, V_o = 48 Volts at 100% and 10% of full load. If the load starts transferring back power to the input, then it is a reverse mode of operation. Case 3: In reverse mode, $V_{battery}$ = 48 Volts, V_o' = 42.5 Volts at 100% and 10% of full load. The comparison values of the theoretical and simulation values are given in Table 3.

Table 2. Designed values of the converter.

Parameter	Specification
Frequency ratio (F)	1.1
Gain (M)	0.95
V_o reflected to the primary side (V_o')	42.75 V
Transformer turns ratio (1:n)	1.12
Load resistance (R_L)	13.6 Ω
Resonant Inductor (L_r)	23.8 μH
Resonant Capacitor (C_r)	0.128 μF
Impedance (Z_{ab})	11.02 + j2.52 Ω
Peak current (I_{Lr})	4.71 A
Peak voltage (V_{cp})	65.5 V
Filter inductance (L_o)	≤403.4 μH
Filter capacitance (C_o)	≥1405 μF

Table 3. Comparison of theoretical and simulation values in forward mode with an input voltage V_{in} = 12V$_{(min)}$ & 18V$_{(max)}$, and reverse mode with an input voltage V_2 or V_o = 48 V.

Parameter	Forward Mode									Reverse Mode					
	$V_{in(min)}$ = 12 V						$V_{in(max)}$ = 18 V				V_2 = 48 V					
	100% Load			10% Load			Full Load		10% Load		Full Load			10% Load		
	Cal.	Sim.	Expt.	Cal.	Sim.	Expt.	Cal.	Sim.	Cal.	Sim.	Cal.	Sim.	Expt.	Cal.	Sim.	Expt.
V_{load}(V)	48	44	48	48	49	50	48	44.8	48	49	42.7	39.8	50	42.7	43	50
I_{load}(A)	2.81	2.57	2.5	0.28	0.3	0.6	2.81	2.62	0.28	0.29	3.1	4.38	2.8	0.31	0.46	0.6
$I_{sr,peak}$	3.92	4.4	5	1.09	1.23	0.5	3.93	4.41	1.29	1.45	3.9	4.38	5	0.41	0.46	0.5
V_{cp}	62.9	56.8	52	5.7	12.9	14	62.9	57.1	5.7	15	62.9	58	55	5.7	6.29	16
δ (°)	180		180		170		170		180		170		179		179		165		165
ZVS/ZCS	All		All		All		All		All		All		All		All		All		All

The voltage at the inverter output terminal voltage (v_{ab}), rectifier bridge input (v_{cd}), primary current (i_p) and the voltage across the resonant capacitor (v_{cr}) are shown in waveforms. The HV-side terminal current i_{load}, load voltage v_{load}, load current i_{Load} and current input to the IBDC converter i_b from the boost stage and the switches Q_1 to Q_8 voltages and currents are shown in Figures 9–11. Simulation waveforms of the CFIBDC for Case 1 are shown in Figure 9, and for Case 2 are shown in Figure 10. Case 3, with the current input to the IBDC converter $i_{discharge}$ from stored energy in a forward mode, which is fed back, is also shown in Figure 11. Soft switching is achieved for the switches Q_1 to Q_8 at full load; the corresponding voltage and current waveforms are offered for both forward and reverse modes of operations. Various losses of the converter are used for the simulation of the losses and efficiency of the converter. The simulation results and theoretical calculations are compared and found to be approximately equal, as can be seen in Table 3.

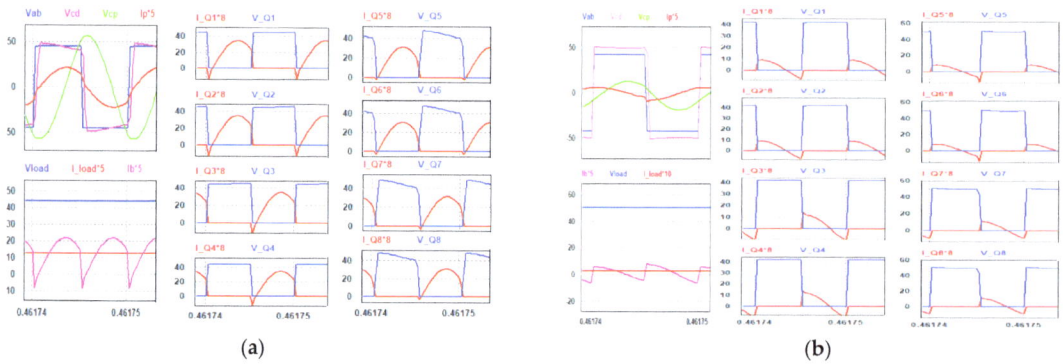

Figure 9. (a). Simulation results in FM mode with 100% Load at $V_{in(min)}$ = 12 V, v_{ab}, v_{cd}, i_p and v_{cr}. Soft-switching characteristics for the primary (Q_1 to Q_4) and secondary side (Q_5 to Q_8) switches are also shown. (b). Simulation results in FM mode with 10% Load at $V_{in(min)}$ = 12 V, v_{ab}, v_{cd}, i_p and v_{cr}. Soft-switching characteristics for the primary (Q_1 to Q_4) and secondary characteristics for the primary (Q_1 to Q_4) and secondary side (Q_5 to Q_8) switches are also shown.

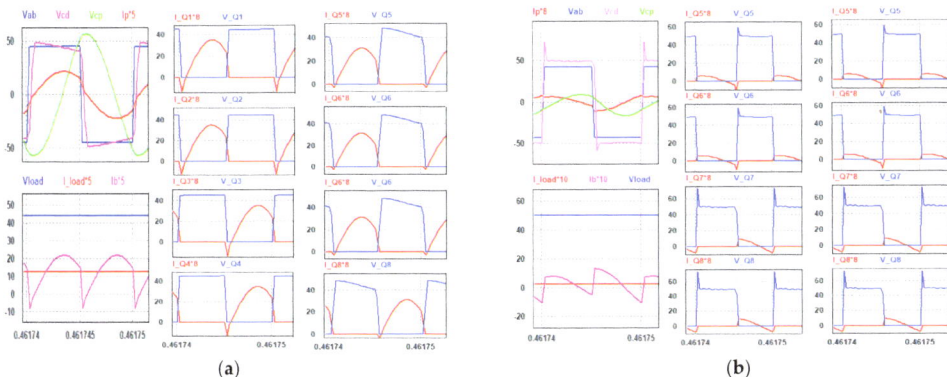

Figure 10. (**a**). Simulation results in FM mode with 100% Load at $V_{in(max)}$ = 18 V, v_{ab}, v_{cd}, i_p and v_{cr}. Soft-switching characteristics for the primary (Q_1 to Q_4) and secondary side (Q_5 to Q_8) switches are also shown. (**b**). Simulation results in forward mode (FM) with 10% Load at $V_{in(max)}$ = 18 V, v_{ab}, v_{cd}, i_p and v_{cr}. Soft-switching characteristics for the primary (Q_1 to Q_4) and secondary side (Q_5 to Q_8) switches are also shown.

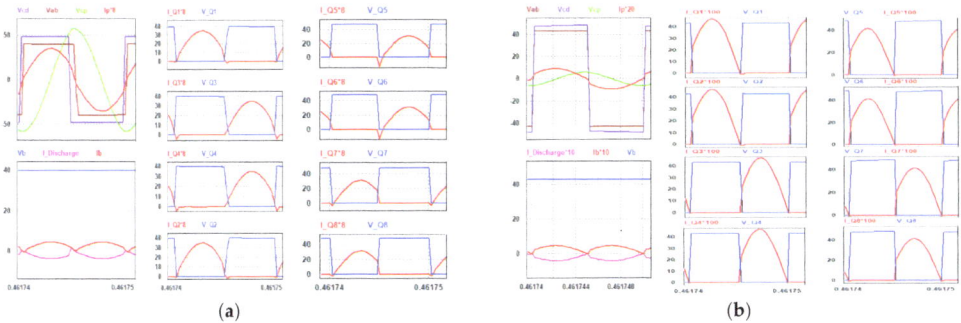

Figure 11. (**a**). Simulation results in reverse mode (RM) with 100% Load at $V_{Battery}$ = 48 V, v_{ab}, v_{cd}, i_p and v_{cr}. Soft-switching characteristics for the primary (Q_1 to Q_4) and secondary side (Q_5 to Q_8) switches are also shown. (**b**). Simulation results in RM mode with 10% Load at $V_{Battery}$ = 48 V, v_{ab}, v_{cd}, i_p and v_{cr}. Soft-switching characteristics for the primary (Q_1 to Q_4) and secondary side (Q_5 to Q_8) switches are also shown.

4.2. Experimental Results

To study the proposed converter's performance, an experiment has been conducted. Experimental results are taken at $V_{in(min)}$ and V_o for a variation in load from 100% to 10% in both forward and reverse modes of operation. The specifications and designed values are shown in Tables 1 and 2. The experimental waveforms are presented in forward and reverse modes. The forward mode has two cases, Case 1 (Full load): square-wave voltages (v_{ab} & v_{cd}) across the terminals 'a', 'b' and 'c', 'd' are shown in Figure 12a along with the primary current and voltage across the resonant capacitor. The voltages v_{ab} & v_{cd} are the resultant of a single-phase shift control technique. The current flowing through the cross-connected switches such as Q_1, Q_2 & Q_3 and Q_4, in a bridge, is the same. The waveforms showing the ZVS for the four switches on the primary side through the complementary switches Q_1 and Q_3 are shown in Figure 12b. Similarly, the currents flowing through the switches Q_5, Q_6 & Q_7 and Q_8 are also the same. The corresponding ZCS for the four switches is shown in Figure 12c through switches Q_5 and Q_7. The load voltage, load current and input current are shown in Figure 12d, along with the voltage at the boost stage of the converter at 100% load. Case 2 (Full Load): The waveforms, as mentioned above, are shown in Figure 12e–h in the same order at 10% of full load. The reverse mode of operation waveforms are shown in Figure 13. Case 1 (10% of full load): Square waveforms v_{ab} & v_{cd} with an input voltage V_2 = 48 V and an output voltage V_o = 42.5 V are shown in Figure 13a along with the primary current i_p and resonant capacitor voltage v_{cr}. The waveforms showing ZVS for the switches (Q_5 to Q_8) on the primary side in the reverse mode of operation through complementary switches are shown in Figure 13b. The ZCS for the secondary side switches (Q_1 to Q_4) is shown in Figure 13c through the voltage and current waveforms of switches Q_1 and Q_4. The output voltage, load current and input current or discharge current in the reverse mode of operation are shown in Figure 13d. Case 2 (10% of full load): As mentioned above in the reverse mode of operation, the waveforms are shown in Figure 13e–h in the same order for 10% of the full load.

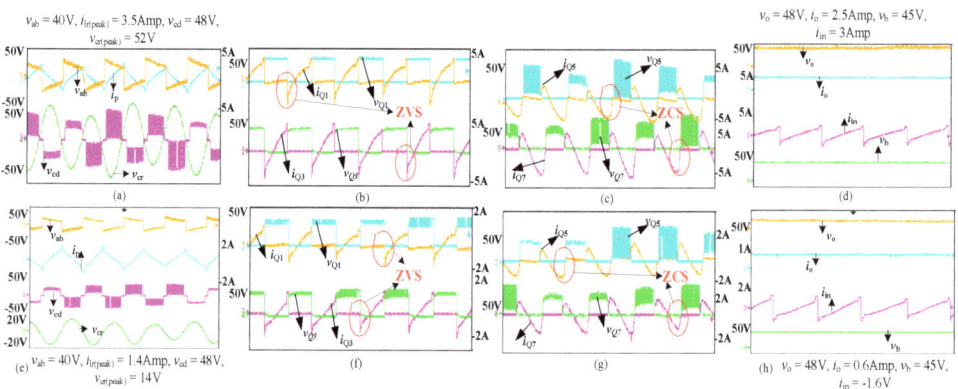

Figure 12. Forward Mode Case 1: At full load (**a**) Waveforms of v_{ab}, i_p, v_{cd}, and v_{cr} (**b**) waveforms showing ZVS for switches Q_1 & Q_3 (**c**) waveforms showing ZCS for switches Q_5 & Q_7 (**d**) Waveforms of v_o, i_o, v_b, and i_{dis}; Case 2: (**e**) Waveforms of v_{ab}, i_p, v_{cd} and v_{cr} (**f**) waveforms showing ZVS for switches Q_1 & Q_3 (**g**) waveforms showing ZCS for switches Q_5 & Q_7 (**h**) Waveforms of v_o, v_b, i_o and i_s.

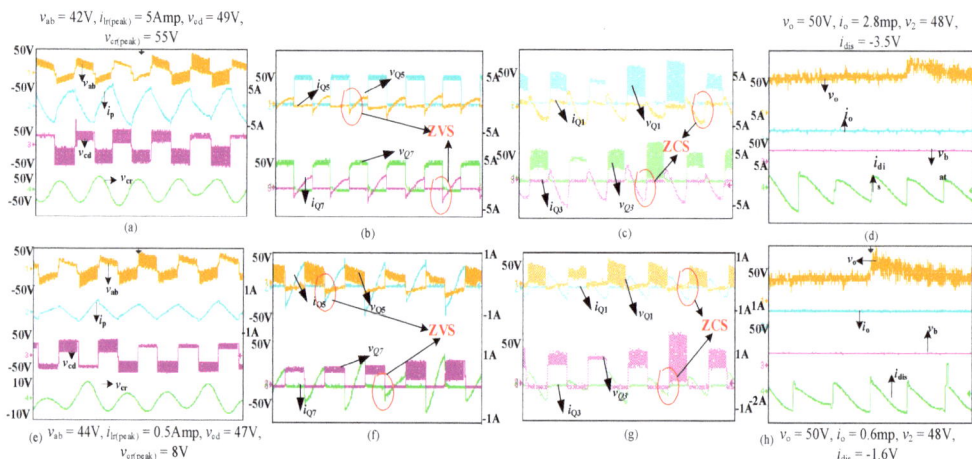

Figure 13. Reverse Mode Case 1: At full load (**a**) Waveforms of v_{ab}, i_p, v_{cd} and v_{cr} (**b**) waveforms showing ZVS for switches Q_5 & Q_7 (**c**) waveforms showing ZCS for switches Q_1 & Q_3 (**d**) Waveforms of v_o, v_b, i_o and i_s; Case 2: (**e**) Waveforms of v_{ab}, v_{cd}, i_p and v_{cr} (**f**) waveforms showing ZVS for switches Q_5 & Q_7 (**g**) waveforms showing ZCS for switches Q_1 & Q_3 (**h**) Waveforms of v_o, v_b, i_o and i_s.

5. Comparison of Results

The performance of the proposed current fed series resonant dual active bridge converter with inductive output filter is analysed by comparing various losses as shown in Figure 14. The efficiecny of proposed converter at various loads is shown in Figure 15, and efficiency comparison of various topologies [7,8,10,11] is shown in Figure 16. The current stress of the proposed converter with the existing dual active bridge (DAB) converters of different topologies [7,8,10,11] is shown in Figure 17. The conventional DAB converter [7] without resonance achieved 93.5% efficiency at full load and 81% efficiency at 10% of full load. However, the converter is operated in discontinuous conduction mode to avoid the circulating current. The voltage-fed dual active-bridge converter with series resonance [11] has been operated with a single-phase shift control technique, and an efficiency of 95% at full load and 77% at 10% of full load is obtained. It has a circulating current at both full load and light load conditions. The non-resonant voltage-fed dual active-bridge (DAB) converter given in [8] has been operated with a zero circulating current modulation scheme. The proposed converter has achieved 94.5% efficiency at full load and 88% efficiency at 10% of full load. This modulation scheme results in low efficiency at light load as the converter is operated in discontinuous mode. The proposed current-fed series-resonant DAB converter has achieved 94.6% efficiency at full load and 83.02% efficiency at 10% of full load. In addition, this converter has achieved 96.31% efficiency at 50% of full load with an input voltage $V_{in(max)}$ = 18 V and V_o = 48 V. The efficiency comparison of various topologies with the proposed current-fed isolated bidirectional DC-DC converter (CFIBDC) is given in Table 4. The proposed converter operated with soft switching for a wide range of loads; the reduced loading conditions are generally due to the peak switch currents (current stress) not reducing with the load.

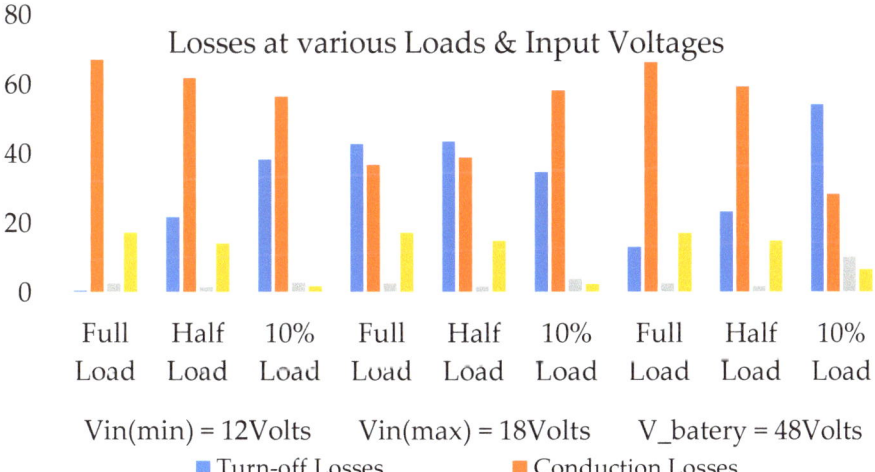

Figure 14. Conduction, turn-off, transformer and body diode losses of the CFIBDC converter at various input voltages ($V_{in(min)}$, $V_{in(max)}$ and V_o) and loads (100%, 50% and 10% of load).

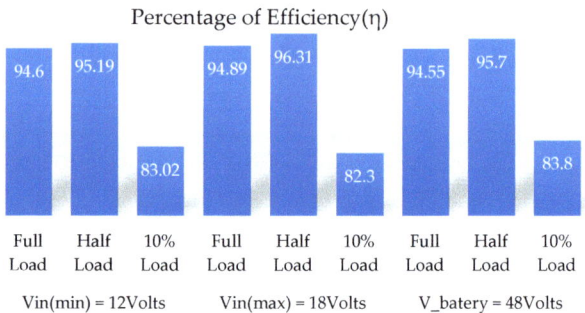

Figure 15. Efficiency of the current-fed isolated bidirectional DC-DC converter (CFIBDC) at various input voltages ($V_{in(min)}$, $V_{in(max)}$ and V_o) and loads (100%, 50% and 10% of load).

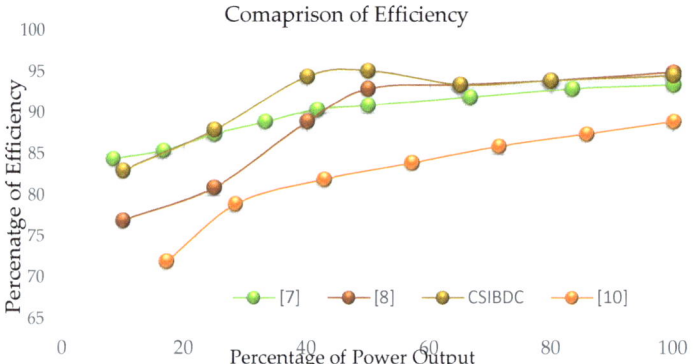

Figure 16. Comparison of efficiencies of various dual active bridge converters with CFIBDC converter.

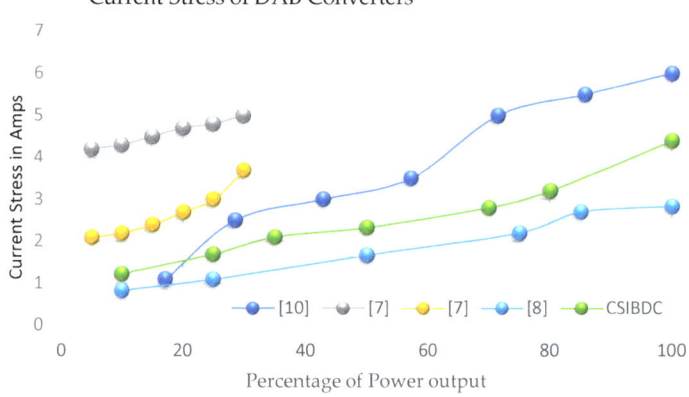

Figure 17. Comparison of current stress of various dual active bridge converters with CFIBDC converter.

Table 4. Comparison of efficiencies of different topologies with proposed current-fed isolated bidirectional DC-DC converter (CFIBDC) (Percentage of output power (% of P_o) Vs. Efficiency of Topologies given with reference number).

% of P_o	[7]	[8]	[10]	CFIBDC
10	84.5	77.5	69	83.02
20	86.5	79.5	74	86.5
30	88	83.5	89.5	90
40	90.5	89	82	94.5
50	91	92.5	83.5	95.19
60	91.5	92.5	84	94.3
70	92	93	86	94
80	93	93.5	87	94.2
90	93.2	94	88	94.3
100	93.5	95	89	94.6

The current stress of various topologies compared to that of the proposed CFIBDC converter is shown in Table 5. High current stress on the switching devices leads to more conduction losses of the switches, which in turn reduces the efficiency of the converter and requires a more oversized heat sink. The current stress for a conventional dual active-bridge (DAB) converter is less during the light load condition, but it increases gradually. The conventional DAB [7] has been operated with two inductors, L_1 and L_2, to increase the light load efficiency of the DAB converter. If this converter uses only one inductor, L_1, it will have higher current stresses, but with two inductors, $L_1 + L_2$, it offers less current stress, giving higher efficiency at light load conditions. The voltage-fed series resonant DAB [11] converter offers less current stress, as shown in Figure 17. As per the percentage of full load current, the proposed current-fed series resonant converter with an inductive output filter offers lower current stresses from light load to full load compared to any other dual active-bridge converter topologies as can be seen in Figure 17.

Table 5. Comparison of current stress in different topologies with current stress of proposed converter CFIBDC (Percentage of output power (% P_o) Vs. Current of Topologies given with reference number).

% of P_o	[7] (Amp)	[8] (Amp)	[10] (Amp)	CFIBDC (Amp)
10	2.2	0.83	0.8	1.23
20	2.7	1	1.5	1.42
30	3.7	1.3	2.3	1.6
40	-	1.45	3	2
50	-	1.67	3.4	2.33
60	-	1.82	3.8	2.5
70	-	2.1	5	2.8
80	-	2.5	5.2	3.2
90	-	2.75	5.8	3.7
100	-	2.83	6	4.4

6. Conclusions

A current-fed series resonant dual active-bridge isolated bidirectional DC-DC converter with an inductive output filter has been proposed, and its steady-state analysis has been carried out. The performance of the proposed converter has been compared with voltage-fed non-resonant and series-resonant isolated bidirectional DC-DC converters with a capacitive filter, and the results are presented. The proposed converter has been operated in continuous conduction mode without loss of duty cycle. It has shown better efficiency and less current stress and zero circulating current at the load than the voltage-fed non-resonant and series resonant dual-active bridge converters. The condition required for soft switching in both forward and reverse modes of operation for both bridges is derived. Simulation and experimental results for a load variation of 100% to 10% of full load are presented. A maximum efficiency of 96.31%, along with the breakdown of losses for the converter, are presented for a 135 W converter.

Strategies can be designed to overcome the disadvantages of current-fed isolated bidirectional converters, such as charging the inductor at the instant of turn-on and the occurrence of voltage spikes across the switches during the turn-off mode. The latter appears due to the mismatch in current flowing through the boost inductor at the input and leakage inductor of the HF isolation transformer. These current-fed IBDC converters can aid in building hybrid microgrid systems by using renewable energy resources with low voltage capacities such as fuel cells, PV systems and wind energy.

Author Contributions: Conceptualization, K.B.; Investigation, D.K.; Resources, D.K.; Writing—original draft, K.B.; Writing—review & editing, N.H. All authors have read and agreed to the published version of the manuscript.

Funding: This research received no external funding.

Institutional Review Board Statement: This study does not require any ethical approval.

Informed Consent Statement: Not applicable.

Conflicts of Interest: The authors declare no conflict of interest.

References

1. Kramer, W.; Chakraborty, S.; Kroposki, B.; Thomas, H. *Advanced Power Electronic Interfaces for Distributed Energy Systems Part 1: Systems and Topologies*; No. NREL/TP-581-42672; National Renewable Energy Lab. (NREL): Golden, CO, USA, 2008.
2. Lund, H.; Munster, E. Integrated energy systems and local energy markets. *Energy Policy.* **2006**, *34*, 1152–1160. [CrossRef]
3. Chakraborty, S.B.; Kramer, K.W. *Advanced Power Electronic Interfaces for Distributed Energy Systems, Part 2: Modeling, Development, and Experimental Evaluation of Advanced Control Functions for Single-Phase Utility-Connected Inverter*; No. NREL/TP-550-44313; National Renewable Energy Lab. (NREL): Golden, CO, USA, 2008.
4. Henao-Bravo, E.E.; Ramos-Paja, C.A.; Saavedra-Montes, A.J.; González-Montoya, D.; Sierra-Pérez, J. Design method of dual active bridge converters for photovoltaic systems with high voltage gain. *Energies* **2020**, *13*, 1711. [CrossRef]
5. Pan, X.; Li, H.; Liu, Y.; Zhao, T.; Ju, C.; Gae, A.K.R. An overview and comprehensive comparative evaluation of current-fed-isolated-bidirectional DC/DC converter. *IEEE Trans. Power Electron.* **2019**, *35*, 2737–2763. [CrossRef]
6. Zhao, B.; Song, Q.; Liu, W.; Sun, Y. Overview of dual-active-bridge isolated bidirectional DC-DC converter for high-frequency-link power-conversion system. *IEEE Trans. Power Electron.* **2013**, *29*, 4091–4106. [CrossRef]
7. Karthikeyan, V.; Gupta, R. Light-load efficiency improvement by extending ZVS range in DAB-bidirectional DC-DC converter for energy storage applications. *Energy* **2017**, *130*, 15–21. [CrossRef]
8. Karthikeyan, V.; Gupta, R. Zero circulating current modulation for isolated bidirectional dual-active-bridge DC-DC converter. *IET Power Electron.* **2016**, *9*, 1553–1561. [CrossRef]
9. Karthikeyan, V.; Gupta, R. FRS-DAB converter for elimination of circulation power flow at input and output ends. *IEEE Trans. Ind. Electron.* **2017**, *65*, 2135–2144. [CrossRef]
10. Prasetya, T.; Wijaya, F.D.; Firmansyah, E. Design of Full-Bridge DC-DC Converter 311/100 V 1kW with PSPWM Method to Get ZVS Condition. *Int. J. Power Electron. Drive Syst.* **2017**, *8*, 59. [CrossRef]
11. Li, X.; Bhat, A.K.S. Analysis and design of high-frequency isolated dual-bridge series resonant DC/DC converter. *IEEE Trans. Power Electron.* **2009**, *25*, 850–862.
12. Bhat, A.K.S.; Zheng, R.L. Analysis and design of a three-phase LCC-type resonant converter. *IEEE Trans. Aerosp. Electron. Syst.* **1998**, *34*, 508–519. [CrossRef]
13. Chen, H.; Bhat, A.K.S. Analysis and design of a dual-bridge series resonant DC-to-DC converter for capacitor semi-active battery-ultracapacitor hybrid storage system. In Proceedings of the 2014 IEEE 23rd International Symposium on Industrial Electronics (ISIE), Istanbul, Turkey, 1–4 June 2014; IEEE: Piscataway, NJ, USA, 2014.

14. Chen, H.; Bhat, A.K. A bidirectional dual-bridge LCL-type series resonant converter controlled with modified gating scheme. In Proceedings of the 2016 IEEE 8th International Power Electronics and Motion Control Conference (IPEMC-ECCE Asia), Hefei, China, 22–26 May 2016; IEEE: Piscataway, NJ, USA, 2016.
15. Lee, S.; Hong, W.; Kim, T.; Kim, G.-D.; Lee, E.S.; Lee, S.-H. Voltage balancing control of a series-resonant DAB converter with a virtual line shaft. *J. Power Electron.* **2022**, *22*, 1347–1356. [CrossRef]
16. Yang, J.; Zhang, Y.; Wu, X. Minimum Current Optimization of DBSRC Considering the Dead-Time Effect. *Energies* **2022**, *15*, 8484. [CrossRef]
17. Biswas, M.; Biswas, S.P.; Islam, R.; Rahman, A.; Muttaqi, K.M. A New Transformer-Less Single-Phase Photovoltaic Inverter to Improve the Performance of Grid-Connected Solar Photovoltaic Systems. *Energies* **2022**, *15*, 8398. [CrossRef]
18. Deng, J.; Wang, H. A hybrid-bridge and hybrid modulation-based dual-active-bridge converter adapted to wide voltage range. *IEEE J. Emerg. Sel. Top. Power Electron.* **2019**, *9*, 910–920. [CrossRef]
19. Hebala, O.M.; Aboushady, A.A.; Ahmed, K.H.; Abdelsalam, I.; Burgess, S.J. A New Active Power Controller in Dual Active Bridge DC-DC Converter with a Minimum-Current-Point-Tracking Technique. *IEEE J. Emerg. Sel. Top. Power Electron.* **2020**, *9*, 1328–1338. [CrossRef]
20. Jiang, C.; Liu, H. A novel interleaved parallel bidirectional dual-active-bridge DC-DC converter with coupled inductor for more-electric aircraft. *IEEE Trans. Ind. Electron.* **2020**, *68*, 1759–1768. [CrossRef]
21. Majmunović, B.; Maksimović, D. 400–48-V Stacked Active Bridge Converter. *IEEE Trans. Power Electron.* **2022**, *37*, 12017–12029. [CrossRef]
22. Yamada, R.; Hino, A.; Wada, K. Improvement of Efficiency in Bidirectional DC-DC Converter with Dual Active Bridge Using GaN-HEMT. In Proceedings of the 2022 International Power Electronics Conference (IPEC-Himeji 2022-ECCE Asia), Himeji, Japan, 15–19 May 2022; IEEE: Piscataway, NJ, USA, 2022.
23. Hou, N.; Zhang, Y.; Li, Y.W. A load-current-estimating scheme with delay compensation for the dual-active-bridge DC-DC converter. *IEEE Trans. Power Electron.* **2021**, *37*, 2636–2647. [CrossRef]
24. Haneda, R.; Akagi, H. Power-Loss Characterization and Reduction of the 750-V 100-KW 16-KHz Dual-Active-Bridge Converter With Buck and Boost Mode. *IEEE Trans. Ind. Appl.* **2021**, *58*, 541–553. [CrossRef]
25. Sha, D.; Wang, X.; Chen, D. High-efficiency current-fed dual active bridge DC-DC converter with ZVS achievement throughout full range of load using optimized switching patterns. *IEEE Trans. Power Electron.* **2017**, *33*, 1347–1357. [CrossRef]
26. Xuewei, P.; Rathore, A.K. Novel interleaved bidirectional snubberless soft-switching current-fed full-bridge voltage doubler for fuel-cell vehicles. *IEEE Trans. Power Electron.* **2013**, *28*, 5535–5546. [CrossRef]
27. Pan, X.; Rathore, A.K. Comparison of bi-directional voltage-fed and current-fed dual active bridge isolated DC-DC converters low voltage high current applications. In Proceedings of the 2014 IEEE 23rd International Symposium on Industrial Electronics (ISIE), Istanbul, Turkey, 1–4 June 2014; pp. 2566–2571.
28. Bathala, K.; Kishan, D.; Harischandrappa, N. Current Source Isolated Bidirectional Series Resonant DC-DC Converter for Solar Power/Fuel Cell and Energy Storage Application. In Proceedings of the IECON 2021–47th Annual Conference of the IEEE Industrial Electronics Society, Toronto, ON, Canada, 13–16 October 2021; IEEE: Piscataway, NJ, USA, 2021.
29. Wu, H.; Sun, K.; Li, Y.; Xing, Y. Fixed-frequency PWM-controlled bidirectional current-fed soft-switching series-resonant converter for energy storage applications. *IEEE Trans. Ind. Electron.* **2017**, *64*, 6190–6201. [CrossRef]
30. Wu, H.; Ding, S.; Sun, K.; Zhang, L.; Li, Y.; Xing, Y. Bidirectional soft-switching series-resonant converter with simple PWM control and load-independent voltage-gain characteristics for energy storage system in DC microgrids. *IEEE J. Emerg. Sel. Top. Power Electron.* **2017**, *5*, 995–1007. [CrossRef]
31. Justo, J.J.; Mwasilu, F.; Lee, J.; Jung, J.-W. AC-microgrids versus DC-microgrids with distributed energy resources: A review. *Renew. Sustain. Energy Rev.* **2013**, *24*, 387–405. [CrossRef]
32. Mun, S.-H.; Choi, S.-W.; Hong, D.-Y.; Kong, S. J.; Lee, J.-Y. Three-phase 11 kW on-board charger with single-phase reverse function. *J. Power Electron.* **2022**, *22*, 1255–1264. [CrossRef]
33. Qin, Z.; Shen, Y.; Loh, P.C.; Wang, H.; Blaabjerg, F. A dual active bridge converter with an extended high-efficiency range by DC blocking capacitor voltage control. *IEEE Trans. Power Electron.* **2017**, *33*, 5949–5966. [CrossRef]
34. Mazumder, S.K.; Rathore, A.K. Primary-side-converter-assisted soft-switching scheme for an AC/AC converter in a cycloconverter-type high-frequency link inverter. *IEEE Trans. Ind. Electron.* **2010**, *58*, 4161–4166. [CrossRef]
35. Harischandrappa, N. High-Frequency Transformer Isolated Fixed-Frequency DC-DC Resonant Power Converters for Alternative Energy Applications. Ph.D. Thesis, University of Victoria, Victoria BC, Canada, 2015.
36. Harischandrappa, N.; Bhat, A.K.S. A 10 kW ZVS Integrated Boost Dual Three-Phase Bridge DC-DC Resonant Converter for a Linear Generator-Based Wave-Energy System: Design and Simulation. *Electronics* **2019**, *8*, 115. [CrossRef]

Disclaimer/Publisher's Note: The statements, opinions and data contained in all publications are solely those of the individual author(s) and contributor(s) and not of MDPI and/or the editor(s). MDPI and/or the editor(s) disclaim responsibility for any injury to people or property resulting from any ideas, methods, instructions or products referred to in the content.

Article

Practical Nonlinear Model Predictive Control for Improving Two-Wheel Vehicle Energy Consumption

Yesid Bello [1,2], Juan Sebastian Roncancio [1,2], Toufik Azib [1], Diego Patino [2], Cherif Larouci [1,*], Moussa Boukhnifer [3], Nassim Rizoug [1] and Fredy Ruiz [4]

[1] Energy and Embedded Systems for Transportation Research Department, ESTACA-LAB, 78066 Montigny-Le-Bretonneux, France
[2] Javeriana Electronics Department, Pontificia Universidad, Bogotá 110231, Colombia
[3] Université de Lorraine, LCOMS, F-57000 Metz, France
[4] Systems and Control Department Italy, Politecnico de Milano, 20158 Milan, Italy
* Correspondence: cherif.larouci@estaca.fr

Abstract: Increasing the range of electric vehicles (EVs) is possible with the help of eco-driving techniques, which are algorithms that consider internal and external factors, like performance limits and environmental conditions, such as weather. However, these constraints must include critical variables in energy consumption, such as driver preferences and external vehicle conditions. In this article, a reasonable energy-efficient non-linear model predictive control (NMPC) is built for an electric two-wheeler vehicle, considering the Paris-Brussels route with different driving profiles and driver preferences. Here, NMPC is successfully implemented in a test bed, showing how to obtain the different parameters of the optimization problem and the estimation of the energy for the closed-loop system from a practical point of view. The efficiency of the brushless DC motor (BLCD) is also included for this test bed. In addition, this document shows that the proposal increases the chance of traveling the given route with a distance accuracy of approximately 1.5% while simultaneously boosting the vehicle autonomy by almost 20%. The practical result indicates that the strategy based on an NMPC algorithm can significantly boost the driver's chance of completing the journey. If the vehicle energy is insufficient to succeed in the trip, the algorithm can guide the minimal State of Charge (SOC) required to complete the journey to reduce the driver energy-related uncertainty to a minimum.

Keywords: NMPC; two wheel; electric vehicle; eco-driving profile; efficiency; optimization; autonomy increasing

1. Introduction

Technology is improving the efficiency and use of electric automobiles due to the increasing number of electric vehicles (EVs) on the market. EVs are becoming increasingly popular since they are considered a clean option in terms of pollution, even though they face various challenges (charging time, range, standardization of the charging process, charging stations, and the recycling process). The balance between an eco-driving profile and performance impacts the autonomy of the electric vehicle. EVs can be charged heavily using renewable energy systems, which means fewer carbon emissions and more use of renewable energy [1]. Switching to EVs can help reduce greenhouse gas emissions and help people in some countries get ahead economically.

Advanced Driver Assistance Systems (ADAS) are algorithms that assist the driver. According to [2,3], the algorithm's main task is to set performance boundaries for a vehicle based on a given criterion. Theoretical models, historical data, or unsupervised learning systems can all be used [4–6]. When the algorithm determines that the current power demand is insufficient to complete the journey, constraints on the use of the driver modes are placed on the vehicle's performance [5]. These settings can be divided into Eco, Normal, and Sports categories. The constraints may change depending on the algorithm. In some cases, it alludes to the maximum motor power, maximum acceleration, or top speed [7,8]. There are also cases when eco-driving is computed when receiving trip time importance and presenting each road point's optimal speed [9].

Several techniques can improve the driving profile and vehicle-based autonomy [10]. The two most common are the driving alert system and the eco-driving characteristics optimizer [11]. The distinction between them is the assessment of energy needs. The driving alert system is used to notify the driver if the energy available in the vehicle is insufficient to continue the trip. In the eco-driving optimizer, energy estimation helps to decide what type of driving mode is necessary to ensure an energy consumption rate for the duration of the trip. Therefore, the bounds provided by the speed profile optimizer are constant and, in some cases, excessive concerning the actual needs of the driver. In particular weather and road conditions, the efficiency of the BLDC engine decreases from 64% to 12%. Moreover, it has been proven that the input parameters affect an electric motorcycle's dynamic characteristics and consumption characteristics [12].

In this study, evaluating autonomy involves using dynamic models and external data to ascertain whether the driver can successfully achieve the route. Considering factors such as traffic, vehicle limits, and driver preferences can determine the ideal speed profile to lower total energy use. Therefore, an eco-driving approach using a non-linear predictive controller has been developed. All-electric vehicles can use the eco-driving strategy. However, this article concentrates on two-wheeled EVs because the driver's activities account for the majority of energy consumption, and the size of the vehicles imposes additional limits that must be taken into account. In [12], a simulation model of the electric motorcycle was used to determine the velocity, propulsion torque, and electric consumption characteristics with variable electric motorcycle mass, driver mass, wheel radius, frontal area, and transmission ratio.

In this article, eco-driving algorithm design incorporates thermal, electrical, mechanical, and theoretical models to predict power usage. To do that, the non-linear model predictive controller (NMPC) is a helpful technique for prediction and optimization, for example, in speed terms, [13]. This work develops a tool that enables users to maximize energy efficiency depending on input data such as state of charge (SOC), distance traveled, elapsed time, and speed. Because of this procedure, the EV will tell the driver the best speed and driving conditions to cover the remaining distance most efficiently and cost-effectively. This research shows how to implement the NMPC and practically compute different optimization problem parameters, such as (i) Weights of the cost function, (ii) Necessary physical bounds in the constraints, and (iii) Efficiency of the motor concerning torque and speed. This article gives a deeper understanding of what the authors proposed in [14] and presents experimental results through a testbed platform. Here, it is shown how the test bed is built and adapted to emulate the route Paris-Brussels. Practical implementation in the test bed is crucial since it validates results from theory. Results indicate that the strategy based on an MPC algorithm can significantly boost the driver's chance of completing the journey. If the vehicle's energy is insufficient to succeed in the trip, the algorithm can guide the minimal State of Charge (SOC) required to complete the journey to reduce the driver's energy-related uncertainty to a minimum. It gives results that can be achieved afterward to integrate different power sources [15].

The rest of the paper is organized as follows. Section 2 presents the mathematical model used in this work. Section 3 explains the closed-loop controller considering perturbations. The experimental platform development that served as the basis for the present research can be detailed in Section 4. The model validation is presented in Section 5, and the experimental results and some findings of this research are presented in Section 6. Finally, conclusions and future studies are given in Section 7.

2. Mathematical Model

This section summarizes the mechanical model used to represent the degrees of freedom of the vehicle motion. All the variables are listed at the beginning of this article. The longitudinal model examines the accumulation of forces along the X-axis. Longitudinal dynamics are set as shown in Figure 1. Other phenomena, such as wheel slip and angular velocity, are excluded from this study [16–18]. According to [16], Newton's second law of motion in the X-axis is stated in Equation (1).

$$m\ddot{x} = \frac{T}{R_{wf}} + F_{roll} + F_{aer_x} + F_w sin(\theta_s) sin(\beta) \tag{1}$$

Considering that the aerodynamic force, rolling resistance force, and road slope are all critical forces in a vehicle's longitudinal motion [19,20], the aerodynamic force and Rolling Resistance Force are stated as in (2) and (3), respectively.

$$F_{aer} = \frac{-1}{2}\rho C_d A_f (V_{wind} * cos(\alpha_{air}))^2 \tag{2}$$

$$F_{roll} = -(\mu_0 + \mu_1 \dot{x}^2) F_z cos(\theta_s) \tag{3}$$

Road friction parameters can take values from 0.001 to 0.00082 according to the road conditions (new, wet, frozen, etc.) [20]. β is assumed used to be near 90°. Road Slope is stated as $F_w = mg$. The total traction force, from the right-hand of Equation (1), is computed based on a power profile associated with θ_s. This process is vital for comparing the performance of different vehicles since the energy demand depends indirectly on speed values and the energy required by the EV to maintain a desired speed, for example, if there are considered different road classifications (urban or rural), climate, and driving skills can also be included in driving profiles.

The driving profile is the representation of vehicle speed vs. time. The shapes are advantageous to designing, calibrating, or improving the test. Different driving cycles exist, such as (EPA - United States Environmental Protection Agency, WLTP, and NEDC—created by UNECE World Forum for Harmonization of Vehicle Regulations, Artemis—cycles created during the Artemis project in EU) [21]. Figure 2 shows six different driving profiles, which is helpful for the research.

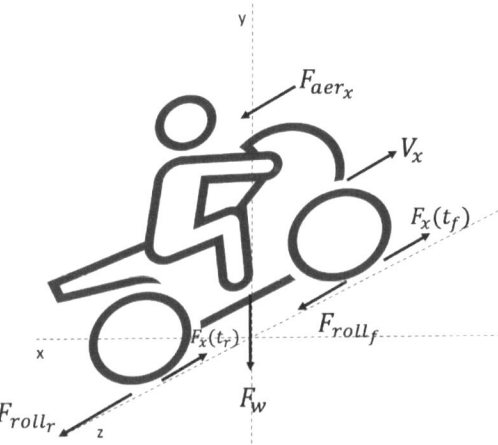

Figure 1. Force Diagram.

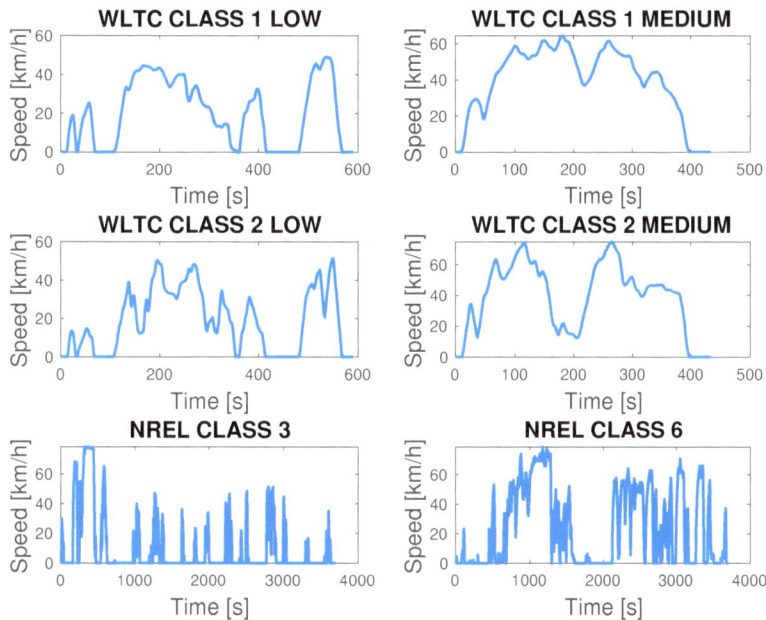

Figure 2. Driving Cycles for Evaluation.

Therefore, the energy used is a function of how a vehicle of 150 kg is affected by outside forces and how well it works electrically. The study's target route is Paris to Brussels, and Google Maps determines the slope profile and wind speed data. Its analysis is required to improve the performance of the controller. Moreover, according to [22], energy losses may exist in various electrical or mechanical devices. Indeed, power losses are caused by the BLDC motor's electrical and magnetic properties. Magnetic losses are caused by temperature fluctuations in the motor's magnetic material. There are also losses which are due to power loss in the winding. When the winding is loaded, core losses result from the unintended magnetization of the core via inductive action [23–26].

Consequently, an efficiency map's geometric representation can orient the optimal controller in the correct direction and reduce the energy estimation error [18,27]. In [14], a "Cauer network" is proposed in each circumstance to alter the resistance and capacitance values anticipated by electrical models under temperature changes. The battery losses are incorporated into the model, allowing for consideration of the internal losses of the battery and their temperature-dependent change. Then, the expected error in energy use can be kept below 5% without making the electrical models challenging to understand [28,29]. Therefore, according to [14], energy E can be modeled as stated in (4).

$$\dot{E} = \frac{T}{R_{wf} Eff(\dot{x}, T)} \dot{x} \qquad (4)$$

where $Eff(\dot{x}, T)$ represents the geometrical abstraction of the motor efficiency that shows how the vehicle can behave in the real world [14].

3. Model Predictive Control

This section shows a proposal for a Non-linear model predictive control (NMPC). Its objective is to determine the proper torque T to drive a given distance using the least feasible energy. Figure 3 depicts the implemented controller. It has a prediction horizon of $N = 10$. These values balance the compilation time and speed/acceleration dynamics.

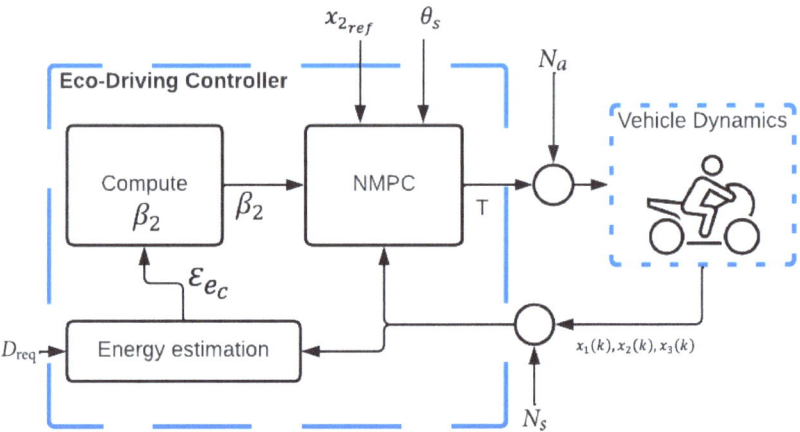

Figure 3. Control Diagram.

The proposed controller Figure 3 calculates the energy required to finish the trip ε_{e_c} based on the remaining distance traveled D_{req} and the current vehicle dynamics. External variables related to noise from the N_a actuators and N_s sensors are also included in the control loop. The vehicle's driver determines the speed $x_{2_{ref}}$. However, based on the driver's behavior and the estimated energy path, the system suggests the optimal driving style β_2 to be adopted by the driver. The algorithm implemented by the NMPC controller is shown in Figure 4. There are both external and internal sources of errors. External factors, such as traffic signals, influence the behavior of a vehicle driver. Internal sources include sensors and actuators and can also alter the operation depending on their performance. The closed-loop controller must also be able to reduce the impact of disturbances on vehicle energy performance.

The estimated speed profile is responsive to the driver's desires but bounds the maximum speed values and tends to bring the average speed closer to the most efficient speed value based on the torque required. The optimal problem solved by the NMPC is stated in Equations (5)–(11).

$$\min \quad J(x, T, k) \tag{5}$$

$$J(x, T, k) = \beta_1 \beta_2 \phi_1(x_3(N)) + \sum_0^N -\beta_2 \phi_2(x_2(k), T(k)) + (x_2(k) - x_{2_{ref}})^2 \tag{6}$$

$$\text{s.t.} \quad T_{min} \leq T(k) \leq T_{max} \tag{7}$$
$$0 \leq x_1(k) \leq x_{1_f} \tag{8}$$
$$x_{2_{min}} \leq x_2(k) \leq x_{2_{max}} \tag{9}$$
$$0 \leq x_3(k) \leq TotEne \tag{10}$$
$$x(k+1) = f(x(k), T) \tag{11}$$

where (5) and (6) represent the objective function of the NMPC, β_1 and β_2 are parameters for the optimization problem. β_2 is computed to be used within the forecast horizon without exceeding the vehicle or system capabilities and the estimation of the energy to finish the journey ε_{e_c} [14]. Constraint (7) is the physical constraint that describes the maximum force mechanical structures can generate, including the maximum torque the system can produce. The vehicle cannot replicate the behavior if the states are outside this feasible region. Constraints (8) and (9) impose the feasible region and avoid an over-damped behavior; constraint (10) is a physical limitation of the quantity of energy available in the battery. (4), $x(k)$ is a vector composed by the discrete variable of x, $\dot{x}$ and E. Finally, ϕ_1 to

ϕ_2 are stated in (12) and (13) and (11) is a simplified discrete representation of the model given in (1).

$$\phi_1(x_3(k)) = x_3(k) \tag{12}$$
$$\phi_2(x_2(k), T(k)) = Eff(x_2, T) \tag{13}$$

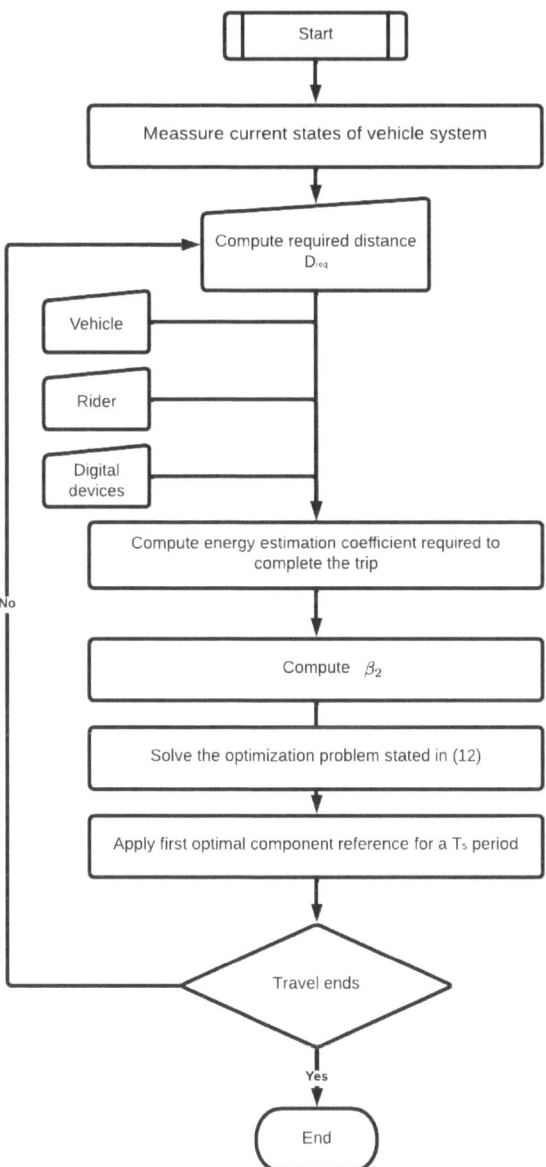

Figure 4. NMPC Algorithm.

4. Experimental Platform

One of the objectives of this research is to propose a viable method for estimating efficiency values in real-time of EV. In addition, it is necessary to ensure the accuracy of the estimation process in the control algorithm. For achieving accuracy and viability, a scale motorcycle model is built with a test bench to validate the proposed method (see Figure 5). Furthermore, the geometric representation can be modified depending on the engine parameters to obtain any efficiency point from the hyperbolic Equation (14). In the test, two main characteristics are altered. First, the engine power is decreased due to safety constraints, and the speed profile used is WLTP in urban, rural, and mixed environments. Thus, this section describes the mechanical specifications of the motorcycle platform and the framework required to reflect internal and external events during operation accurately. Experiments are conducted to determine the controller parameters indicated in the previous section.

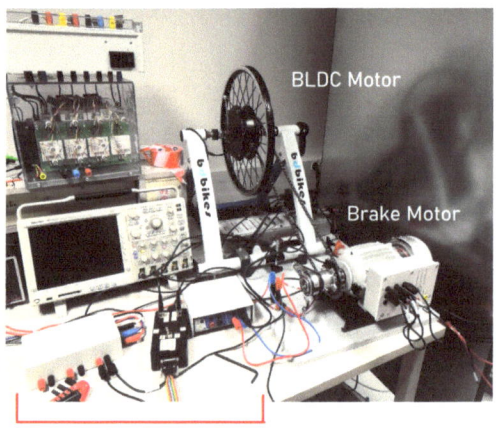

Figure 5. Practical Test bench image.

4.1. Specifications

The test bench consists of a BLDC motor that replicates the motorcycle's traction (traction motor). The traction motor is supported by a framework that ensures its integrity even at maximum torque and speed. In addition, a system that may be adjusted to define several torque profiles due to wind, slope, or weight variations is incorporated. In this method, opposing forces are simulated by a torque generator by a 300-Watt auxiliary engine connected to the traction motor through a generator-equipped bicycle chain. Both electric machineries (traction motor and generator) are coupled to a 32-toothed cycling disc. A Zenon load AL3008BLDC-200V-10KW MD2-1.06 is used to change the torque of the generator. The Semikron 08753450/309 inverter interacts with the generator and the load. This component rectifies the current profile of the asynchronous motor to supply the changeable load with the current. Figure 5 depicts the experimental setup.

A BLDC1200 W from the OZO Speed Donkey kit with a maximum power rating of 1200 W, rated at 1000 W, and bounded to 25 A, is used. This kit converts conventional motorcycles to electricity, and the maximum torque is 98 Nm and the maximum top speed is 30 km/h (due to the power supply of 36 V). It includes a field-oriented control mechanism (FOC) and the option to contain or exclude Hall effect sensors. Here, dSPACE DS1104 real-time controller board is the hardware and software interface. The traction is measured using the encoder XCC1514TSM02Y with a constraint of 6000 rpm as the maximum revolution speed. Figure 6 shows the electric integration between different elements.

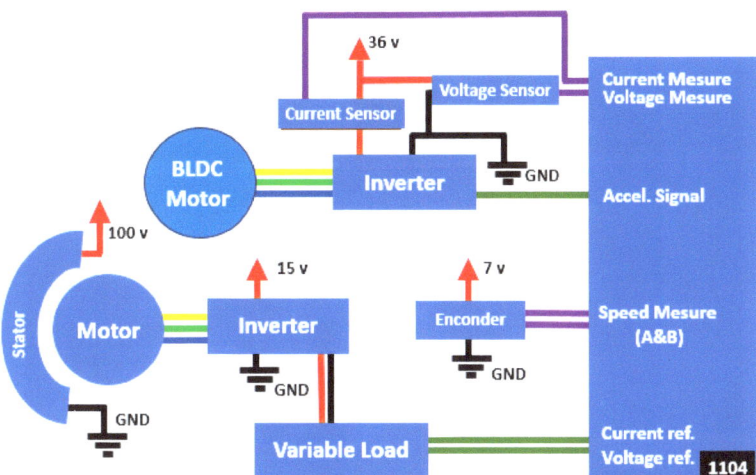

Figure 6. Electrical diagram of the test bench.

In determining the relationship between the voltage of the throttle signal and the speed of the traction motor, the signal range is determined, and the throttle signal and traction motor speed are compared. The measured voltage range for the unaltered throttle signal is between 1.35 V and 2.65 V. This value means that the traction motor speed can be controlled with a V_{ref} from 0 to 12 m/s. This values determines v_{min} and v_{max}. The equation which describes the data behavior is $V_{ac} = V_{ref} * (1/9.23) + (12.4/9.23)$, where V_{ac} and V_{ref} are the acceleration in volts and velocity in meters per second, the behavior is showed in Figure 7.

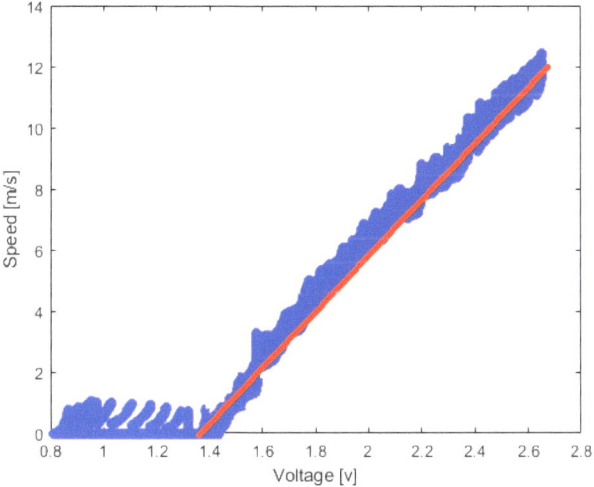

Figure 7. Accelerator vs. motor speed voltage behavior.

It is important to consider noise due to hardware and software components. There are two significant noise sources, the interaction between the sensor and the analog-to-digital converter on the DSpace board and the exchange between the actuator and the digital-to-analog converter. Furthermore, they can produce system oscillation.

The characteristics of the test are shown in Table 1 and in [30]. The sensors have errors of 0.12 and 0.02, respectively. Moreover, the actuator error resulting from throttle and driver variations has an RMS error of 0.4, which is incorporated into the ADC resolution, which can affect the open loop motor controllability is essential.

Table 1. Parameters related to Test bed [30].

Device	Parameter	Value
Motor System	Technology	BLDC
	Nominal power	1200 [W]
	Max power	1365 [W]
	Max speed	40 [km/h]
	Max torque	98 [Nm]
	Wheel diameter	21 [cm]
Braking System	Technology	Brushless
	Nominal power	300 [W]
	Max current	0.95 [A]
	Max speed	1500 [rpm]
BLDC inverter	Control technique	full wave rectifier
	Nominal voltage	15 [V]
	Max voltage	56 [V]
	Max output current	25 [A]
Brake inverter	Control technique	Full wave rectifier
	Nominal control voltage	15 [V]
	Nominal voltage	200 [V]
	Max output current	25 [A]
Encoder	Encoder type	Incremental encoder
	Shaft diameter	14 [mm]
	Resolution	256 to 4096 [points]
	Max speed	60,006 [rpm]
	Voltage Supply	5 to 30 [V]- DC
Voltage sensor	Range	500 [V]
	Resolution	100 [mV]
	Voltage Supply	220–240 [V]-AC
Current sensor	Range	50 [A]
	Resolution	10 [mA]
	Supply voltage	220–240 [V] - AC
Variable load	Voltage range	10–1000 [V]
	Voltage resolution	100 [mV]
	Max input power	14 [kW]
	Current resolution	100 [mA]
	Voltage Supply	220–240 [V] tri phases
DSpace card	I/O range	−10 to 10 [V]
	I/O resolution	1 [mV]
	I/O max current	0.005 [A]
	Voltage Supply	12 [V]

Speed profile NREL class 3 described in Section 2 gives the least sensitive energy behavior for external perturbations, with a 16% increase over the projected energy per kilometer. In addition, the WLTC class 2 velocity profile is most influenced by disturbances present on the experimental platform, with an increase in power per kilometer of 25%.

4.2. Efficiency and Bounds

The mechanical limits and efficiency maps were derived by logical programming, co-simulation, and the magnetic simulation software ANSYS/Maxwell©), such as in [31]. The assessed parameters of the electric motor are shown in Table 1 as in [30]. This initial stage allows the estimation of subsequent phases' and predicts outcomes.

Then, perform rotor blocking testing. This test attempts to impede the rotor with a progressive load. Throughout each trial, the load must exceed the motor's maximum load at maximum speed. Consequently, the rotor's speed must decrease, and the mechanical limits of the engine will be identified. This process is crucial to prevent motor damage and to remember that the maximum torque obtained during this test must be less than 90 percent of the maximum torque motor limitation. Next, the values of the efficiency maps inside the area identified by the rotor blockage test were calculated. The data are extracted from the datasheet, and the traction motor parameters are shown in Table 1. Then, construct a linear representation of the mechanical motor restrictions determined in the preceding phase. Finally, Check the maximum efficiency level. Since an initial efficiency estimate is made in step 1, verifying the maximum value by working as near as feasible to that point and measuring the efficiency value is necessary. After the co-simulation process shown in Figure 8, the rotor blocking test gradually obstructs the motor's rotor. In addition, because torque cannot be measured, the link between speed and torque shown in the depiction of simulated efficiency is used. A method for preventing exceeding 90 percent of the maximum torque is keeping the speed value at no less than 30 percent of the maximum speed. From Figure 8, it is around 160 rpm.

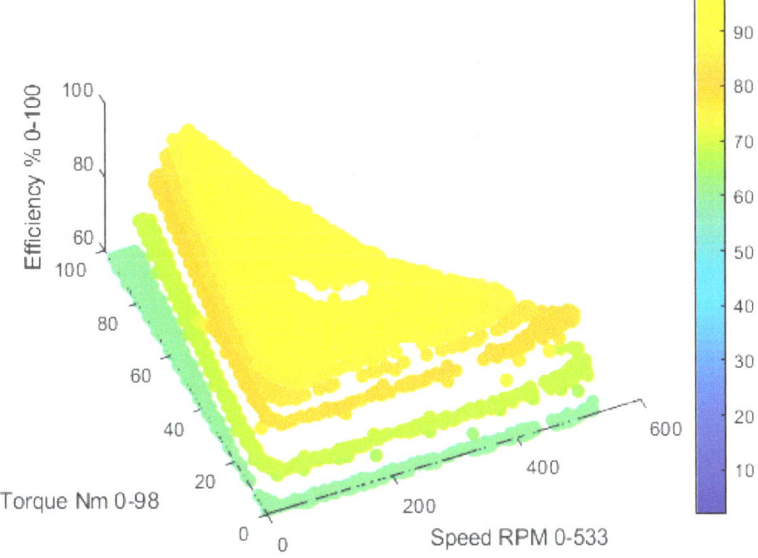

Figure 8. Efficiency map estimation result $Eff(x_2, T)$.

Due to the rotor blocking test, the speed restriction shown in Figure 9 is determined. This representation is oriented and positioned again in a plane (speed-torque) using the following information: minimum and maximum speed ($x_{2_{min}} = 0$ and $x_{2_{max}} = 12$), minimal and maximal torque ($T_{min} = 0$ and $T_{max} = 95$), and co-simulation results as shown in Figure 10.

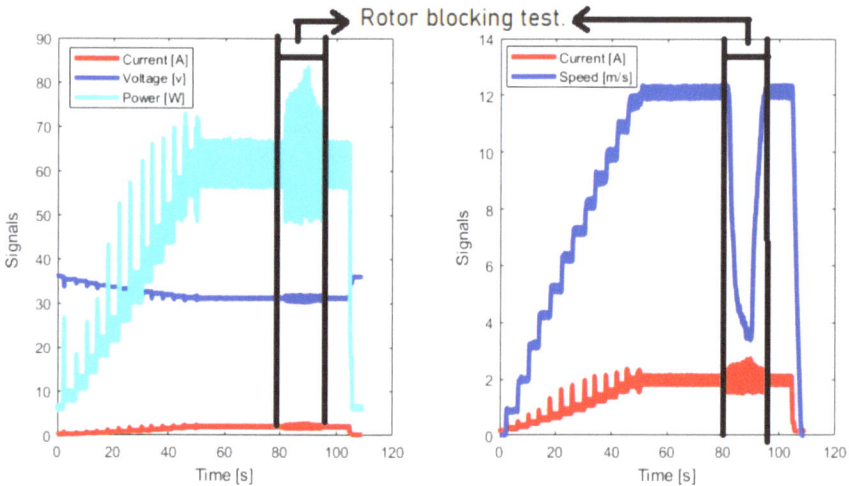

Figure 9. Experimental result of speed curve during rotor blocking test.

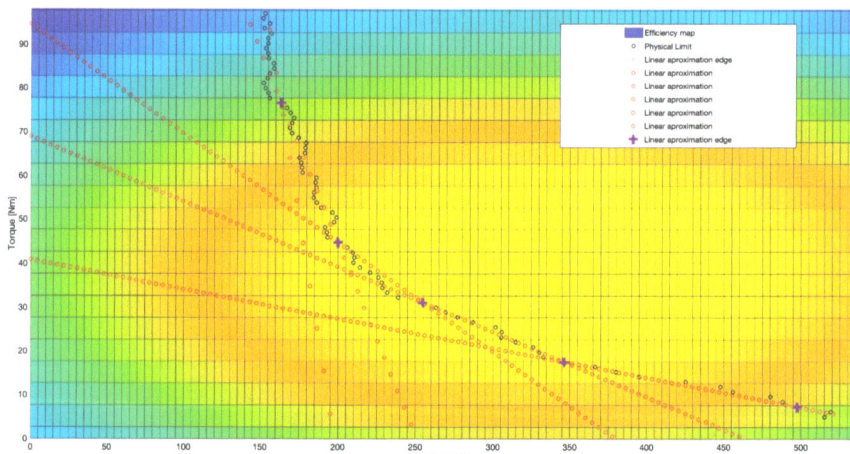

Figure 10. Simplified efficiency representation.

After obtaining the constraint speed curve, linear representations are incorporated and integrated into the estimation process. The linear expressions defined by the expression: $Y_i = m_i X_i + b_i$. Where Y_i is the torque, is X_i is the velocity i-st, and b_i, m_i are the Y-axis intercept and the slope. The obtained values are presented in Table 2, which shows the linear representations of the physical limits of the system. It is essential to mention that this approximation is made to avoid the overburden in computation time due to nonlinearity. Figure 10 shows the speed-torque constraints of the motor with the efficiency map and the linear approximations with the points where these limits start and end.

Finally, the geometrical representation needs a suitable efficiency value to orient the optimization issue from NMPC in Equations (5)–(11). To verify convergence, the bike training roller's brake is used to measure torque near maximal efficiency. Zenon variable electrical load makes modest modifications. The geometrical representation's proposed total efficiency point is inaccurate by 67 rpm in speed and 7.1 Nm in torque. The Speed-

Torque limitations curve and simulated efficiency map from Figure 8 are used to calibrate this number. Thus, the geometrical representation shifts to the maximum point. The simulation and calibration resulted in Figure 10 and Equation (14).

$$Eff(T, x_2) = B_0 - B_1(T - x_c)^2 - B_2(x_2 - y_c)^2 \tag{14}$$

where $B_0 = 92.6$, $B_1 = 1.2859e^{-4}$, $B_2 = 0.0011$, $x_c = 204$ and $y_c = 32.6$.

Table 2. Linear approximation of torque-speed constraints.

Slope	Value	Intercept	Value
m_1	−2.92	b_1	40.86
m_2	−6.45	b_2	68.96
m_3	−10.81	b_3	94.57
m_4	−37.97	b_4	219.74
m_5	−97.18	b_5	443.03

Figure 10 represents the physical constraints of the experiment. The absolute power limits of the torque speed curve, which has a non-linear nature, are depicted in black "o". Because of the non-linear limitations, an approximation to the natural limits is required; the approximation is replicated by the five red "o" curves. This recreation aims to develop a series of linear functions "function by parts" to linearize the physical limits to ensure the shortest possible computation time in the MPC structure. Each linear limit's change limits are denoted by an "x" in red.

4.3. Speed and Energy Coefficient

Once the efficiency function has been computed, the information must be included in the block "energy estimation" of Figure 3. This parameter allows you to evaluate the battery's energy quantity concerning the required distance to perform a more stringent or light estimation. The first test estimates the energy parameter along the velocity signal in the form of steps. This test allows investigation of the energy coefficient at various values to check the suggested efficiency function and demonstrate the minimal energy coefficient the controller platform can get in constant speed profiles. The current, velocity, power, and voltage signals are displayed in Figure 11 are acquired due to the test.

The power is integrated along the time corresponding to each step and divided by the distance covered to calculate the energy used in each speed signal phase in stairs shapes. Torque area test from 0 to 10 Nm is also essential. This torque spans all speeds. The driver's necessary distance and battery SOC are used to calculate the highest practicable energy coefficient needed to complete the journey. The ε_{e_c} coefficient is determined from the total energy consumed over the trip's distance and represents the energy used per kilometer. In addition, the estimation block looks at past traffic data to estimate the speed profile. Estimating the profile is vital for the Eco-Driving Controller to work in real-time and to take into account the randomness of the actual driving cycle. Figure 12 depicts the relationship between EFF and β_2. In this figure, while the β_2 increases, the energy required to achieve the trip decreases. Figure 12 shows the energy coefficient data and the ε_{e_c} energy estimation comparison with β_2.

Now that the whole efficiency function and parameters have been calculated, a test to achieve the β_2 value may be performed. This test estimates the speed profile along various variables to determine the controller's optimal performance. This test is performed without any extra disturbance (actuator and sensor disturbances proper of test bed), and the velocity profile estimates are assured of avoiding a perturbation that an open-loop controller cannot fix. As a result of the tests, the most efficient velocity in this torque range is roughly 4 m/s, which translates to 204.6 rpm. Then, increasing β_2 always ensures that the energy coefficient goes down. Lastly, Figure 12 shows the result of this test.

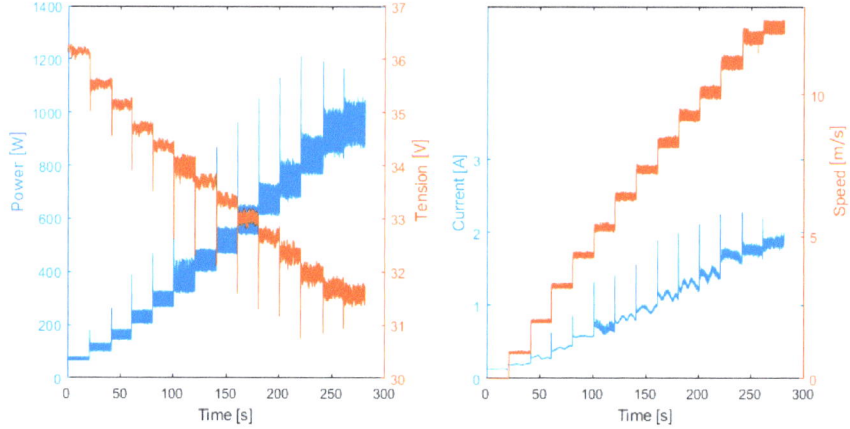

Figure 11. Speed vs. Power.

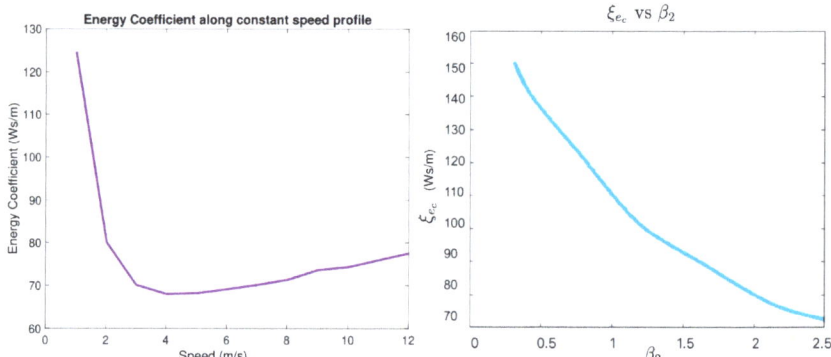

Figure 12. Energy coefficient behavior.

5. Model Validation

The approach used to validate the mathematical model with the experimental model under specific operation settings is described in the following paragraphs. In this section, it is vital to note that, according to [17], lateral dynamics are not required to indicate the vehicle's energy condition. This assumption is only valid if the speed is less than 60 km/h, and as demonstrated below, all tests are only valid for velocities less than 60 km/h; indeed, the testbed has a max speed of 40 km/h. The proposed road (see Figure 13) consists of two 50-meter-radius curves with a bank angle of 30% and a junction with an elevation of 2 meters and a slope angle of 5°. A virtual test scenario is incorporated into a vehicle dynamic simulation tool to validate the model's behavior. Two tests are conducted: one with constant and varying speeds to determine the yaw reaction at different speeds and another with variable speeds within the same range to determine the error's dynamic response.

Figure 13. Designed Road.

5.1. Constant Speed Test

This test verifies the behavior shown in Equation (1) and identifies discrepancies in the model. Using the linear pneumatic friction model, the inaccuracy of angle yaw increases rapidly. As shown in Figure 14, the estimation is deemed acceptable if the longitudinal and lateral speed has an error less than 5%.

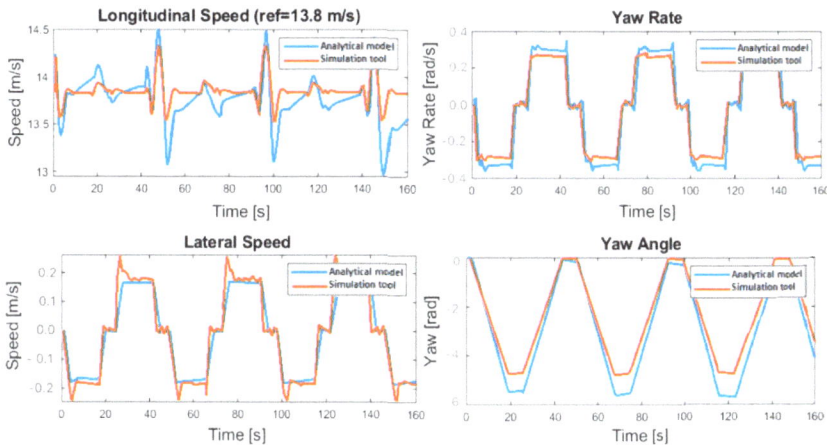

Figure 14. Static speed profile results (error percentage).

Therefore, a speed profile estimator cannot use yaw information as a constraint under these conditions. Figure 15 displays the test's error rate.

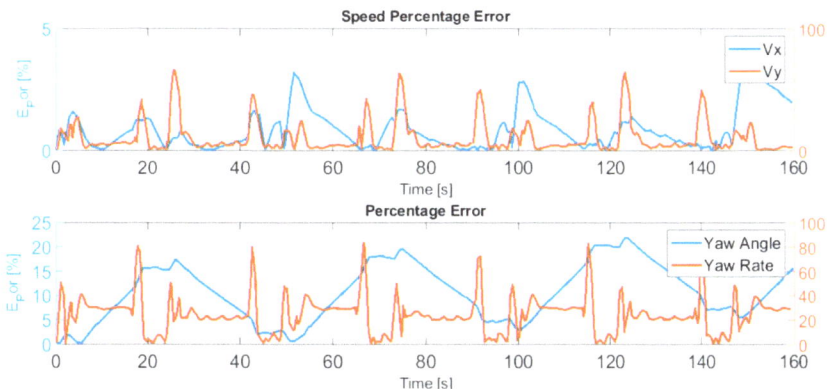

Figure 15. Static speed profile results (error percentage).

5.2. Dynamic Speed Test

Due to changing speed references, this test's error rate may rise. As illustrated in Figure 16, under dynamic speed conditions, a forecast horizon of fewer than 10 s is reasonable.

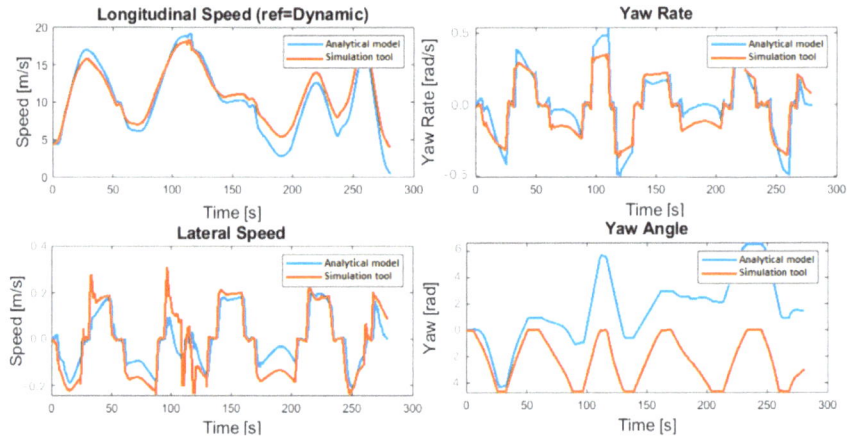

Figure 16. Dynamic speed profile results (error percentage).

Figure 17 depicts the prediction horizon for the model's present accuracy. These results demonstrate the prediction horizon in 8 s for errors in longitudinal velocity and yaw angle of less than 10%.

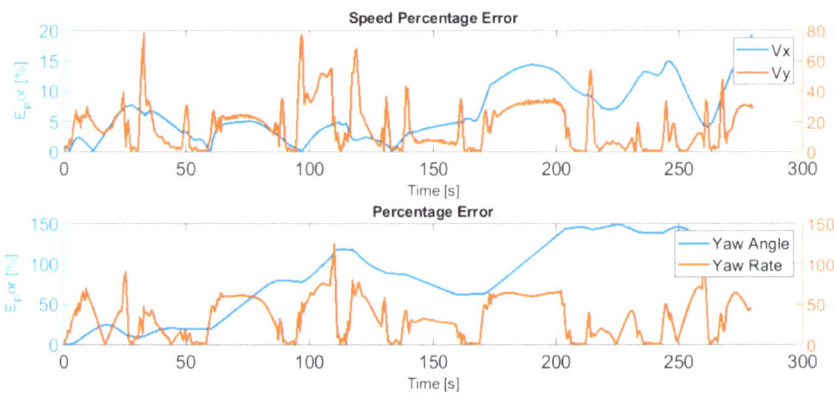

Figure 17. Dynamic speed profile results (error percentage).

To fully understand the impact of the current findings, a sensitivity analysis must be conducted to see how the error % varies as a function of the friction coefficients. However, the Pacejka equation represents the friction coefficients' non-linear behavior in the simulator, despite the model's assumption that they are constant. (see Figure 18 [17]), better results are expected with a more accurate estimation of these parameters for a 0 to 60 km/h urban drive profile.

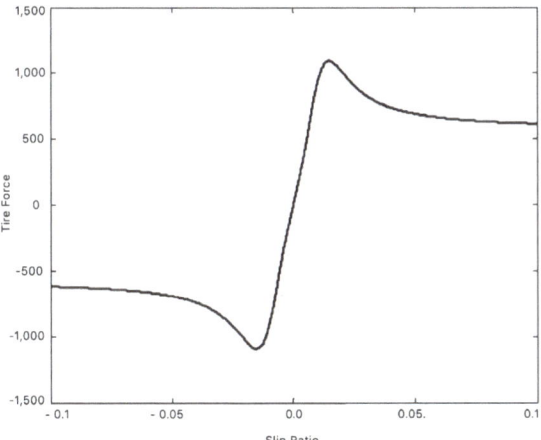

Figure 18. Tire Pressure Curve Calculator (with Magical Formula- Pacejka equation).

5.3. Simulation Results after Sensitivity Test

The sensitivity test is based on how each adjustment in friction coefficient impacts the average percentage error between the analytical model and the simulated model. The pneumatics industry has recognized physical limits to pneumatic dynamics and friction coefficients, which are described by minimum and maximum test results. It is crucial to note that in the test with an a priori estimation value, only one of the four parameters changes during the experiment. The selected delta values guarantee that the model's state does not fluctuate by more than half a percent from iteration to iteration. The average yaw error fell by 73% after new friction coefficients were implemented, as expected by the sensitivity analysis. According to the results of the static speed test (shown in Figures 19 and 20, the average error percentage is 1.15 percent for longitudinal speed and 3.40 percent for yaw angle. Even if the mean errors increase during the test, the ratio stays around 1/100 s for a longitudinal speed and 1/40 s for a yaw angle.

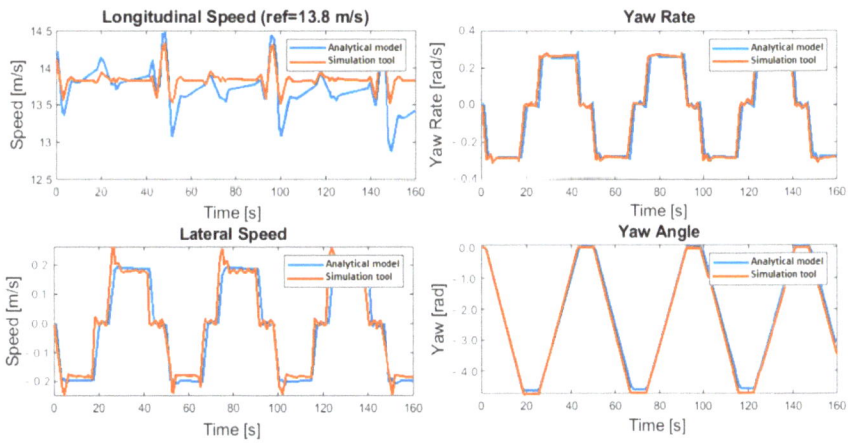

Figure 19. Static speed profile results after sensitivity test.

Moreover, the dynamic speed test in Figures 21 and 22 indicate a mean error percentage of 4.47% for longitudinal speed and 23.12% for yaw angle, with the sensitivity changes. The revised friction coefficients determined by the sensitivity test drop to 71.97% of the mean error percentage during the dynamic speed test compared to the previously reported yaw mean error percentage.

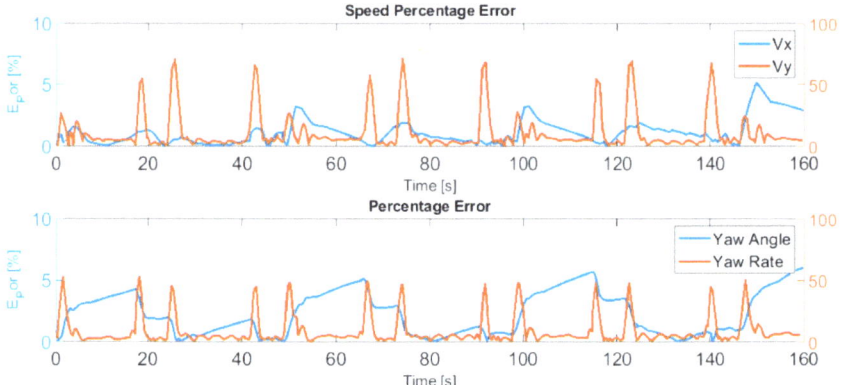

Figure 20. Static speed profile results after sensitivity test (error percentage).

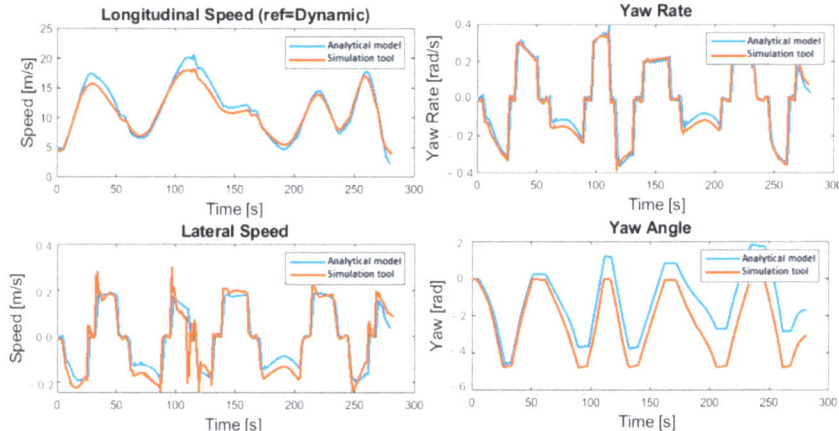

Figure 21. Dynamic speed results after sensitivity test.

As shown in Figure 22, even though the yaw angle implies the error percentage is still significant, the improved results allow for a forecast horizon of 66 s when an acceptable yaw error is less than 10%. Results may be satisfactory up to a speed of 60 km/h (urban speed profile); however, the lateral dynamic model can reproduce the mechanical behavior of a motorcycle in the presence of roll, slope, and bank angles with a maximum error of 10% on time around 1 min; this error represents an unacceptable energetic representation in the electric model. The lateral dynamic model's maximum speed values will be used as constraints in estimating behavior since their primary function is to ensure the driver's safety and comfort over the proposed speed profile. Because of its ability to provide an energy estimate that is both fast and precise enough for use in the control phases, the longitudinal model is chosen.

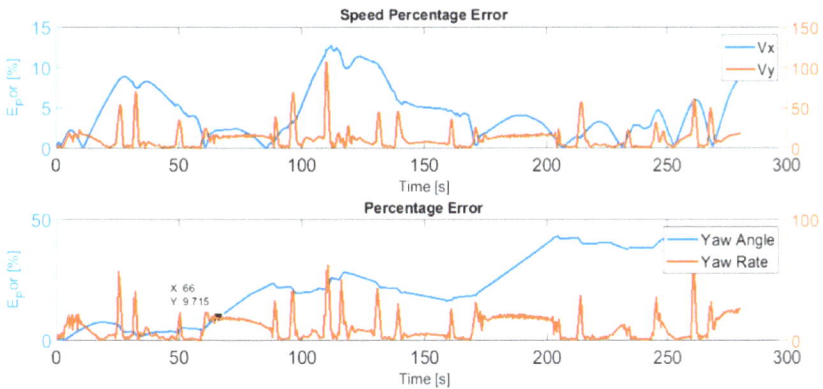

Figure 22. Dynamic speed profile results after sensitivity test (error percentage).

6. Test Results

This section covers the elements needed to determine the eco-driving strategy's operating torque and speed range. The purpose of the exam is to:

- Expected autonomy exploration and comparison with autonomy attained using speed and position state feedback.
- To contrast predicted autonomy with autonomy attained in a particular scenario with energy feedback.
- The effect of the velocity estimator on the control signal is investigated. Because the velocity estimator has a 28% possibility of misinterpreting, this inaccuracy generates an initial energy estimate error that the closed-loop controller must fix. To understand the controller capabilities appropriately, the estimation results must address the worst-case scenario.

The first test consists in running the same distance many times at different speed profiles with a specific Set on β_2 value to establish the energy consumption throughout the path. Then, β_2 is varied to assess its impact on all velocity profiles, sensor, and actuator errors. Additionally, β_2 is adjusted based on the maximum normalized energy usage (kW/km) for each WLTC category (class 6, class 3, class 2, and class 1). Each criterion assesses the maximum normalized energy usage based on a rural and urban trajectory. Figure 23 illustrates the relationship between β_2 and the normalized peak energy consumption (kW/km). In Figure 24 is presented the autonomy behavior of the test.

The estimated autonomy in the test bench findings is reduced from 30% in simulation to 20%. The energy coefficient exhibited similar behavior to that observed in the simulation but with notable changes in autonomy. While the speed profiles have equal autonomy with low β_2 values, the simulations demonstrate that the "Class 3" speed profile is the most battery-consuming. Furthermore, rural speed profiles are more likely to improve autonomy than urban speed profiles. Both variances are due to motor power and projected efficiency function. The speed range of the motor is limited. It results in a distinct behavior of the efficiency function, leading the stop times and lower speed trajectories to be more efficient than in a larger motor, even when the speed profiles are normalized to be coherent between both tests. At the same time, there are striking parallels. The ranges of autonomy estimation are still consistent across the two studies. Except for the "NREL Class 3" speed profile in its urban and rural approaches, the autonomy variation of β_2 from minimum to maximum is roughly 20% (in the simulation experiment, it was 30%). Because of the halted time in each speed profile, the autonomy augmentation recorded by the "Class 3" speed profile is more significant than 30% in both circumstances (simulation and test bench). The controller's primary function is to assure the trip distance required by the driver under various beginning conditions. For these reasons, it is important to investigate the dynamic of energy estate and how β values can correct miscalculations caused by the speed profile estimator.

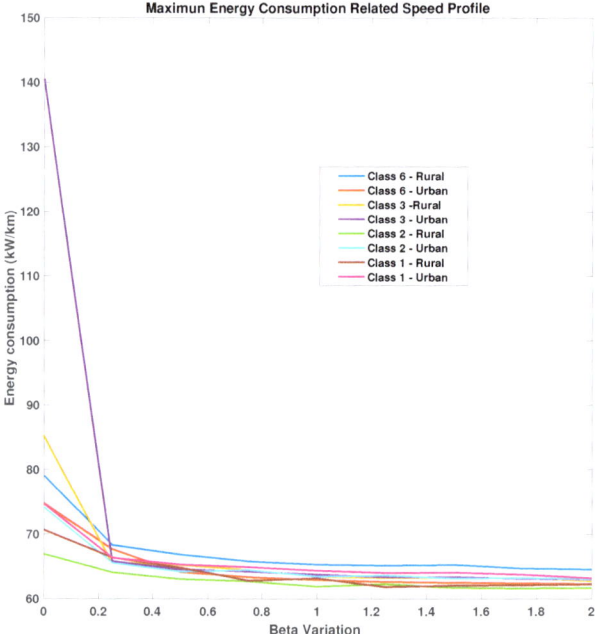

Figure 23. Energy coefficient along each speed profile.

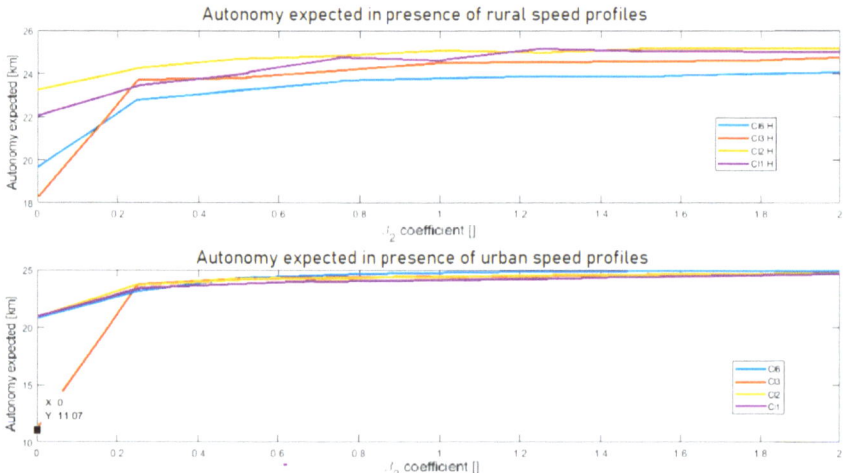

Figure 24. Autonomy comparison.

In the second test, the energy state is sent to the controller so that the energy coefficient can be fixed through the β_2 parameter. This value is dynamically varied from 0 to 5 $\beta_{2_{maxA}}$ without traffic restrictions and from 0 to 2 with traffic constraints $\beta_{2_{maxNA}}$. The test results are provided in Table 3.

Table 3. Closed-Loop distance error values.

	Maximum $\beta_2 \to \beta_{2_{maxA}}$			
SOC [%]	Expected by Estimator [km]	Reference [km]	Computed [km]	Distance Error [%]
100	79.70	71.73	71.05	−0.95
50	39.84	35.86	35.48	−1.06
20	15.93	14.34	14.07	−1.92
	Middle $\beta_2 \to \beta_{2_{maxNA}}$			
SOC [%]	Expected by Estimator [km]	Reference [km]	Computed [km]	Distance Error [%]
100	70.5	70.5	70.61	0.16
50	35.25	35.25	35.25	0.54
20	14.1	14.1	14.08	0.09
	Minimum $\beta_2 \to \beta_{2_{min}}$			
SOC [%]	Expected by Estimator [km]	Reference [km]	Computed [km]	Distance Error [%]
100	60.7	60.7	64.07	5.49
50	30.36	30.36	32	5.4
20	12.1	12.14	12.24	0.8

As can be seen, distance error (difference between driver request and vehicle location) is always less than 5.5%. This indicator can improve if only cases in which the trip still needs to be completed are considered. These produce error decreases to approximately 2% in the absence of traffic conditions.

Under traffic conditions where β_2 cannot be rapidly saturated due to unplanned stops and a brief period where traffic is not restricted (the vehicle is stopped), distance error drops to a minimum of 0.6%, which means the eco-driving approach can guarantee 99.4% of the distance required by the driver under realistic traffic, weather, and road circumstances. This result shows an optimum speed profile with a correlation of approximately 0.87 with the driver-proposed initial speed profile. The speed profile provided by the NMPC can be seen in Figure 25. The maximum and minimum values are avoided in experimental findings and simulation, except for unplanned stops. The resulting speed profile is sensitive to speed changes proposed by the driver. It maintains a soft constraint over the speed values to ensure the safety and comfort of passengers and dynamically controls the intended energy consumption rate by integrating the speed constraint. All reported results do not account for estimated errors. It indicates that the speed profile calculation was accurate in every instance. The estimator's precision is 72%. The impact of this inaccuracy is examined in the following section.

This evaluation uses a variety of speeds and accelerations to try to guess how much energy will be needed to finish the trip, whether in a city or a rural setting. Appropriately approximating the speed profile (rural or urban) yields accurate estimates. The estimation error is 72% when the data is less than 120 s and 88% when it is 500 s. Then, from this concept, Figures 26 and 27 show two situations of correct estimation via the behavior of the signal $beta_2$ and its effect on the estimated range along the route. The anticipated speed profile utilized in instance A is "class 3 urban", with SOC in 100% for the start and 70.5 km of distance trip. The estimation is correct in this situation, but due to the inaccuracy produced by the amplitude of the disturbance relative to the velocity signal, β_2 must fix the energy parameter to demand the suitable energy required and reduce the distance error to 0.82%. However, in some circumstances, the perturbation magnitude is modest enough that the β_2 value does not need to be corrected. Figure 27 depicts an example of this circumstance. The estimation and used velocity profile are equaled, except because the initial SOC is 20%, and the distance is 14.1 km. The distance inaccuracy in this situation is roughly 0.18 m.

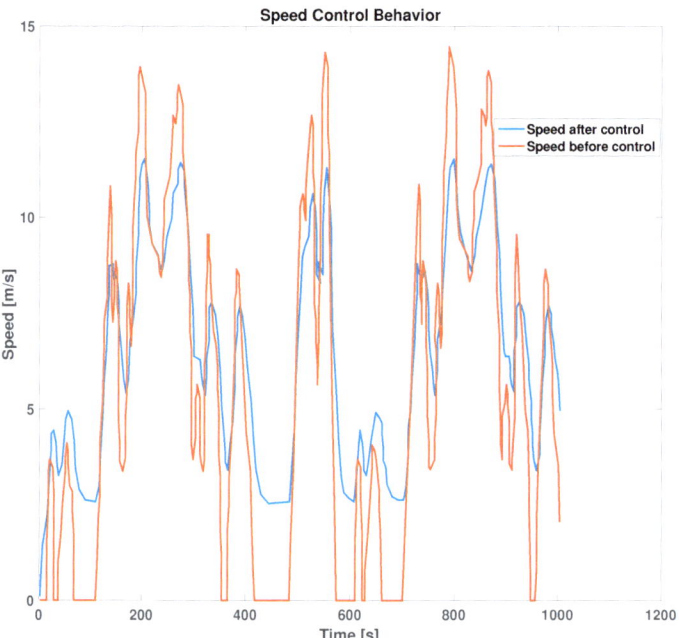

Figure 25. Effect of speed estimation.

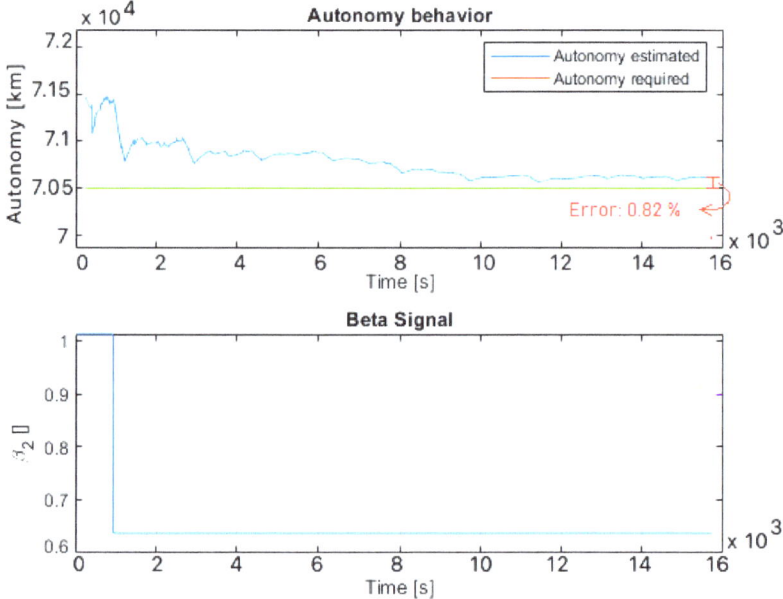

Figure 26. Case A, Good estimation.

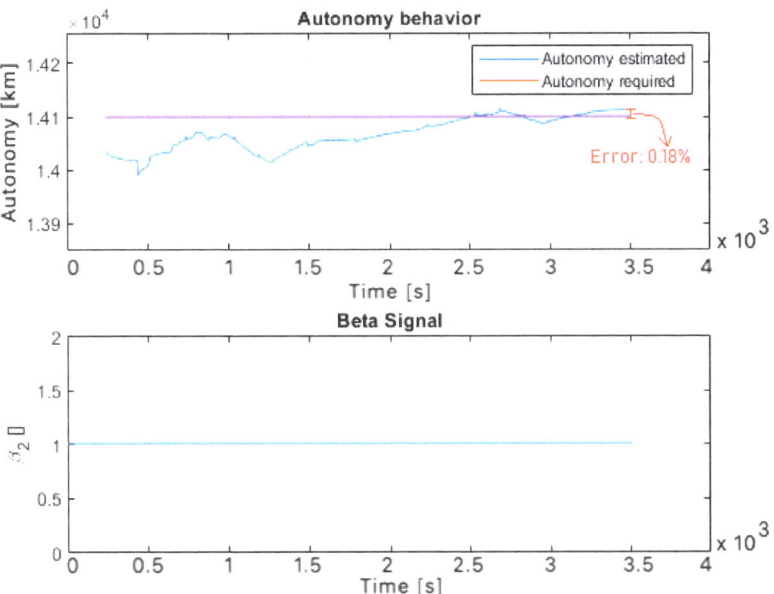

Figure 27. Case B, Good estimation.

Between proper and inaccurate categorization, a medium category will be designated (wrong estimation). An accurate velocity estimation seeks to decrease the distance error, but the random behavior of the disturbances accounts for a significant portion of the error compared to the speed data. The proportional energy controller can manage the energy coefficient inaccuracy with some time delays if the speed profile is incorrect, but the approach (rural or urban) is right. The outcomes of these cases are depicted in Figure 28. In this scenario, the starting SOC is 50% and is configured at 35.25 km for the trip. The inaccuracy induced by incorrect velocity profile estimation generates an inaccuracy corrected by the controller, but the control signal is sluggish due to the significant variation in sampling time. In those circumstances, though, the distance inaccuracy remains at approximately 1.3 percent. This final distance mistake is determined by the amount of distance traveled and the misunderstood speed profile.

A "completely erroneous estimation" is achieved when the urban/rural approach and velocity profile are misconstrued or the anticipated velocity profile has a high energy coefficient compared to the genuine one. As seen in Figure 29, the control signal saturates itself to minimize the energy coefficient error as much as possible. If the required distance is short, the inaccuracy can be as low as 4% if enough time is allowed. The inaccuracy is even reduced to 2.12%. (as in the presented example). In certain circumstances, the control signal climbs to the maximum speed variable limitations, resulting in a trip that behaves more like a driver-controlled experience than an autonomous experience.

Furthermore, the distance error is negative, indicating that the vehicle will not be able to complete the journey. The likelihood of encountering this circumstance within a measurable period of 120 s is approximately 20%. This reduces the possibility of completing the trip with a distance error of less than 1.5% from 98.4% to 78.4%. As a result, an additional estimating tool is used. When the difference between the predicted and observed energy coefficient is more than 10 W/m, the rural/urban approach estimates are replaced by the hypothetical rural/urban approach inferred from the road type used (highway, urban road, etc.). Using this approach, if the distance error is about 1.3 percent and the error is positive, it is considered "totally incorrect estimation", in addition to "bad estimation" (It means the trip is completed with the remaining energy in the battery).

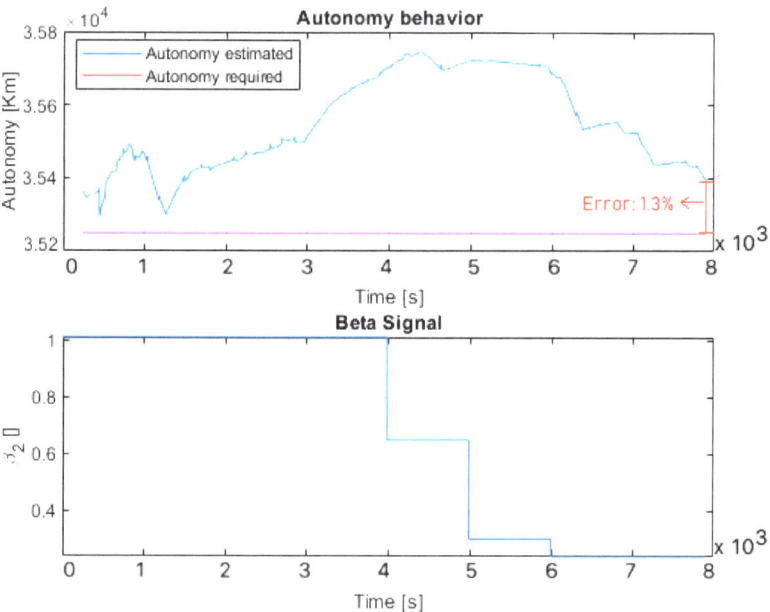

Figure 28. Correct urban/rural estimation.

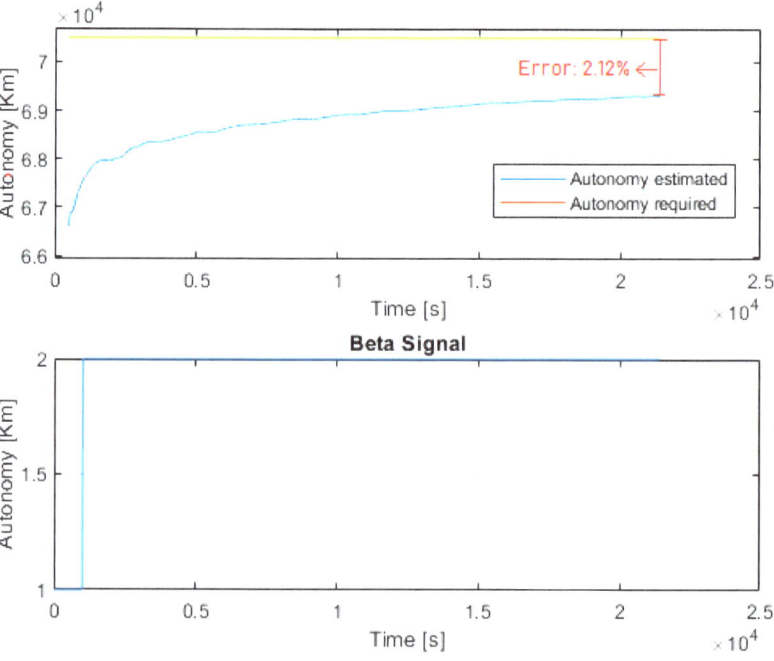

Figure 29. Bad speed profile estimation.

7. Conclusions

This article presented a practical energy-efficient non-linear model predictive for an electric two-wheeler vehicle with different driving profiles and driver preferences. In the proposal, the feedback loop allows energy computation and estimates the parameters in the cost function of the problem. Moreover, it is shown how to implement the NMPC in a test bed with its specifications and the different test. The proposed method has been successful regarding the energy consumption of the vehicle. Indeed, the presented implementation concludes that the NMPC can be used under realistic conditions with specific configurations well adapted to real-time operation. Comparisons with simulations showed an excellent approach. Even if the speed profile estimator has 72% reliability, the supplemental information on the rural/urban component of the speed profile derived from historical data was enough to guarantee a 98% probability of finishing the trip with various initial SOC or driving behaviors. The gap between the absolute needed and the final required distance was less than 1.5 percent. The distance error is positive in 98% of the cases, which indicates that the distance traveled is always more significant than the amount required by the driver. Finally, the tighter speed profile restrictions allow for a 20% increase in autonomy.

Future research could compare and improve new speed profile estimators to mitigate this effect. Lastly, estimating energy use increases autonomy by about 20%, even when the estimated speed profile keeps a 68% correlation with the driver's most demanding speed profile. The eco-driving technique suggested and validated in this work effectively reduces the energy requirements for transportation. It is a tool for increasing the presence of electric vehicles in the transportation sector due to the driver's confidence in the vehicle's autonomy capacity and the reduction in energy needs.

Author Contributions: Methodology Y.B., T.A., J.S.R. and D.P.; Implementation Y.B.; validation, Y.B., T.A., J.S.R. and D.P.; formal analysis, Y.B., T.A., J.S.R. and D.P.; research, Y.B., T.A., J.S.R. and D.P.; physical resources, Y.B., T.A., C.L., N.R., M.B. and F.R.; writing—original draft preparation, Y.B., T.A., J.S.R. and D.P.; writing—review and editing, C.L., N.R. and M.B., F.R., Y.B., T.A., J.S.R. and D.P.; design, Y.B., J.S.R., D.P. and T.A.; supervision, C.L., N.R., M.B., F.R., T.A. and D.P.; project administration, C.L., N.R., M.B., F.R., T.A. and D.P.; funding acquisition, T.A. and D.P. All authors have read and agreed to the published version of the manuscript.

Funding: This research was funded by: Minciencias, ECOSNORD project 890-2019, and grant number 779 of 2017 by Boyaca government.

Institutional Review Board Statement: Not applicable.

Informed Consent Statement: Not applicable.

Data Availability Statement: Not applicable.

Acknowledgments: Authors want to thank Minciencias, ECOSNORD for the sponsorship to the project "Control asistido para la conducción de una motocicleta eléctrica orientado a la eficiencia energética" with contract 890-2019, and Boyacá government grant number 779 of 2017.

Conflicts of Interest: The authors declare no conflict of interest.

Abbreviations

ADAS	Advance Driver Assistance System
BLDC	Brushless DC
EV	Electric Vehicles
FOC	Feld-Oriented Control
GUI	User Graphic Interface
MPC	Model Predictive Controller
NMPC	Nonlinear Model Predictive Controller
NREL	National Renewable Energy Laboratory Drive Cycle
PI	Proportional–Integral
SOC	State of Charge
WLTC	Worldwide harmonized Light-duty vehicles Test Cycles

Nomenclature

$\ddot{x}$	Acceleration
F_{aer_x}	Aerodynamic Force in X-axis
ρ	Air Density
α_{air}	Angle Between Air and Vehicle Direction
β	Bank Angle
N_c	Control Horizon
$\dot{x}_1$	Distance
C_d	Drag Coefficient
R_{wf}	Effective Radius of Rear Wheel
EVs	Electric Vehicles
E	Energy
ϵ_{e_c}	Energy necessary to end the trip
t_f	Final Time
A_f	Front Area
g	Gravity
t_0	Initial Time
m	Mass
T_{max}	Maximum Torque
T_{min}	Minimum Torque
$E_{ff}(X_2, T)$	Motor Efficiency Abstraction
F_z	Normal Force
β_{11}	Cost Function Normalized coeff
β_{13}	Cost Function Normalized coeff
β_{12}	Cost Function Normalized coeff
N_p	Prediction Horizon
$\mu_{0,1,2}$	Road Parameters
$\mu_{1,2}$	Road Parameters
F_{roll}	Roll Resistance
s	Seconds
θ_s	Slope
$\dot{x}$	Speed
t	Time
t_s	Time Period (1000 s)
T	Torque
TotEne	Total Energy
d(t)	Travel Distance
F_l	Uncontrolled Torque Inputs (losses)
$\dot{x}_3$	Vehicle energy
$\dot{x}_2$	Vehicle Speed
F_w	Weight
V_{wind}	Wind Speed

References

1. Bugaje, A.; Ehrenwirth, M.; Trinkl, C.; Zörner, W. Electric two-wheeler vehicle integration into rural off-grid photovoltaic system in Kenya. *Energies* **2021**, *14*, 7956. [CrossRef]
2. Xue, Q.; Wang, K.; Lu, J.J.; Liu, Y. Rapid driving style recognition in car-following using machine learning and vehicle trajectory data. *J. Adv. Transp.* **2019**, *2019*, 9085238. [CrossRef]
3. Musa, A.; Pipicelli, M.; Spano, M.; Tufano, F.; De Nola, F.; Di Blasio, G.; Gimelli, A.; Misul, D.A.; Toscano, G. A review of model predictive controls applied to advanced driver-assistance systems. *Energies* **2021**, *14*, 7974. [CrossRef]
4. Gao, G.; Wang, Z.; Liu, X.; Li, Q.; Wang, W.; Zhang, J. Travel behavior analysis using 2016 Qingdao's household traffic surveys and Baidu electric map API data. *J. Adv. Transp.* **2019**, *2019*, 6383097. [CrossRef]
5. Rahimi-Eichi, H.; Chow, M.Y. Big-data framework for electric vehicle range estim. In Proceedings of the IECON 2014-40th Annual Conference of the IEEE Industrial Electronics Society, Dallas, TX, USA, 29 October–1 November 2014; pp. 5628–5634.
6. Zhang, Y.; Wang, W.; Kobayashi, Y.; Shirai, K. Remaining driving range estimation of electric vehicle. In Proceedings of the 2012 IEEE International Electric Vehicle Conference, Greenville, CA, USA, 4–8 March 2012; pp. 1–7.
7. Lin, X.; Görges, D.; Liu, S. Eco-driving assistance system for electric vehicles based on speed profile optimization. In Proceedings of the 2014 IEEE Conference on Control Applications (CCA), Juan Les Antibes, France, 8–10 October 2014; pp. 629–634.

8. Kato, H.; Ando, R.; Kondo, Y.; Suzuki, T.; Matsuhashi, K.; Kobayashi, S. Comparative measurements of the eco-driving effect between electric and internal combustion engine vehicles. In Proceedings of the 2013 World Electric Vehicle Symposium and Exhibition (EVS27), Barcelona, Spain, 17–20 November 2013; pp. 1–5.
9. Farzaneh, A.; Farjah, E. a novel smart energy management system in pure electric motorcycle using COA. *IEEE Trans. Intell. Veh.* **2019**, *4*, 600–608. [CrossRef]
10. Koch, A.; Teichert, O.; Kalt, S.; Ongel, A.; Lienkamp, M. Powertrain Optimization for Electric Buses under Optimal Energy-Efficient Driving. *Energies* **2020**, *13*, 6451. [CrossRef]
11. Chen, C.; Zhao, X.; Yao, Y.; Zhang, Y.; Rong, J.; Liu, X. Driver's eco-driving behavior evaluation modeling based on driving events. *J. Adv. Transp.* **2018**, *2018*, 1–12. [CrossRef]
12. Hieu, L.T.; Khoa, N.X.; Lim, O. An Investigation on the Effects of Input Parameters on the Dynamic and Electric Consumption of Electric Motorcycles. *Sustainability* **2021**, *13*, 7285. [CrossRef]
13. Benotsmane, R.; Vásárhelyi, J. Towards Optimization of Energy Consumption of Tello Quadrotor with MPC Model Implementation. *Energies* **2022**, *15*, 9207. [CrossRef]
14. Bello, Y.; Azib, T.; Larouci, C.; Boukhnifer, M.; Rizoug, N.; Patino, D.; Ruiz, F. Eco-Driving Optimal Controller for Autonomy Tracking of Two-Wheel Electric Vehicles. *J. Adv. Transp.* **2020**, *2020*, 1–15. [CrossRef]
15. Kamal, E.; Adouane, L. Optimized EMS and a Comparative Study of Hybrid Hydrogen Fuel Cell/Battery Vehicles. *Energies* **2022**, *15*, 738. [CrossRef]
16. Zhang, X.; Mi, C. *Vehicle Power Management: Modeling, Control and Optimization*; Springer Science & Business Media: Berlin/Heidelberg, Germany, 2011.
17. Rajamani, R. *Vehicle Dynamics and Control*; Springer Science & Business Media: Berlin/Heidelberg, Germany, 2011.
18. Bello, Y.; Azib, T.; Larouci, C.; Boukhnifer, M.; Rizoug, N.; Patino, D.; Ruiz, F. Two wheels electric vehicle modelling: Parameters sensitivity analysis. In Proceedings of the 2019 6th International Conference on Control, Decision and Information Technologies (CoDIT), Paris, France, 23–26 April 2019; pp. 279–284.
19. Jazar, R.N. *Vehicle Dynamics*; Springer: Berlin/Heidelberg, Germany, 2008; Volume 1.
20. Cossalter, V.; Lot, R.; Massaro, M. Motorcycle dynamics. *Modelling, Simulation and Control of Two-Wheeled Vehicles*; Wiley: Hoboken, NJ, USA, 2014; pp. 1–42.
21. Sun, Z.; Wen, Z.; Zhao, X.; Yang, Y.; Li, S. Real-world driving cycles adaptability of electric vehicles. *World Electr. Veh. J.* **2020**, *11*, 19. [CrossRef]
22. Jape, S.R.; Thosar, A. Comparison of electric motors for electric vehicle application. *Int. J. Res. Eng. Technol.* **2017**, *6*, 12–17. [CrossRef]
23. Atallah, K.; Howe, D.; Mellor, P.H.; Stone, D.A. Rotor loss in permanent-magnet brushless AC machines. *IEEE Trans. Ind. Appl.* **2000**, *36*, 1612–1618.
24. Bianchi, N.; Bolognani, S.; Frare, P. Design criteria for high-efficiency SPM synchronous motors. *IEEE Trans. Energy Convers.* **2006**, *21*, 396–404. [CrossRef]
25. Song, L.; Li, Z.; Cui, Z.; Yang, G. Efficiency map calculation for surface-mounted permanent-magnet in-wheel motor based on design parameters and control strategy. In Proceedings of the 2014 IEEE Conference and Expo Transportation Electrification Asia-Pacific (ITEC Asia-Pacific), Beijing, China, 31 August–3 September 2014; pp. 1–6.
26. Fasil, M.; Mijatovic, N.; Jensen, B.B.; Holboll, J. Nonlinear dynamic model of PMBLDC motor considering core losses. *IEEE Trans. Ind. Electron.* **2017**, *64*, 9282–9290. [CrossRef]
27. Neacă, M.I.; Neacă, A.M. Determination of the power loss in inverters which supplies a BLDC motor. In Proceedings of the 2016 International Symposium on Fundamentals of Electrical Engineering (ISFEE), Bucharest, Romania, 30 June–2 July 2016; pp. 1–6.
28. Rong, Y.; Yang, W.; Wang, H.; Qi, H. SOC estimation of electric vehicle based on the establishment of battery management system. In Proceedings of the 2014 IEEE Conference and Expo Transportation Electrification Asia-Pacific (ITEC Asia-Pacific), Beijing, China, 31 August–3 September 2014; pp. 1–5.
29. Lee, K.Y.; Lai, Y.S. A novel magnetic-less bi-directional dc-dc converter. In Proceedings of the 30th Annual Conf. of IEEE Industrial Electronics Society IECON 2004, Busan, Republic of Korea, 2–6 November 2004; Volume 2, pp. 1014–1017.
30. Cristhian Yesid, B.C. Eco-Driving Planification Profile for Electric Motorcycles. Ph.D. Thesis, Paris-saclay, Pontficia Universidad Javeriana, Bogota, Colombia, 2020.
31. Bello, Y.; Azib, T.; Larouci, C.; Boukhnifer, M.; Rizoug, N.; Patino, D.; Ruiz, F. Motor efficiency modeling towards energy optimization for two-wheel electric vehicle. *Energy Effic.* **2022**, *15*, 1–14. [CrossRef]

Disclaimer/Publisher's Note: The statements, opinions and data contained in all publications are solely those of the individual author(s) and contributor(s) and not of MDPI and/or the editor(s). MDPI and/or the editor(s) disclaim responsibility for any injury to people or property resulting from any ideas, methods, instructions or products referred to in the content.

Article

Interaction among Multiple Electric Vehicle Chargers: Measurements on Harmonics and Power Quality Issues

Andrea Mazza [1,*], Giorgio Benedetto [1], Ettore Bompard [1], Claudia Nobile [1], Enrico Pons [1], Paolo Tosco [2], Marco Zampolli [2] and Rémi Jaboeuf [2]

1. Dipartimento Energia "Galileo Ferraris", Politecnico di Torino, 10129 Turin, Italy; giorgio.benedetto@polito.it (G.B.); ettore.bompard@polito.it (E.B.)
2. Edison SpA, 20121 Milano, Italy
* Correspondence: andrea.mazza@polito.it

Citation: Mazza, A.; Benedetto, G.; Bompard, E.; Nobile, C.; Pons, E.; Tosco, P.; Zampolli, M.; Jaboeuf, R. Interaction among Multiple Electric Vehicle Chargers: Measurements on Harmonics and Power Quality Issues. *Energies* **2023**, *16*, 7051. https://doi.org/10.3390/en16207051

Academic Editor: Byoung Kuk Lee

Received: 31 July 2023
Revised: 25 September 2023
Accepted: 26 September 2023
Published: 11 October 2023

Copyright: © 2023 by the authors. Licensee MDPI, Basel, Switzerland. This article is an open access article distributed under the terms and conditions of the Creative Commons Attribution (CC BY) license (https://creativecommons.org/licenses/by/4.0/).

Abstract: The electric vehicle (EV) market is growing rapidly due to the necessity of shifting from fossil fuel-based mobility to a more sustainable one. Smart charging paradigms (such as vehicle-to-grid (V2G), vehicle-to-building (V2B), and vehicle-to-home (V2H)) are currently under development, and the existing implementations already enable a bidirectional energy flow between the vehicles and the other systems (grid, buildings, or home appliances, respectively). With regard to grid connection, the increasingly higher penetration of electric vehicles must be carefully analyzed in terms of negative impacts on the power quality; and hence, the effects of electric vehicle charging stations (EVCSs) must be considered. In this work, the interactions of multiple electric vehicle charging stations have been studied through laboratory experiments. Two identical bidirectional DC chargers, with a rated power of 11 kW each, have been supplied by the same voltage source, and the summation phenomenon of the current harmonics of the two chargers (which leads to an amplification of their values) has been analyzed. The experiment consisted of 100 trials, which considered four different combinations of power set-points in order to identify the distribution of values and to find suitable indicators for understanding the trend of the harmonic interaction. By studying the statistical distribution of the Harmonic Summation Index, defined in the paper, the impact of the harmonic distortion caused by the simultaneous charging of multiple electric vehicles has been explored. Based on this study, it can be concluded that the harmonic contributions of the electric vehicle charging stations tend to add up with increasing degrees of similarity of the power set-points, while they tend to cancel out the more the power set-points differ among the chargers.

Keywords: electric vehicles (EVs); vehicle-to-grid (V2G); EV charging stations (EVCSs); harmonics; power quality

1. Introduction

The increasing share of renewable energy sources (RESs) in the electrical infrastructure is a mandatory consequence of the European set of proposals, designed to make the EU's climate, energy, transport and taxation policies fit for reducing net greenhouse gas emissions by at least 55% by 2030, compared to 1990 levels [1]. As a result, in the last few years, the Electric Vehicle (EV) market has experienced a rapid growth, as the world is moving toward more sustainable mobility systems, sustained by the increasing electrification of the road transport sector, with a consequent reduction in its carbon emissions. According to the Global EV Outlook 2023 [2], the sales of EVs, including both Battery Electric Vehicles (BEVs) and Plug-in Hybrid Electric Vehicles (PHEVs), reached USD 10.5 million, and this rising trend seems to be continuing. Moreover, according to the IEA report [3], by 2030, EVs will represent more than 60% of vehicles sold globally, so the the number of public (or publicly accessible) charging points (2.7 million reached in 2022) will continue to expand. This will be reflected in the number of products available on the market for charging vehicles, which need to be evaluated as in the review [4]. In the context of a rising number of EVs, the Vehicle-to-X (V2X) paradigm, including both Vehicle-to-Grid (V2G) and Vehicle-to-Building (V2B), is becoming increasingly interesting, both for the grid and for different energy community configurations. In these applications, a high number of Electric Vehicle Charging Stations (EVCSs) will be connected to a grid infrastructure (either public or part of an energy community) close to each other. The storage capabilities of EVs have led to the development of smart charging approaches, enabling to use of EV batteries in the most cost-effective way. In fact, thanks to the bidirectional electricity flow between vehicles and the network, the power can be sent back to the grid when needed, thus providing benefits both to the vehicle owners (in terms of additional revenues) and to grid operators (in terms of the offered services that may improve quality, reliability, and sustainability of the grid itself) [5]. The harmonic current content has to adhere to the limits established by the IEEE Standards [6] or to those fixed by the European Standards [7]. The widespread adoption of EVs may however have negative effects on the power quality of the electrical grid. Therefore, it is crucial to consider the impact of EVCSs on the grid side. According to [8], integrating several EVs into the grid could lead to voltage imbalances and to a decrease in the transformer efficiency. Moreover, when multiple vehicles are charging or discharging simultaneously, as in large parking lots, the individual harmonics could add up and hence approach the standard limits, resulting in strong harmonic injections into the power grid. This behavior can negatively impact the energy supplied to the relevant electrical node, potentially hindering proper equipment functioning. This paper will try to study the disturbances injected from multiple EV chargers connected to the same electrical node, as in a realistic environment. Unlike other works, the results of this paper have been obtained by analyzing the currents coming from two identical real chargers, by studying the contribution of each of them to the selected frequencies and by comparing the theoretical sum of the current harmonics with the measured harmonics. The paper explores the known effects of the EV charger aggregation in Section 2 and introduces the setup of the tests in Section 3. Then, Section 4 reports the approach used to carry out to the tests. In Section 5, the results for the considered harmonic orders are reported. Finally, some concluding remarks are given.

2. Overview of the Effects of Aggregating Several EV Charging Points

The expression "power quality" indicates the "characteristics of the electric current, voltage and frequency at a given point in an electric power system, evaluated against a set of reference technical parameters" [9]. In Europe, the EN 50160 Standard specifies the main characteristics that the grid voltage should meet at the public low-, medium-, and high-voltage AC supply terminals [10]. Any deviation in the voltage or current waveforms that can degrade the performance of a device, equipment, or system, or adversely affect living or inert matter is an "electromagnetic quality disturbance", as stated in [11]. In more detail, the definition of a power quality disturbance is generally accepted as any change in voltage, current, or frequency that interferes with the normal operation of electrical equipment [12] or in the quality of power while supplying an electrical equipment [13].

The integration of several EVs can have a potential impact on the power quality of the grid they are connected to: the magnitude of the impact depends on the number of EVs being charged at the same time, their location, and their charging rate [14]. The paper [15] examines the impact of V2G operation when multiple vehicles are connected. Various scenarios and EV penetration levels are analyzed to study both the harmonic distortion and stability effects. The results indicate that the primary concern for power quality is the harmonic distortion; in fact, the higher the EV share, the higher the Total Harmonic Distortion (THD). Consequently, the number of equipment affected by the reduced power quality and absorbing distorted current from the grid increases. Some studies, such as [16], have analyzed the impact of equipment diversity within a EVCS. The study tested four different fast chargers and recorded full charging cycles four times for each charger, analyzing the amplitude and phase angles of each harmonic. The research found that the phase angles of the current harmonics varied within a preferential range that could potentially lead to an increase in the current THD. In [17], one EV charger was analyzed both in AC and DC. Current harmonics emissions and conductive electromagnetic disturbances were considered, while the THD was used to evaluate the global power quality level. This analysis, however, was based on a single EV charger, so it was suggested that it should be repeated by including different Devices Under Test (DUT) to highlight possible differences. Another case study which considers both the harmonics and supra-harmonics content is presented in [18]. Supra-harmonics are defined in [19] as waveform distortions in the frequency range from 2 to 150 kHz. This range is still only partially standardized, while the amount of devices emitting in this range is increasing [20], so it could lead to undesirable effects. In [18], the emissions caused by nine different BEV models have been studied and eigth out of nine have proven to be the source of supra-harmonics. The tests have also been conducted for power levels differing by the nominal one: they show a variation of the fundamental reactive power and lead to the recommendation to execute future tests at non-nominal power in order to account for possible differences. With reference to the aggregation of EV charging points, the effects of multiple EVs connected to electrical grids have already been studied in order to assess how the disturbances injected from the aggregate propagate in the low- and medium-voltage networks. For example, the JRC report [21] studied the grid harmonic impact of multiple EVs. Focusing on the phase summation or cancellation of the harmonics, it was revealed that the phase angles between the same harmonic order tend to be lower than 90°, leading to a summation of the harmonics and therefore suggesting that there could be a maximum acceptable number of chargers connected to the same infrastructure. In contrast, the authors of [22] had previously pointed out how adding chargers from different manufacturers may result in a notable harmonic cancellation. Some of the tests, however, showed how the chargers failed to comply to standard limits, suggesting that the harmonics added up until reaching a maximum tolerable value. What seems to be a recurrent element in these published works is that when multiple chargers are operating together on the same electrical node (e.g., a parking lot), their disturbances would sum-up, approaching the standard limits, potentially amplifying their values and hence causing issues to the other customers.

3. Laboratory Set-Up

The authors of previous work established a laboratory test bed using real-time simulation and Power Hardware-in-the-Loop (PHIL) layout to study bidirectional EVCSs [23,24]. In this setup, the currents measured on a single EV charging point under test were multiplied to emulate 20 charging points connected to the same electrical node. However, this resulted in a harmonic spectrum summation, leading to instabilities and preventing the equipment from operating correctly. Starting from those results, the current work studies the behavior of the charger in depth without the simulated environment, evaluating the interaction of multiple chargers by employing two identical EV DC chargers with a rated power equal to 11 kW. The chargers were supplied by the same voltage source: in *Case 1*, the voltage source was a linear power amplifier; in *Case 2*, the voltage source was the electrical grid. The latter case enabled us to achieve the maximum power for both chargers simultaneously. Figure 1 shows a simplified layout of the test bed, with the current measurement points highlighted.

The main components used to carry out the tests are as follows:

- *Power amplifier:* A three-phase linear power amplifier with a nominal power of 7 kVA per phase has been used. This technology has been chosen because of its performance. In fact, the linear amplifier has a maximum distortion on the generated voltages equal to 0.7% at the maximum output power [25]. Moreover, the introduced short time delay enables the use of a simple interface topology and guarantees high stability.
- *Chargers and electric vehicles:* The cars used are two Nissan Leaf, with battery capacities of 62 kWh and 40 kWh, respectively, and equipped with the DC CHAdeMO plug.

The model of charger (which cannot be disclosed for confidentiality reasons) is a bidirectional WB, with rated power 11 kW in G2V and 10 kW in V2G operation.
- *Data acquisition:* The HBM GEN7tA is a transient recorder and data acquisition system. It has been used to visualize, monitor, record, and post-process the electrical quantities involved in the tests. Three identical Hioki 9018-50 clamp probes have been employed for capturing the currents during the tests. The clamps have a range from 10 A to 500 A AC, for a total of six ranges, with the amplitude accuracy equal to $\pm 1.5\%$ rdg $\pm 0.1\%$ f.s. (45 to 66 Hz) and the phase accuracy equal to ± 2.5 for frequencies from 40 Hz to 3 kHz.

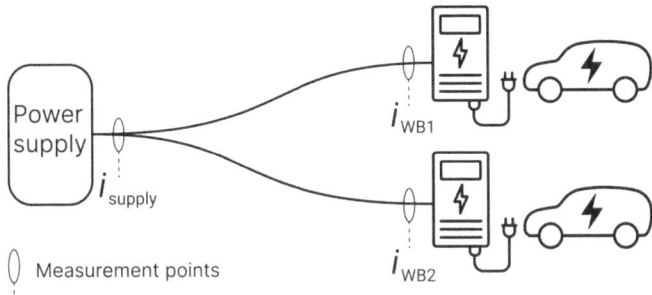

Figure 1. Representation of the experimental setup with measurement points.

4. Test Description

The WBs were supplied with a standard voltage $V_{nom} = 230$ V by the linear power amplifier in order to have low disturbances injected from the amplification stage. The testing was conducted using different charging power set-points. Initially, the vehicles were charged at power $P_1 = 4$ kW, and then the power set-point was changed for the two chargers. To ensure that the tests were independent from each other, the WBs were turned in the stand-by state after each measurement. Different power scenarios were studied as suggested by the authors of [18], which indicated that there may be slight variations in the disturbance during a charging with power differing by the nominal one. The preliminary measurements and comparisons were carried out by using the power amplifiers (*Case 1*); after that, we proceeded by using as a voltage source the real network, in order to reach the maximum rated power for both the WBs (*Case 2*). The harmonic content was similar in both cases, as the THD of the network supply was very low. The power set-point combinations and the relative cases are listed in Table 1.

Table 1. Set-point combinations.

Cases	WB1 (kW)	WB2 (kW)
	4	4
Case 1	8	6
	10	2
Case 2	11	11

4.1. Procedure

The current measurements, as depicted in Figure 1, have been taken on the first phase of each WB and on the output of the power amplifier (in *Case 1*), or grid supply (in *Case 2*), in order to acquire the sum of the currents of the two WBs. The observed behavior was the same for the other phases, so, the results are reported for the first phase only. The data gathering consisted of acquiring the current waveforms 100 times for each power set-point combination, with a sampling rate of 20 kSample/s. This sampling frequency will be sufficient to avoid the aliasing phenomenon till the 40th order, as the frequency of the acquisition is ten times higher than the highest measured order. The acquisitions have been made with the recorder presented above. The Fast Fourier Transform (FFT) has been calculated in Matlab from the acquired current recordings, based on 10 cycles of the target currents, as stated in the IEEE Standard [26], in order to achieve a 5 Hz resolution. The harmonic component magnitude is then calculated by taking the RMS value of the center frequency combined with the values at the two adjacent ±5 Hz frequency bins as calculated in Equation (1) for the example in Figure 2.

$$G_h = \sqrt{\sum_{i=-1}^{1} X^2_{(10h+i)\Delta f}} \qquad (1)$$

where:
- Δf = Bins width (5 Hz in this case);
- h = Harmonic order;
- G_h = Group value at the order h.

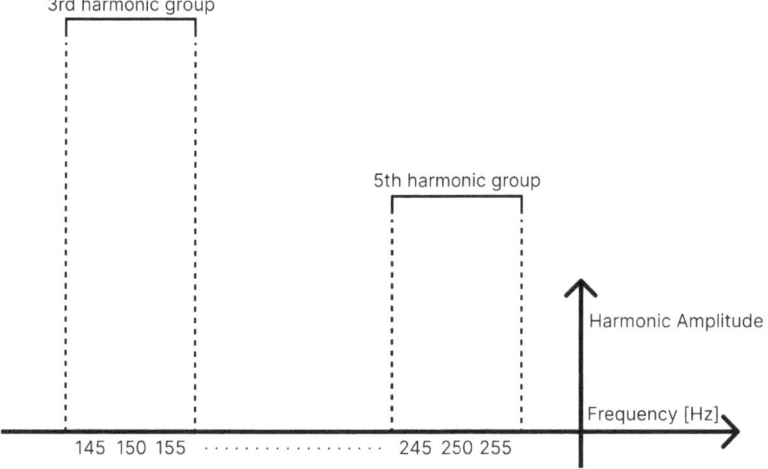

Figure 2. Main order grouping.

4.2. THD Evaluation

Harmonics can be evaluated in two ways:
- Individually, by comparing their amplitude to the fundamental frequency;
- Globally, using the THD.

The THD% has been computed for each one of the 100 tests. Tables 2 and 3 contain the mean values of the THD for the 100 tests for each set-point combination under test. Currents and voltages of the THD have been measured both on the power supply (as shown in Figure 1) and on the two WBs. The THD% is always below the 8% limit, so the disturbances are acceptable in all tests. Based on the previous measurements, it was noted that on the grid side, the THD was under the limits, so the impact of the grid voltage harmonics on the tests was considered negligible. It is visible how the total THD can be lower than the one on each WB; nevertheless, as the THD averages on all the harmonic orders, the contribution on each frequency needs to be investigated due to possible problems caused by some specific harmonics.

Table 2. Current THD %.

Cases	WB1 Power	WB2 Power	Power Supply	WB1	WB2
Case 1	4 kW	4 kW	5.26%	5.63%	5.25%
	8 kW	6 kW	4.05%	4.30%	4.57%
	10 kW	2 kW	3.79%	3.52%	7.30%
Case 2	11 kW	11 kW	3.01%	3.05%	3.10%

Table 3. Voltage THD %.

Cases	WB1 Power	WB2 Power	Power Supply	WB1	WB2
Case 1	4 kW	4 kW	2.08%	2.09%	2.07%
	8 kW	6 kW	2.31%	2.32%	2.30%
	10 kW	2 kW	2.22%	2.26%	2.22%
Case 2	11 kW	11 kW	2.08%	2.11%	2.07%

4.3. Harmonic Sum Evaluation

After collecting all the measurements for the 100 tests, a dataset has been extracted, whereby an example of the structure (for the case $P_1 = 10\,\text{kW}, P_2 = 2\,\text{kW}$) is reported in Table 4. Each column shows the harmonic order, whereas the rows represent the data collected in terms of the value of HSI as defined in Equation (2). Then, in Tables 5–7, the harmonic current components used to calculate the HSI for the same case are reported. A similar approach, which defines an index to evaluate the summation phenomenon called the diversity factor, was adopted by the authors of [27] for the study of harmonics summation or cancellation at an industrial facility. This index is the ratio between the RMS current value extracted from the FFT of the grouping described in Equation (1) and is directly measured on the first phase of the network (namely $I_{h,SUPPLY}$), and is the sum of the same quantities measured on the first phase of each one of the two WBs (i.e., $I_{h,WB1}$ and $I_{h,WB2}$). This ratio is the only cause of error propagation due to the elaboration of the data that will reflect on the HSI and, from the current clamp specification, will be around the 3% of the HSI index itself.

$$HSI = \frac{I_{h,SUPPLY}}{I_{h,WB1} + I_{h,WB2}} \qquad (2)$$

This index can be:

- HSI $\geq$ 1: The harmonics summed;
- HSI $<$ 1: The harmonics canceled.

The lower the HSI, the highest is the cancellation effect on a certain harmonic order.

Table 4. HSI dataset structure.

Test No.	Harmonic Order				
	1	2	...	39	40
0	0.9976	0.6073	...	0.7367	1.0349
1	0.9973	0.7611	...	0.8793	0.6176
2	0.9975	0.4813	...	1.0236	0.9997
...	...	...	...	...	...
98	0.9977	0.3999	...	0.8607	0.7854
99	0.9974	0.4660	...	1.0150	0.3912
100	0.9976	0.8801	...	0.5548	0.9968

Table 5. I_{WB1} harmonic components.

Test No.	Harmonic Order				
	1	2	...	39	40
0	14.5776	0.0249	...	0.0082	0.0027
1	14.3311	0.0224	...	0.0082	0.0048
2	14.4597	0.0187	...	0.0048	0.0042
...	...	...	...	...	...
98	14.4420	0.0201	...	0.0097	0.0059
99	14.4745	0.0175	...	0.0069	0.0037
100	14.5909	0.0154	...	0.0079	0.0025

Table 6. Current I_{WB2} harmonic components.

Test No.	Harmonic Order				
	1	2	...	39	40
0	4.2347	0.0123	...	0.0047	0.0042
1	4.2324	0.0192	...	0.0062	0.0049
2	4.2647	0.0204	...	0.0030	0.0035
...	...	...	...	...	...
98	4.2683	0.0167	...	0.0045	0.0041
99	4.2781	0.0128	...	0.0032	0.0046
100	4.2731	0.0106	...	0.0049	0.0045

Table 7. Current I_{SUPPLY} harmonic components.

Test No.	Harmonic Order				
	1	2	...	39	40
0	18.7676	0.0226	...	0.0095	0.0071
1	18.5142	0.0317	...	0.0127	0.0060
2	18.6780	0.0188	...	0.0079	0.0077
...	...	...	...	...	...
98	18.6665	0.0147	...	0.0122	0.0078
99	18.7041	0.0141	...	0.0102	0.0032
100	18.8189	0.0229	...	0.0071	0.0070

4.4. Data Cleaning

Based on the experience gained from previous studies [23,24] on the analyzed DUT, it has been observed that the power set-point of the charger exhibits high variability. To obtain a more accurate representation of this phenomenon, a statistical approach has been used. This approach does not only focus on increasing the number of measurements taken, but also enables both removing possible outliers and analyzing the probability distribution of the index defined for each acquired harmonic. Firstly, the instantaneous currents acquired have been elaborated through the FFT in order to obtain the RMS value of the current for each frequency. The first outcome was a spectrum of the absorbed current harmonics from the two chargers for each of the one hundred tests and for each of the four power set-points. In Figure 3, the FFT spectrum of one charger at the power of 10 kW is shown. The harmonic spectra of the same charger for the other five power set-points are presented in Appendix A. The amplitudes of each frequency have been collected in a dataset and used to compute the index defined in Equation (2).

Then, the point cloud plot of each frequency for the four different power set-points has been created to identify possible outliers in order to find non-reliable points. The outlier identification was firstly conducted with the box plots of each frequency, but the number of non-feasible points was not reasonably high, so a second approach, based on the distance calculation, has been used. For this purpose, the Mahalanobis distance **m** is calculated for all the observations as in Equation (3). The Mahalanobis distance is an effective multivariate distance metric that measures the distance between a point and a distribution. It has excellent applications in multivariate anomaly detection and classification on highly imbalanced datasets.

$$\mathbf{m} = \sqrt{(\mathbf{x} - \overline{x}) \cdot \mathbf{S}^{-1} \cdot (\mathbf{x} - \overline{x})^T} \qquad (3)$$

where:
- $\mathbf{x}$ is the measured sample (taken as column);
- $\overline{x}$ is the mean value of the measured sample;
- $\mathbf{S}$ is the variance–covariance matrix of the transposed measured sample;
- ^{T} represents the transpose operator.

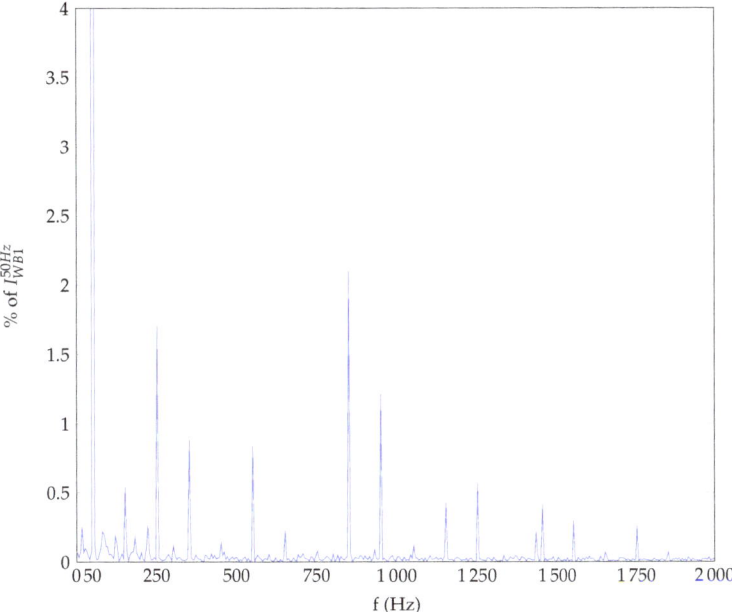

Figure 3. FFT spectrum of I_{WB1} at 10 kW.

The confidence ellipse, whose dimensions depend on the distance calculated using Equation (3), is then used to graphically distinguish the possible outliers, as shown in Figure 4, where, for the third harmonic order taken as an example, the point cloud plot is reported and the points outside the confidence ellipse are the possible outliers. The found distances are then compared with a cut-off value based on a chi-squared distribution with the probability of rejecting the null hypothesis when it is true and equal to 0.02.

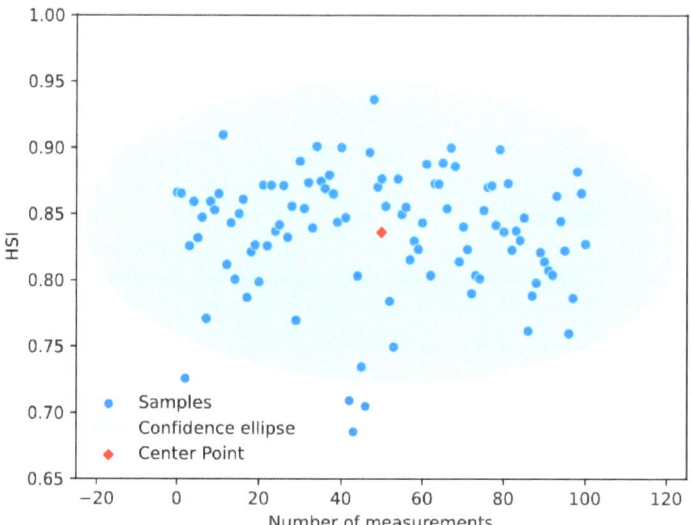

Figure 4. Third harmonic HSI scatter plot with confidence ellipse in the 10/2 kW case.

5. Test Results

The main hypothesis, as mentioned in other works like [16,23,24], was that the harmonics will add-up due the common supply voltages. The HSI represents how much each harmonic tends to sum up with the frequency of the same order injected by the other chargers connected nearby. For this purpose, each frequency multiple of the fundamental, up till the 40th harmonics, has been individually analyzed based on the valid measures obtained from the data cleaning process previously described in Section 4.4. Figures 5–8 show the third harmonic HSI distribution for each tested power set-point in the form of both a continuous distribution and a bar plot. On the the x-axis, the index is reported, while on the y-axis, the number of occurrences is shown. The third order was chosen due to the importance noted on this specific hardware; however, the behavior is the same for the other studied frequencies. Is is clear that varying the power set-point impacts the distribution of the index, moving its average values.

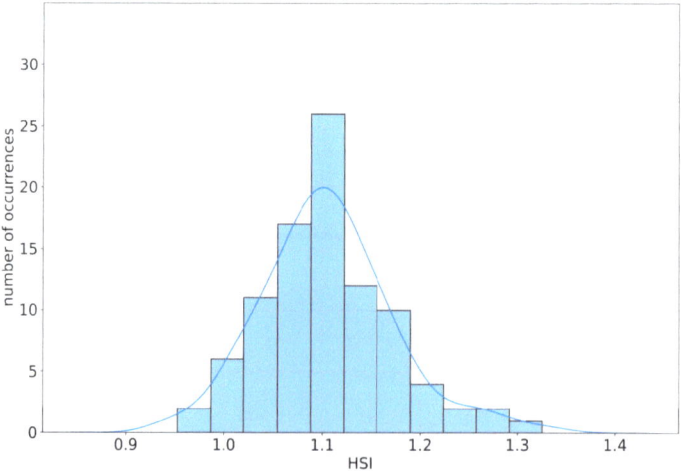

Figure 5. Third harmonic HSI distribution for the combination WB1 = 11 kW, WB2 = 11 kW.

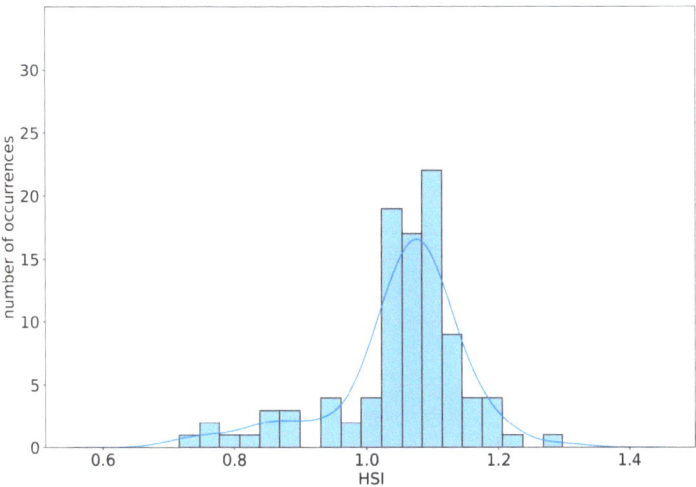

Figure 6. Third harmonic HSI distribution for the combination WB1 = 4 kW, WB2 = 4 kW.

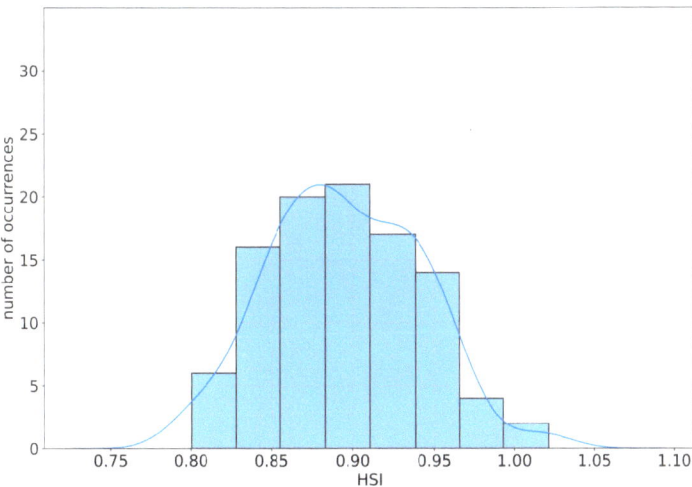

Figure 7. Third harmonic HSI distribution for the combination WB1 = 8 kW, WB2 = 6 kW.

Figures 9 and 10 and Tables 8 and 9 summarize the phenomenon, showing the two most different scenarios in terms of the power level of the tested chargers' set-points for two of the analyzed cases. The y-axis represents the amplitude of a certain harmonic order expressed in percentage with respect to the fundamental. The shown frequencies are the nine with a higher amplitude in percentage with respect to the fundamental, since real-life applications could benefit more from their cancellation. The x-axis represents the HSI as computed using Equation (2). The values presented are the average values of the 100 tests, after the removal of the outliers. It is clear how the summation index increases as the difference between power set-points decreases. In certain cases, the HSI can reach values slightly larger than 1. This can be due to the combination of the measurement errors of the current probes and data acquisition system.

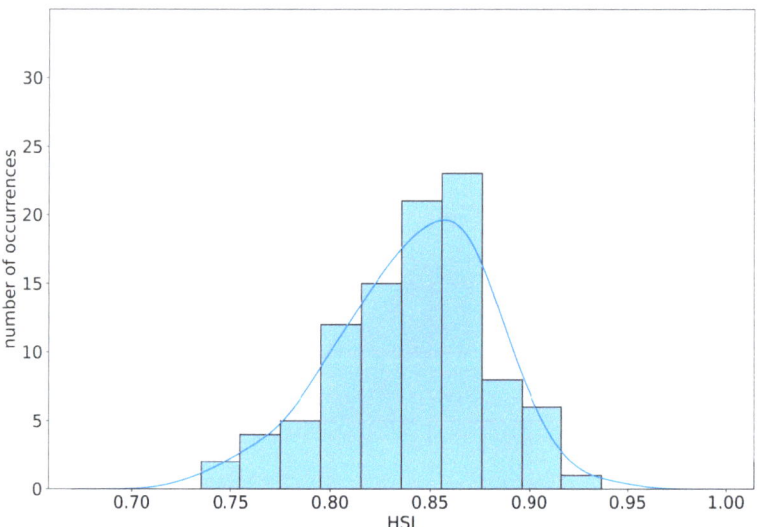

Figure 8. Third harmonic HSI distribution for the combination WB1 = 10 kW, WB2 = 2 kW.

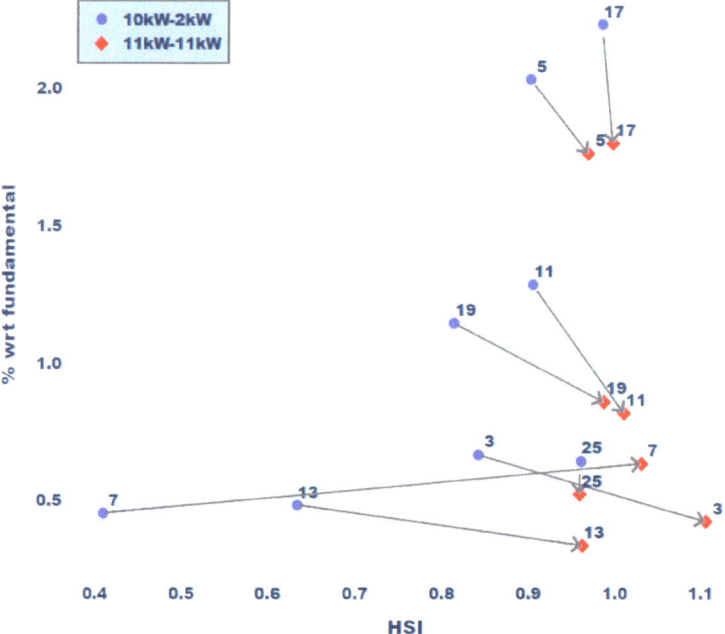

Figure 9. Difference between power set-points: 10/2 kW and 11/11 kW.

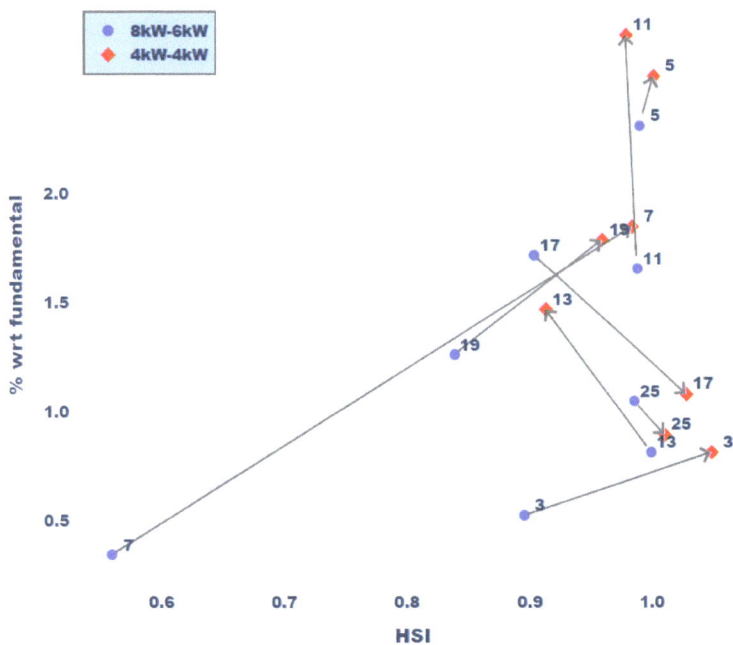

Figure 10. Difference between power set-points: 4/4 kW and 8/6 kW.

Table 8. Difference between power set-points: 10/2 kW and 11/11 kW.

Harmonic Order	Index		Values	
	P1 = 10 kW P2 = 2 kW	P1 = 11 kW P2 = 11 kW	P1 = 10 kW P2 = 2 kW	P1 = 11 kW P2 = 11 kW
17	0.98	1	2.23	1.8
5	0.9	0.97	2.03	1.76
11	0.91	1.01	1.29	0.82
19	0.81	0.99	1.15	0.86
3	0.84	1.11	0.67	0.42
25	0.96	0.96	0.64	0.53
13	0.63	0.96	0.48	0.34
7	0.41	1.03	0.46	0.64

Table 9. Difference between power set-points: 4/4 kW and 8/6 kW.

Harmonic Order	Index		Values	
	P1 = 8 kW P2 = 6 kW	P1 = 4 kW P2 = 4 kW	P1 = 8 kW P2 = 6 kW	P1 = 4 kW P2 = 4 kW
11	0.99	0.98	1.65	2.73
5	0.99	1	2.31	2.54
7	0.56	0.98	0.34	1.85
19	0.84	0.96	1.26	1.79
13	1	0.91	0.81	1.47
17	0.9	1.03	1.72	1.08
25	0.99	1.01	1.05	0.89
3	0.9	1.05	0.52	0.81

6. Conclusions

This paper presents an analysis of the harmonics interaction between two electric vehicle chargers based on measurements on real devices. These results are applicable on the specific tested hardware, but in future works, different charging converter topologies will be studied, both in terms of technology and rated charging power. The experimental distributions show how the summation phenomenon appears on the tested hardware for some specific harmonic orders. It is clearly visible how the power set-points impact the cancellation of certain harmonics. The results are presented in detail for the third harmonic; even though other frequencies have slightly different behavior, the trend of the harmonic cancellation was essentially repeated. The phenomenon is all the more evident the more distant the setpoints are in terms of power. From these results, it can be stated that to decrease the impact of the harmonics on the low-voltage grid, in the case of a high number of chargers connected to the same electrical node, it is possible to elaborate a control strategy of the power set-points of the WB, in order to keep the power of the chargers as different as possible.

Author Contributions: Conceptualization, E.P.; Software, G.B. and C.N.; Validation, M.Z.; Formal analysis, A.M.; Investigation, G.B.; Resources, P.T., M.Z. and R.J.; Data curation, G.B. and C.N.; Writing—original draft, A.M., G.B. and E.P.; Writing—review & editing, A.M. and E.P.; Supervision, E.P.; Project administration, R.J.; Funding acquisition, E.B. and P.T. All authors have read and agreed to the published version of the manuscript.

Funding: This publication is part of the project NODES, which has received funding from the MUR—M4C2 1.5 of PNRR, with grant agreement no. ECS00000036, and of the project PNRR-NGEU, which has received funding from the MUR—DM352/2022.

Data Availability Statement: The data presented in this study are available on request from the corresponding author. The data are not publicly available due to confidentiality reasons.

Conflicts of Interest: The authors declare no conflict of interest. The funders had no role in the design of the study; in the collection, analyses, or interpretation of data; in the writing of the manuscript; or in the decision to publish the results.

Abbreviations

The following abbreviations are used in this manuscript:

BEV	Battery Electric Vehicle
DUT	Device Under Test
G2V	Grid-to-Vehicle
EV	Electric Vehicle
EVCS	Electric Vehicle Charging Station
FFT	Fast Fourier Transformation
HSI	Harmonic Summation Index
PHEV	Plug-in Hybrid Electric Vehicle
PHIL	Power Hardware-in-the-Loop
RES	Renewable Energy Source
RMS	Root Mean Square
THD	Total Harmonic Distortion
V2X	Vehicle-to-Everything
V2B	Vehicle-to-Building
V2G	Vehicle-to-Grid
WB	WallBox

Appendix A

In Figures A1–A5, the harmonic spectra of the same charger for the other five power setpoints are shown; note that these signatures are only one of the one hundred measurements taken in the study. The behavior among the tests are similar, but some differences can be found; this aspect has been considered in the present work to find the distribution of the HSI.

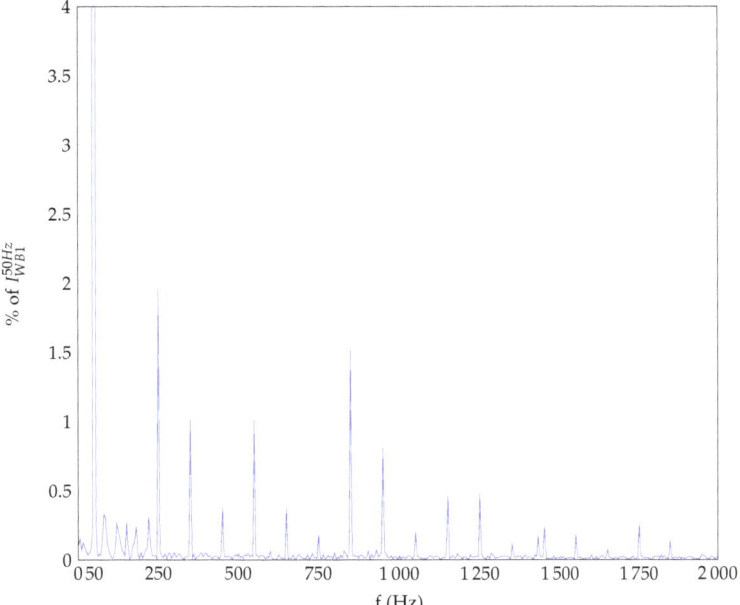

Figure A1. FFT spectrum of I_{WB1} at 11 kW.

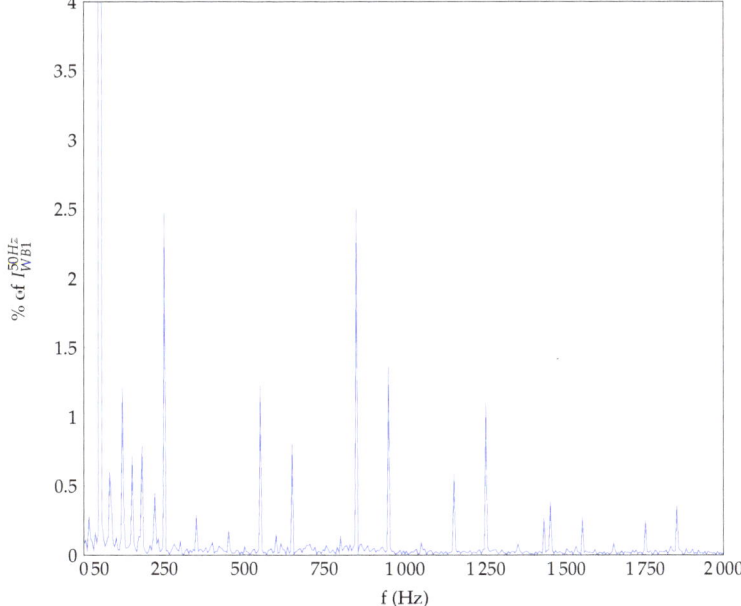

Figure A2. FFT spectrum of I_{WB1} at 8 kW.

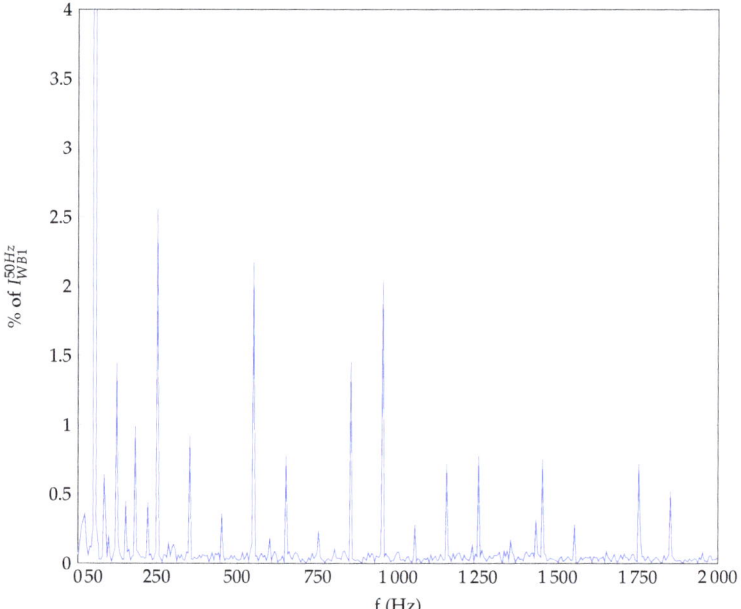

Figure A3. FFT spectrum of I_{WB1} at 6 kW.

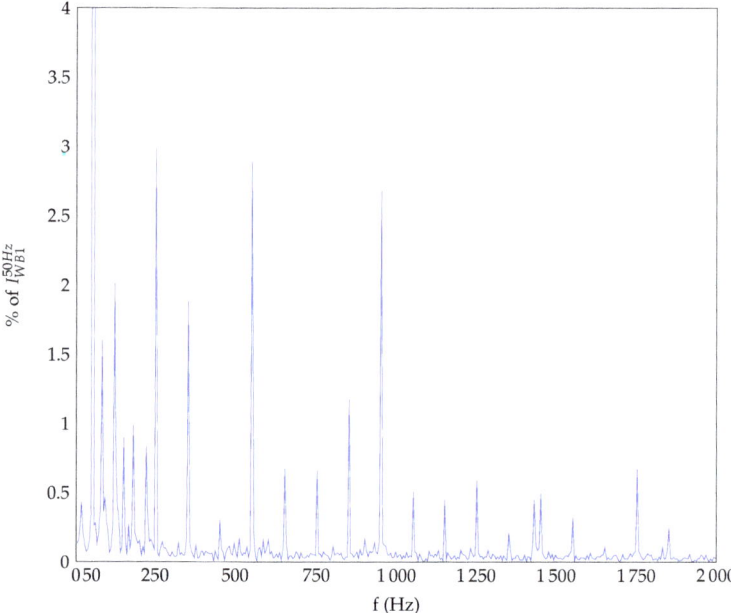

Figure A4. FFT spectrum of I_{WB1} at 4 kW.

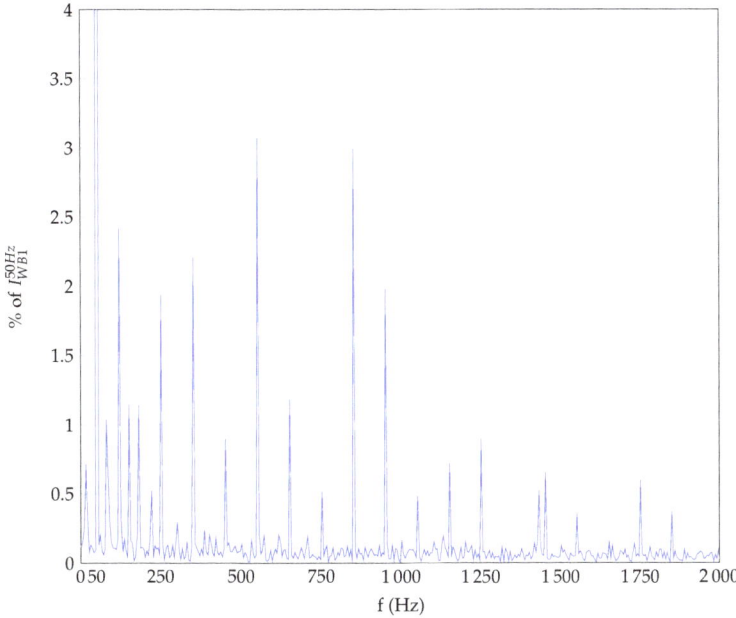

Figure A5. FFT spectrum of I_{WB1} at 2 kW.

References

1. EU. European Green Deal. Available online: https://commission.europa.eu/strategy-and-policy/priorities-2019-2024/european-green-deal_en (accessed on 28 July 2019).
2. IEA. *Global EV Outlook 2023–Analysis*; IEA: Paris, France, 2023.
3. IEA. *Technology and Innovation Pathways for Zero-Carbon-Ready Buildings by 2030–Analysis*; IEA: Paris, France, 2022.
4. Savari, G.F.; Sathik, M.J.; Raman, L.A.; El-Shahat, A.; Hasanien, H.M.; Almakhles, D.; Abdel Aleem, S.H.E.; Omar, A.I. Assessment of charging technologies, infrastructure and charging station recommendation schemes of electric vehicles: A review. *Ain Shams Eng. J.* **2023**, *14*, 101938. [CrossRef]
5. Barreto, R.; Faria, P.; Vale, Z. Electric Mobility: An Overview of the Main Aspects Related to the Smart Grid. *Electronics* **2022**, *11*, 1311. [CrossRef]
6. *IEEE Std 519-2014 (Revision of IEEE Std 519-1992)*; IEEE Recommended Practice and Requirements for Harmonic Control in Electric Power Systems. IEEE: New York, NY, USA, 2014; pp. 1–29. [CrossRef]
7. *IEC 61000-4-30*; Standard IEC 61000-4-30 Electromagnetic Compatibility (EMC). Testing and Measurement Techniques. Power Quality Measurement Methods. International Electrotechnical Commission: Geneva, Switzerland, 2015.
8. Nour, M.; Chaves-Avila, J.P.; Magdy, G.; Sanchez-Miralles, A. Review of Positive and Negative Impacts of Electric Vehicles Charging on Electric Power Systems. *Energies* **2020**, *13*, 4675. [CrossRef]
9. IEC. *International Electrotechnical Vocabulary*; International Electrotechnical Committe (IEC): Geneva, Switzerland, 2016.
10. *EN 50160*; Standard EN 50160 Voltage Characteristics of Electricity Supplied by Public Electricity Networks. CENELEC: Brussels, Belgium, 2022.
11. *TR 61000-1-1*; IEC TR 61000-1-1 Electromagnetic Compatibility (EMC)-Part 1-1: General-Application and Interpretation of Fundamental Definitions and Terms. International Electrotechnical Commission: Geneva, Switzerland, 2023.
12. Dougherty, J.; Stebbins, W. Power quality: A utility and industry perspective. In Proceedings of the 1997 IEEE Annual Textile, Fiber and Film Industry Technical Conference, Greenville, SC, USA, 6–8 May 1997; pp. 5–10. 598528. [CrossRef]
13. Pandya, R.; Bhavsar, F. An Overview on Power Quality Issues In Smart Grid. *IOSR J. Electr. Electron. Eng.* **2018**, *13*, 1–4.
14. Awadallah, M.A.; Singh, B.N.; Venkatesh, B. Impact of EV Charger Load on Distribution Network Capacity: A Case Study in Toronto. *Can. J. Electr. Comput. Eng.* **2016**, *39*, 268–273. [CrossRef]
15. Alghsoon, E.; Harb, A.; Hamdan, M. Power quality and stability impacts of Vehicle to grid (V2G) connection. In Proceedings of the 2017 8th International Renewable Energy Congress (IREC), Dead Sea, Jordan, 21–23 March 2017; pp. 1–6. [CrossRef]
16. *Fast Charging Diversity Impact on Total Harmonic Distortion Due to Phase Cancellation Effect: Fast Charger's Testing Experimental Results*; Publications Office of the European Union: Brussels, Belgium, 2017. [CrossRef]
17. Mazurek, P.; Chudy, A. An Analysis of Electromagnetic Disturbances from an Electric Vehicle Charging Station. *Energies* **2022**, *15*, 244. [CrossRef]

18. Slangen, T.M.H.; van Wijk, T.; Ćuk, V.; Cobben, J.F.G. The Harmonic and Supraharmonic Emission of Battery Electric Vehicles in The Netherlands. In Proceedings of the 2020 International Conference on Smart Energy Systems and Technologies (SEST), Istanbul, Turkey, 7–9 September 2020; pp. 1–6. [CrossRef]
19. Bollen, M.H.J.; Olofsson, M.; Larsson, A.; Ronnberg, S.K.; Lundmark, M. Standards for supraharmonics (2 to 150 kHz). *IEEE Electromagn. Compat. Mag.* **2014**, *3*, 114–119. [CrossRef]
20. Rönnberg, S.; Bollen, M. *Propagation of Supraharmonics in the Low Voltage Grid. Report 2017:461*; Energiforsk: Stockholm, Sweden, 2017.
21. Lucas, A.; Bonavitacola, F.; Kotsakis, E.; Fulli, G. Grid harmonic impact of multiple electric vehicle fast charging. *Electr. Power Syst. Res.* **2015**, *127*, 13–21. [CrossRef]
22. Dharmakeerthi, C.; Mithulananthan, N.; Saha, T. Overview of the impacts of plug-in electric vehicles on the power grid. In Proceedings of the 2011 IEEE PES Innovative Smart Grid Technologies, Perth, WA, Australia, 13–16 November 2011; pp. 1–8. [CrossRef]
23. Mazza, A.; Pons, E.; Bompard, E.; Benedetto, G.; Tosco, P.; Zampolli, M.; Jaboeuf, R. A Power Hardware-In-the-Loop Laboratory Setup to Study the Operation of Bidirectional Electric Vehicles Charging Stations. In Proceedings of the 2022 International Conference on Smart Energy Systems and Technologies (SEST), Eindhoven, The Netherlands, 5–7 September 2022; pp. 1–6. [CrossRef]
24. Benedetto, G.; Bompard, E.; Mazza, A.; Pons, E.; Jaboeuf, R.; Tosco, P.; Zampolli, M. Impact of bidirectional EV charging stations on a distribution network: a Power Hardware-In-the-Loop implementation. *Sustain. Energy Grids Netw.* **2023**, *35*, 101106. [CrossRef]
25. *4Q POWER AMPLIFIER PCU-3x7000-AC/DC-400V-54A-4G with Different Options-User Manual*; Spherea Puissance Plus: Colomiers, France, 2019.
26. *IEEE Std 519-2022 (Revision of IEEE Std 519-2014)*; IEEE Standard for Harmonic Control in Electric Power Systems. IEEE: New York, NY, USA, 2022; pp. 1–31. https://doi.org/10.1109/IEEESTD.2022.9848440.
27. Ćuk, V.; Cobben, J.F.; Kling, W.L.; Ribeiro, P.F. Analysis of harmonic current summation based on field measurements. *IET Gener. Transm. Distrib.* **2013**, *7*, 1391–1400. [CrossRef]

Disclaimer/Publisher's Note: The statements, opinions and data contained in all publications are solely those of the individual author(s) and contributor(s) and not of MDPI and/or the editor(s). MDPI and/or the editor(s) disclaim responsibility for any injury to people or property resulting from any ideas, methods, instructions or products referred to in the content.

Article

Enhancing Grid Operation with Electric Vehicle Integration in Automatic Generation Control

Zahid Ullah [1], Kaleem Ullah [2], Cesar Diaz-Londono [1,*], Giambattista Gruosso [1] and Abdul Basit [3]

[1] Dipartimento di Elettronica, Informazione e Bioingegneria, Politecnico di Milano, Piazza Leonardo da Vinci, 32, 20133 Milano, Italy; zahid.ullah@polimi.it (Z.U.); giambattista.gruosso@polimi.it (G.G.)
[2] US-Pakistan Center for Advanced Studies in Energy, University of Engineering and Technology Peshawar, Peshawar 25000, Pakistan; kaleemullah@uetpeshawar.edu.pk
[3] Manager R&D, National Power Control Center, National Transmission and Dispatch Company, Islamabad 44000, Pakistan; abdul.basit@ntdc.com.pk
* Correspondence: cesar.diaz@polimi.it

Abstract: Wind energy has been recognized as a clean energy source with significant potential for reducing carbon emissions. However, its inherent variability poses substantial challenges for power system operators due to its unpredictable nature. As a result, there is an increased dependence on conventional generation sources to uphold the power system balance, resulting in elevated operational costs and an upsurge in carbon emissions. Hence, an urgent need exists for alternative solutions that can reduce the burden on traditional generating units and optimize the utilization of reserves from non-fossil fuel technologies. Meanwhile, vehicle-to-grid (V2G) technology integration has emerged as a remedial approach to rectify power capacity shortages during grid operations, enhancing stability and reliability. This research focuses on harnessing electric vehicle (EV) storage capacity to compensate for power deficiencies caused by forecasting errors in large-scale wind energy-based power systems. A real-time dynamic power dispatch strategy is developed for the automatic generation control (AGC) system to integrate EVs and utilize their reserves optimally to reduce reliance on conventional power plants and increase system security. The results obtained from this study emphasize the significant prospects associated with the fusion of EVs and traditional power plants, offering a highly effective solution for mitigating real-time power imbalances in large-scale wind energy-based power systems.

Keywords: electric vehicle area; automatic generation control; forecasting errors; power dispatch strategies; modern power grid

1. Introduction

Among other renewable energy technologies, wind energy technology has made significant progress globally, with interconnections to various voltage levels of power systems. However, the inherent intermittency of wind speed makes wind farms stochastic, yielding inaccurate predictions that can cause mismatches between generation and load demand, affecting power system operations and leading to deviations from scheduled values. Power system schedulers use various strategies to balance generation and load throughout the day. However, the uncertainty of wind power often results in an energy imbalance between supply and load demand that necessitates the deployment of additional operational reserves. These reserves are usually provided by conventional power plants, resulting in higher operational costs and increased CO_2 emissions [1]. Extensive research on large-scale wind power integration has pertinently increased the use of operating reserves to sustain an active power balance in the system. This emphasizes the influence of wind power's uncertain behavior on reserve requirements. To optimize the utilization of wind power resources effectively, it is essential to instill flexibility into electric vehicles (EVs), allowing them to actively engage in the demand–supply equilibrium as needed [2–4].

The concept of flexibility in the smart grid context has been explained in detail using mathematical models for flexibility [5]. Moreover, real-time flexibility is ensured using peer-to-peer energy trading. Implementing superior coordination control strategies is vital for the optimal utilization of EVs, leading to substantial reductions in operating costs and carbon emissions [6–8].

1.1. Related Work

Over the past decade, extensive research has focused on the vehicle-to-grid (V2G) technology of EVs, driven by their significant potential to provide grid ancillary services actively [6–13]. By adopting the V2G mode, EVs can operate as battery storage systems, enabling bi-directional power flow with the power grid. This enhanced capability supports grid flexibility and resilience, paving the way for more efficient energy utilization and demand response management. EVs are not confined to a single location but are dispersed across regions and utilized for commuting or long-distance travel [9]. A study indicated that the average roundtrip driving distance in the U.S. is approximately 50 km, with an average driving time of ~52 min, although there is significant variability. A survey of U.S. drivers nationwide reveals that 60% of commuters travel distances less than 80 km [9]. Importantly, EVs employed for daily commuting remain idle for approximately 22 h per day, accumulating surplus energy stored in their batteries during travel. This excess energy presents an opportunity to support the grid and can be used to recharge EV batteries. Over the past decade, extensive research has explored the contribution of EVs to secondary frequency response and conceptually frame worked the EVs integration into bulk power systems, considering technical grid operation and the electricity market [9–11]. The challenges and benefits of the proposed integrated framework have been examined, focusing on mitigating anticipated errors. Regarding distributed system management, energy communities' growth driven by cheaper storage and economy-driven energy exchange has been explored [12,13]. Further, novel transactive control frameworks were introduced, optimizing energy scheduling between prosumers and storage providers and offering two game-theory-based algorithms adaptable to grid communication.

Meanwhile, the system response has been thoroughly analyzed at the inertial and primary control levels [14,15], employing a finely tuned adaptive mechanism to ensure utmost system reliability even under arduous conditions. Hence, the efficient harnessing of EVs' capabilities for grid regulation holds great potential for significantly augmenting the proportion of renewable energy in future power systems. Giordano et al. [16] investigated the impact of increasing EVs on grids, focusing on aggregator-led scheduling for grid stability. They proposed automated logic for day-ahead EV fleet charging, maintaining grid balance, and successfully testing it on three EVs without grid disruption. Diaz-Londono et al. [17] proposed two optimal strategies; one focused on lower energy prices and the other on providing flexible grid capacity, aimed at integrating EVs efficiently and avoiding transformer overloads. Moreover, the same group discussed how evolving energy practices impact power grid regulations and the role of aggregators in connecting flexible loads, like EVs, to the grid [18]. Based on a financial perspective and methodology, the benefits for aggregators and end-users were assessed, highlighting scenarios where aggregation is advantageous, and revealing potential conflicts of interest, with numerical results demonstrating varied consumer benefits and situations where intermediaries may not be beneficial.

Mignoni et al. [19] presented a novel control strategy for optimizing the scheduling of an energy community comprising prosumers with unidirectional V1G and V2B capabilities. Long-term parked EVs served as temporary storage systems for prosumers, while prosumers offered V1G services to EVs at charging stations. To handle the framework's stochastic nature, EVs shared their parking and recharging time distributions with prosumers, enhancing energy allocation. Prosumers and EVs, acting as self-interested agents, engage in a rolling horizon control framework to reach operating strategy agreements, framed as a generalized Nash equilibrium problem solved in a distributed manner. Meanwhile, Hosseini et al. [20] present a resilient, decentralized charging approach for extensive EV fleets, aiming to reduce energy expenses and battery degradation while addressing fluctuations in power costs and inelastic loads. A robust optimization based on uncertainty sets formulated the challenge as a manageable quadratic programming problem with restrictions on grid resource sharing. The practicality of utilizing a commercially available EV to offer grid flexibility in real distribution networks has been examined [21]. More specifically, the employed controller who adheres to IEC 61851 and SAE J1772 standards [22] and a Nissan Leaf was assessed in a Danish distribution grid to deliver congestion management, voltage support, and frequency regulation. Performance metrics, including EV response time and precision, were appraised to validate smart grid concepts using standard-compliant equipment. EVs provide frequency regulation services in renewable energy-rich power systems [23], employing a leader–follower game between EVs and their aggregator to optimize charging and regulation scheduling while addressing signal uncertainty. The aggregator incentivizes EV participation through pricing, and EVs aim to balance consumption costs and regulation revenues. Moreover, Tushar et al. [24] discuss the importance of microgrid technology and integrating electric vehicles, energy storage, and renewables for efficient electricity management. They introduce a real-time decentralized demand side management system that optimizes residential electricity consumption and improves microgrid planning for enhanced power delivery quality.

While a substantial body of literature has addressed EVs integration challenges, it is imperative to note that significant considerations and gaps persist, awaiting further exploration and resolution. For instance, a closed-loop control methodology implemented in the context of EV participation in the AGC system [25] accomplished bidirectional power flow for charging and frequency regulation. However, the study's assumed time delay of 1–2 s contrasts with actual turbine and EV responses, which suggests a longer delay time of 7–8 s. Moreover, a robust frequency regulator was devised for a power grid comprising multiple interconnected regions, considering EVs involvement in load frequency control services, thus enhancing the resilience of Automatic Generation Control (AGC) services [26]. However, this study lacked consideration for practical constraints, such as higher time delays and dead bands, and did not thoroughly assess realistic EV capacity. Meanwhile, Sanki et al. [27] integrated plug-in EV services into the AGC system to tackle the grid stability challenges of integrating highly intermittent solar and wind technologies into grid operations. Khezri et al. [28] incorporated EVs in the AGC regulation process utilizing a consolidated model of EVs governed by a fractional order-PID controller to manage the discharging state of EVs. However, this study omitted the EV contribution over 24 h when assessing EV availability from the consumer side. Therefore, the responsive involvement of the EVs in AGC necessitates a more thorough examination of extensive power grid models. This requires careful consideration of practical constraints, including delays, parametric uncertainties, and dead bands. The approach used in this study will offer valuable perspectives into the dynamic performance of EVs and their influence on grid stability, ultimately contributing to establishing more resilient and reliable AGC-based power systems.

1.2. Our Contributions

This study investigates utilizing the storage capabilities of EVs to reinforce future power systems, particularly in controlling operations involving massive wind power integration. The primary aim is to create a simple yet sturdy and responsive AGC system for a real power system network to regulate the system frequency efficiently and cost-effectively. By effectively combining the capacities of EVs with thermal energy systems (TES), the AGC model offers improved active power regulation services. The developed model introduces an enhanced allocation of regulating reserves from EVs while considering their power threshold levels. The dispatch strategy formulated for the AGC system prioritizes the utilization of reserves from EVs over those from TESs in grid balancing procedures. Integrating EVs ensures greater flexibility, cost-effectiveness, and reduced environmental strain, leading to a more resilient and eco-friendly energy landscape. Moreover, the DigSILENT Power Factory software (2019 SP3) assesses the proposed AGC dispatch strategy. The proposed model integrates detailed models of various generating units, including wind energy systems (WES), TES, gas turbine energy systems (GTES), and an electric vehicle area (EVA). Additionally, a specific windy day in 2023 was chosen to study and analyze forecasting errors in a large-scale wind energy-based power system network.

The primary contributions of this research are as follows:

- A comprehensive power system model has been developed, incorporating key generating units like TES, GTES, and WES. Furthermore, a comprehensive EVA model is developed, harnessing frequency control capabilities utilizing the concepts of positive and negative regulation capacities.
- A centralized AGC model for the proposed power system is developed to facilitate secondary frequency response and ensure power balancing operations.
- A real-time dynamic dispatch strategy is formulated for the AGC model to efficiently integrate reserve capacities from the EVA model and prioritize its utilization over TES.

1.3. Paper Outline

This paper is structured systematically, with Section 2 focusing on the detailed modeling of TES, GTES, WES, and the EVA model. In Section 3, the AGC system modeling is outlined, incorporating the model of the power plant units and EVA system. Section 4 delineates the proposed power dispatch approach and validates its performance. Section 5 concludes the paper, drawing insightful conclusions from the investigation and providing future recommendations.

2. Generating Units and EVA Modelling

This section comprehensively overviews EVs' modeling process and various power plant units, such as TES, GTES, and WES. Governors are accurately engineered and strategically positioned on each generating unit to ensure a highly efficient primary frequency response. Their vital role is effectively regulating and stabilizing the system's power output. Additionally, we developed a sophisticated EVA model that plays a key role in this integrated system. The EVA model receives dispatch orders from the AGC system and employs an advanced inbuilt algorithm to efficiently route the instructions to individual EVs. This intelligent routing system ensures that the required secondary regulating reserves are promptly provided, contributing to the overall grid stability and efficient energy management.

2.1. EVA Modelling for Grid Support

The integration of EVs offers substantial potential in effectively managing the system frequency and maintaining a harmonious equilibrium between demand and generation. This is accomplished by employing the EVs as both load and source, controllable by the AGC regulator. The AGC controller is crucial in supporting grid operations, as it promptly responds to any fluctuations in the system frequency. An EVA designates a specific zone where many EVs are assembled and overseen by a dedicated control center. These EVAs are entrusted with dispatch orders from the AGC and utilize an intrinsic algorithm, illustrated in Figure 1, to allocate the orders to individual EVs. The algorithm functions in actual time, perpetually calculating the controlling capability of EVAs for the existing dispatch interval. Consequently, the aggregator must comprehend each EV condition and conduct during this duration.

This research proposes a novel EVA model designed specifically for the AGC system. The formulation of the model follows a first-order transfer function, incorporating two pivotal parameters: frequency gain (K_{EV}) and the time constant for charging and discharging (T_{EV}). To attain maximum precision and pragmatic feasibility, we carefully integrated the inherent time lag reaction of the EVA model into the AGC system, covering a span of 0 to 3 s. This time delay encompasses two critical factors significantly influencing the overall system dynamics. The first is the duration for the aggregator to transmit the received orders to individual EVs within the system. This step introduces a certain degree of time delay in the overall response. The second factor contributing to the time delay is the inherent latency arising from communication channels, typically on the order of milliseconds. By acknowledging these time delay elements, we aims to construct a model that closely emulates real-world conditions. Our study delves into analyzing the response of EVs in the AGC system at the power system level, considering various important aspects, such as the time delay, dead band, and dynamic response characteristics. This comprehensive analysis provides valuable insights into the functioning and performance of the AGC system in conjunction with the proposed EVA model.

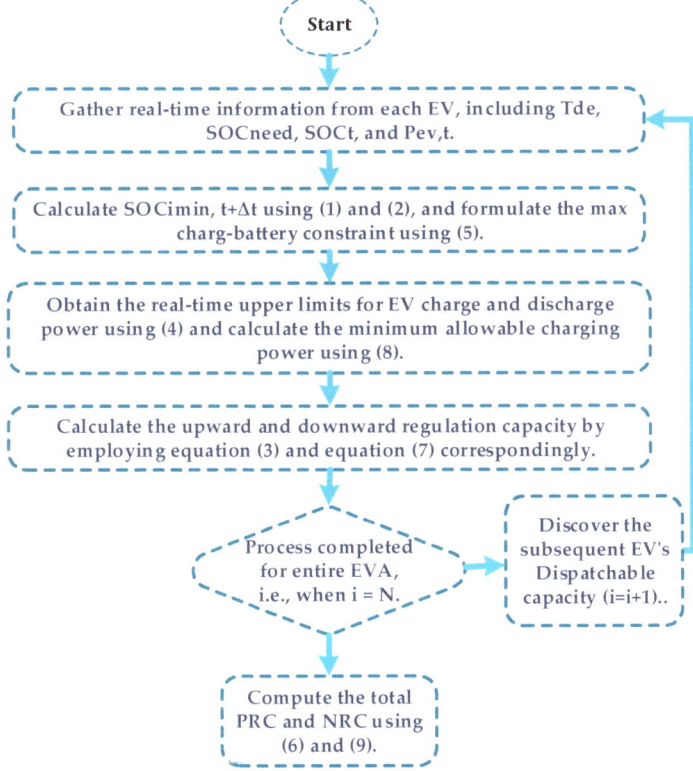

Figure 1. Flow chart for calculation of PRC and NRC.

The EVA model provides positive regulation capacity (PRC) during positive imbalances in the system and negative regulation capacity (NRC) during negative imbalances. We consider a group of 17,000 EVs, each with an average battery capacity of 60 KWh (C_i). The installed inverters have an average capacity of 7.5 KW. Consequently, the cumulative peak power accessible for regulation intentions is ±127.5 MW. Figure 2 illustrates the calculations for PRC and NRC. In the case of a single EV, within a specific time interval Δt, PRC can be defined as the discrepancy between the present charging power ($P^i_{EV,t} < 0$) and the maximum discharging power ($P^i_{\Delta t} > 0$). Conversely, NRC is determined by the difference between the current discharging power ($P^i_{EV,t} > 0$) and the maximum charging power ($P^i_{\Delta t} < 0$). To execute the PRC process, the loads linked to EVs are restricted, or the accumulated energy in their batteries is transmitted back to the grid using sophisticated V2G technology. In contrast, the EVs' power demand is augmented for NRC operation to facilitate power absorption from the grid to charge their batteries. The role of the aggregator is paramount in efficiently managing the collective operation of all EVs, seamlessly orchestrating their contributions during specific time intervals, thereby ensuring a highly effective and harmonized regulation response.

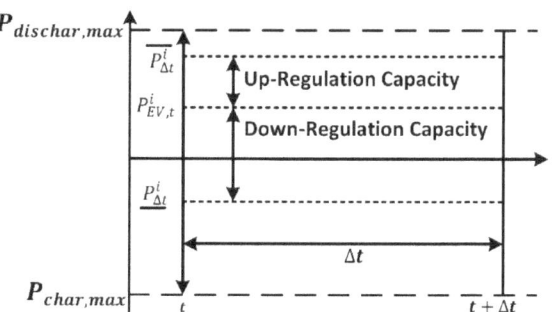

Figure 2. Calculation of PRC and NRC of EVs.

Regulation Capacities

Determining PRC and NRC entails a scrupulous procedure, as depicted in Figure 2, where specific parameters are carefully measured. Subsequently, precise calculations are carried out for each dispatch interval to accurately evaluate the PRC and NRC. The realization of the PRC operation entails two key methods: mitigating the load effect or facilitating the transfer of EV battery stored power back to the grid through precise controls. For PRC computation, this study incorporates two important constraints. First, the current state of charge (SoC) represented as $\left(SoC_t^i\right)$ must align with the user's specific requirements SoC_{need}^i. It is employed for regulation purposes at time $(t + \Delta t)$ as formulated in (1):

$$SoC_{min,\, t+\Delta t}^i \geq \frac{\left(SoC_{need}^i - \left(-P_{charg,\, max}\right) \times \eta \times \left(T_{dep,\, i} - (t + \Delta t)\right)\right)}{Ci} \quad (1)$$

where η represents the coefficient signifying the battery's discharge and charge effectiveness, Ci indicates the battery's capability, and $T_{dep,\, i}$ specifies the time of departure for the ith electric vehicle.

The second constraint pertains to battery deterioration, predominantly induced by charge cycles. Consequently, to balance achieving sufficient regulatory capacity and safeguarding battery health, careful attention is paid to the depth of discharge. This research establishes a limit of 60% for the depth of discharge (DoD) power, ensuring that the battery operates within a controlled DoD, mitigating the detrimental effects of excessive cycling while providing the necessary regulatory capacity.

$$SoC_{min,\, t+\Delta t}^i \geq 40 \quad (2)$$

Figure 2 depicts the charging or discharging power of EVs that can be precisely adjusted or increased within a specified Δt when $P_{\Delta t}^i$ surpasses $P_{EV,t}^i$. This observation highlights the ability to enhance the PRC of each EV for the interval Δt, enabling an active contribution to grid stabilization and power regulation. The PRC of an EV within Δt is determined by considering the dynamic interrelationship between the EV's charging or discharging power and its maximum available discharging power. This calculation effectively quantifies the capacity of each EV to participate in the positive regulation process, optimizing the overall power management system.

$$P_{PRC}^i = P_{\Delta t}^i - P_{EV,t}^i \quad (3)$$

$$
\text{Here, } P^i_{\Delta t} = \begin{cases} min(P_{charg,max}, \dfrac{(\Delta SoC^i_t \times C_i)}{\Delta t \times \eta}) & \text{if } \Delta SoC^i_t > 0 \\ max(P_{dicharg,max}, \dfrac{(\Delta SoC^i_t \times C_i)}{\Delta t \times \eta}) & \text{if } \Delta SoC^i_t < 0 \end{cases} \qquad (4)
$$

where P^i_{PRC} represents the PRC capacity of the ith vehicle, and the variation in SoC represented as ΔSoC^i_t indicates the potential for increasing the maximum capacity of the EVs within Δt.

$$\Delta SoC^t_t = SoC^i_{EV,min,\ t+\Delta t} - SoC^i_{EV,t} \qquad (5)$$

Derived from this, the entire PRC of EVAs can be computed in the following manner:

$$\Delta P^i_{PRC(total)} = \sum_{i=1}^{N} P^i_{PRC} \qquad (6)$$

Analogously, the charge or discharge rate of EVs can be curtailed for a specific time interval Δt when $P^i_{\Delta t}$ surpasses P^i_t as shown in Figure 2. By implementing this measure, the NRC of each EV for the time interval Δt can be effectively ascertained. The NRC capacity of an EV during Δt is determined through a comprehensive assessment of its charging or discharging power in comparison to the maximum available discharging power.

$$P^i_{NRC} = (P^i_{EV,t} - P^i_{\Delta t}) \qquad (7)$$

$$P^i_{\Delta t} = min(P_{charg,\ max},\ \dfrac{(\Delta SoC^i_t \times C_i)}{\Delta t \times eff}) \qquad (8)$$

where P^i_{NRC} represents the ith vehicle participating in the negative regulation capacity. Given this premise, the cumulative NRC of EVA can be computed as follows:

$$\Delta P^i_{NRC(total)} = \sum_{i=1}^{N} P^i_{NRC} \qquad (9)$$

In this context, it is essential to note that the constraints governing NRCs primarily revolve around the SoC and the maximum charging power of the charger.

$$SoC^i_{min,\ t+\Delta t} \leq 100\% \qquad (10)$$

2.2. Modelling of the Thermal Energy System (TES)

This research extensively analyses the aggregated TES model concerning active power balancing control. A particular emphasis is placed on the boiler response time, a crucial parameter affecting the overall plant's reaction and system stability. The TES model, depicted in Figure 3, is derived from previously described models [29,30]. The TES model underwent simplification to facilitate long-term dynamic simulations. Two essential inputs, main steam pressure (P_t) from the boiler and control block, and the control valve (cv) from the governor block, determine the mechanical output power (P_{mech}) of the steam turbine block. When load fluctuations (L_R) arise, the boiler model quickly computes the suitable (P_t) value to offset the load changes, considering the turbine's output limitations and steam energy storage delays.

This all-encompassing approach guarantees superior precision and dependability in the dynamic simulation studies of the TPS model. The L_R signal assumes a pivotal role, functioning as a forward signal to the boiler and a controller for the turbine valve. The model combines the influence of steam temperature regulation and generator reference current for accuracy and steadiness. The ramp-rate limit is maintained at 30 MW/min for a controlled response. The time lags (T_{b1}, T_{b2}, and T_{b3}) in the boiler model profoundly affect frequency and power time reactions. The boiler response severely affects the overall turbine response, taking about 5 to 6 min to settle. The power-to-mechanical conversion relies on the boiler model's response (P_t) and the governor's output (cv), contributing to system dynamic behavior and stability. The steam turbine's response depends on four-time constants (T_1, T_2, T_3, and T_4) representing different volumes. Coefficients (K_1–K_8) determine power contributions from turbine sections. The speed governor regulates the turbine's speed valve, considering generator speed and droop settings as inputs for primary frequency response, ensuring stability with a dead zone to prevent unnecessary adjustments.

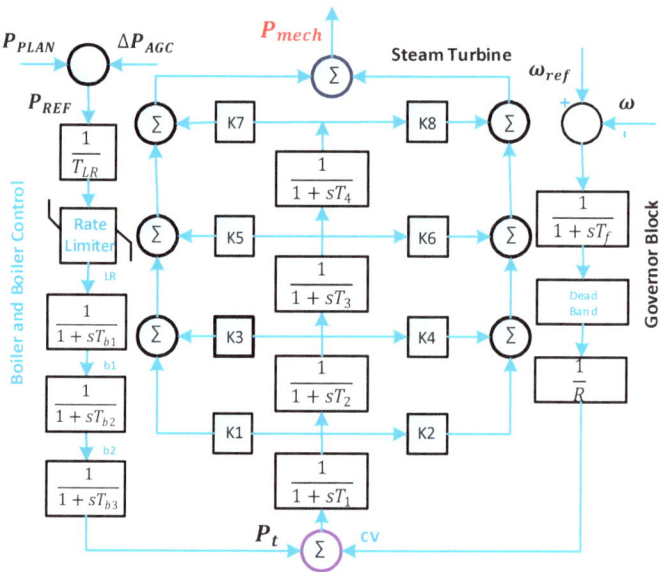

Figure 3. Thermal energy system model.

2.3. Modelling of Gas Turbine Energy System (GTES)

A detailed GTES model was developed (Figure 4). The GTES is the primary response source, achieved through the governor linked to the generator's turbine. The GTES governor incorporates a dead band and a low-pass filter, showcasing droop characteristics. The dead band ensures stability by disregarding low-frequency deviations, while the low-pass filter stabilizes rotor speed against high-frequency deviations, further promoting system stability. Any power deviation beyond the dead band's limits leads to frequency deviations, which activates the droop characteristic signal, ultimately generating a power demand signal (ΔP_c). This signal drives the necessary adjustments in the power generation process, allowing the GTES to promptly and effectively respond to power fluctuations within the system. The ΔP_c signal holds utmost significance in the functioning of the GTES, which includes the power limitation block (PLB), power distribution block (PDB), and gas turbine dynamics block (GTDB) (Figure 4). The PLB imposes physical constraints on the turbine's response, enforcing upper and lower power level restrictions (P_{max} and P_{min}) based on combustion technology limitations. To comply with combustion constraints, the set points L_{max} and L_{min} act as maximum and minimum load limits. Additionally, a rate limiter block carefully regulates the rate of change for the ΔP_c signal to optimize gas turbine performance while ramping up and down processes. The PLB produces a CLC signal that acts as an input to the PDB. Two sequential combustion chambers are included in the PDB block that skillfully blend compressed air with fuel to initiate efficient combustion processes. Initially, the environment incineration chamber receives compressed air, subjecting it to warming, and deftly blends it with 50% of the overall fuel. Subsequently, the mixture is forcibly expelled through a high-pressure turbine, provoking its rapid rotation. The resultant mixture is directed into the SEV chamber. Here, an additional 50% of the distributed gasoline is intricately amalgamated with a measured quantity of supplementary air, guaranteeing an optimal combustion process.

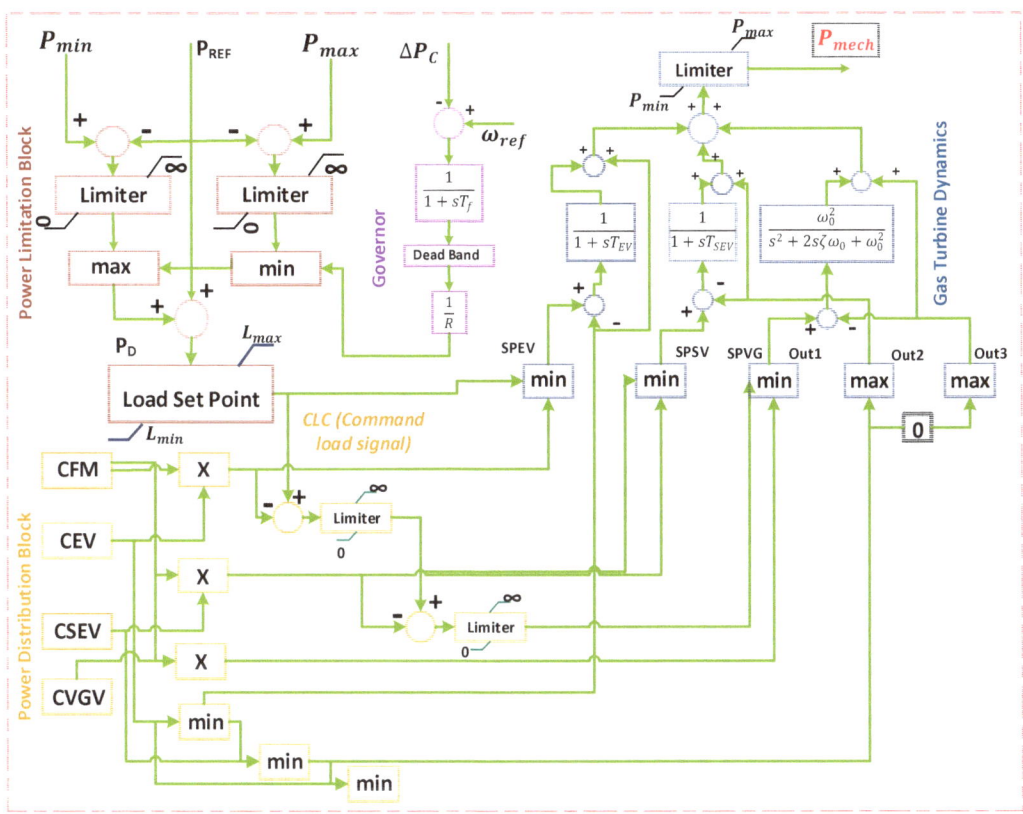

Figure 4. Gas turbine energy system model (GTES).

The setup includes a low-pressure turbine with operational adaptability, low emissions, and high efficacy. Power contribution coefficients such as CVGV, CFM, CEV, and CSEV are precisely adjusted to mirror system traits. The capabilities of the combustors, air compressor, and CLC signal from the power limitation block determine the output variables SPEV, SPSV, and SPVG. The configuration ensures optimal performance and environmental consciousness. The sophisticated interaction of factors significantly influences power generation, providing resource optimization. Gas turbine dynamics are closely linked with compressor and combustor dynamics. First-order lcg functions elegantly represent the environment and sequential environment combustor dynamics, while second-order functions aptly describe VIGV dynamics. The mechanical power output (P_{mech}) of the GTES relies on CFM, CEV, CSEV, CVGV, and CLC, ensuring optimal performance within the specified limits. The GTES exhibits a response time of 30 to 40 s when subjected to a step change in input power, primarily due to turbine ramp rates impacting its overall response time.

2.4. Modelling of Wind Energy System (WES)

Figure 5 provides a comprehensive investigation into the dynamic behavior of WES and its capacity to contribute to grid balancing through active power control. WES emphasizes the overall performance of aggregated wind energy systems (WESs) within the power system. The WES model is expertly streamlined and tailored for active power regulation and long-term dynamic simulation studies [7]. WES comprises essential blocks, such as wind turbine active power controller (WTAPC), WES active power controller (WESAPC), and generator reference current block. The frequency droop block is crucial in providing primary frequency response (ΔP_c) contingent on available wind power and systemic frequency droop parameters. This intricate relationship ensures the power plant responds actively and effectively, maintaining grid stability and balance through active power control. The efficacy of this reaction is contingent upon the magnitude of wind power accessible ($P_{WES_{avail}}$) and the intricacies of the power system's frequency droop parameters, amplifying the plant's dynamic interplay with the power grid.

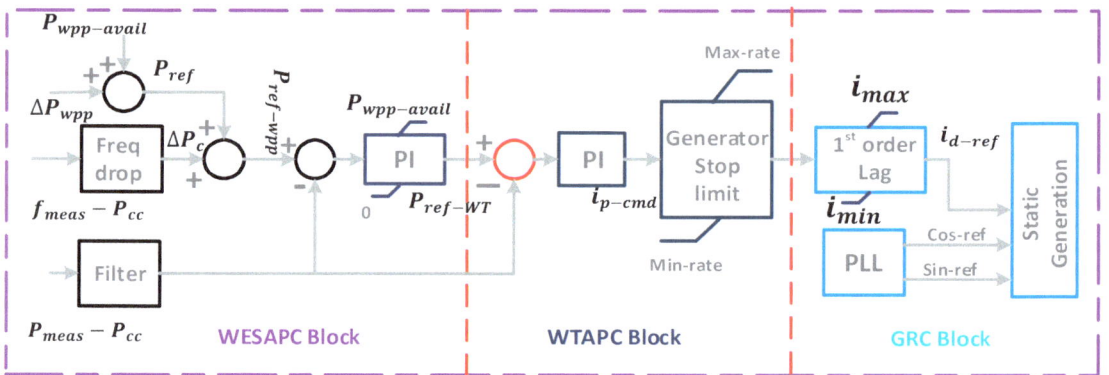

Figure 5. Wind energy system model.

When the reference power (P_{ref_WES}) of the WES is modified, the WESAPC block swiftly generates a novel turbine allusion power P_{ref_WT}. The calculation for P_{ref_WES} within the WESAPC block is reliant on the allusion power P_{ref}, the prime frequency response signal ΔP_c, and the gauged power at the point of shared coupling P_{meas_PCC}. The PI regulator in the WESAPC block maintains the regulation of the allusion power signal for the WTAPC block. This regulation is established through an error comparison between P_{ref_WPP} and P_{meas_PCC}. To avert extra power output, the available power signal $P_{WES_{avail}}$ is employed to restrict the PI regulator's output. Meanwhile, the WTAPC block generates the current active component (I_{Pcmd}) of the generator as its output, calculated by the PI regulator within the WTAPC block. This calculation relies on the discrepancy arising between the wind turbine reference power P_{ref_WT} and P_{meas_PCC}. The investigation involves the intelligent development of the wind turbine's generator type IV model, which confers unmatched operational flexibility compared to other generator models. This advanced design significantly boosts the wind turbine's performance and adaptability in various operational situations, rendering it a beneficial option for wind power plants. The wind turbines are equipped with distinct machine-side and grid-side inverters, operating autonomously. The machine-side converter facilitates seamless generator rotation at the optimal rotor speed, while the grid-side converter independently regulates the flow of active and reactive power. The wind turbine generator is a static generator utilizing the model based on the current sources technology. The reference current and input from the phase-locked loop determine the dynamic response of the stationary generator. Controlled operation is ensured by limiting the ramp rate of the available wind power. The WES showcases a remarkable response time, swiftly adapting to fluctuations in system load within 2 to 4 s.

3. AGC Modelling

An effective, reliable, and consistent electricity supply within a cohesive power grid is achieved through AGC, which continuously monitors load oscillations and adapts generator output accordingly. The efficient operation of the AGC service requires continuous monitoring of frequency fluctuations to establish the area control error (ACE). The essential step in AGC control involves $P_{ACE,i}$ as expressed in (11):

$$P_{ACE,i} = \sum_{j \in \mathcal{A}_n} \beta_i \Delta f + (P_{ij}^{Sch} - P_{ij}^{Act}) \quad (11)$$

where $P_{ACE,i}$ signifies the total discrepancy, and P_{ij}^{Sch} and P_{ij}^{Act} represent the prearranged and real data flows in the line; the difference between these is represented by ΔP_{tie}. In this context, the parameter β_i takes on significance as it embodies the frequency bias constant specific to the ith area. Its computation involves the ratio $D_i + \frac{1}{R_i}$. Notably, Δf symbolizes the frequency deviation from the present value, making it a crucial indicator in AGC operations. During power supply–demand discrepancies, the speed governor initiates the frequency containment reserve (FCR) to address the imbalance. Simultaneously, the AGC system detects alterations in $P_{ACE,i}$ and triggers the frequency regulation reserves to stabilize ACE error and safeguard the pre-activated reserves. AGC, while acquiring the data input from the ACE, regulates the load operating points ($\Delta P_{ref,i}$) of all the power plant units to efficiently operate the system. This study incorporates the characteristics of the PI regulator to regulate P_{ACE}, as defined in (12):

$$\Delta P_{Sec} = K \cdot \Delta P_{ACE} + KT \int \Delta P_{ACE} dt \quad (12)$$

The attainment of the network's original frequency and the restoration of tie-line power to its pre-determined value necessitate the determination of suitable parameters, T and K. These parameters are crucial in governing the secondary control system. To ensure effectiveness and adherence to industry standards, selecting K and T values follows a widely recognized guideline. The K constant typically ranges from 0 to 0.5, providing a spectrum of options for fine-tuning the control response. Meanwhile, T (time constant) spans from 50 s to 200 s, enabling flexibility in adjusting the system's response time. These ranges are widely considered to strike an optimal balance between speed and stability, promoting efficient regulation of the P_{ACE}.

The time constant calculates the tracking speed of the regulator in activating the operating reserves from the power units, which contributes to the AGC regulation process. The resultant generated error of the AGC is divided into the power units and the EVA system as per the defined dispatch strategy in Figure 6. This study incorporates a diverse mix of resources in the AGC system, comprising TES, WES, and EVs, all contributing to the provision of regulation reserves. The dispatching section of the AGC, upon receiving inputs, such as ΔP_{Sec}, the EV aggregator data, and the P_{TES}, P_{WES}, $P_{wind, Avail}$, effectively calculates the necessary adjustments in the load reference of the power producing units, denoted as ΔP_{TES}, and ΔP_{EVs}, respectively.

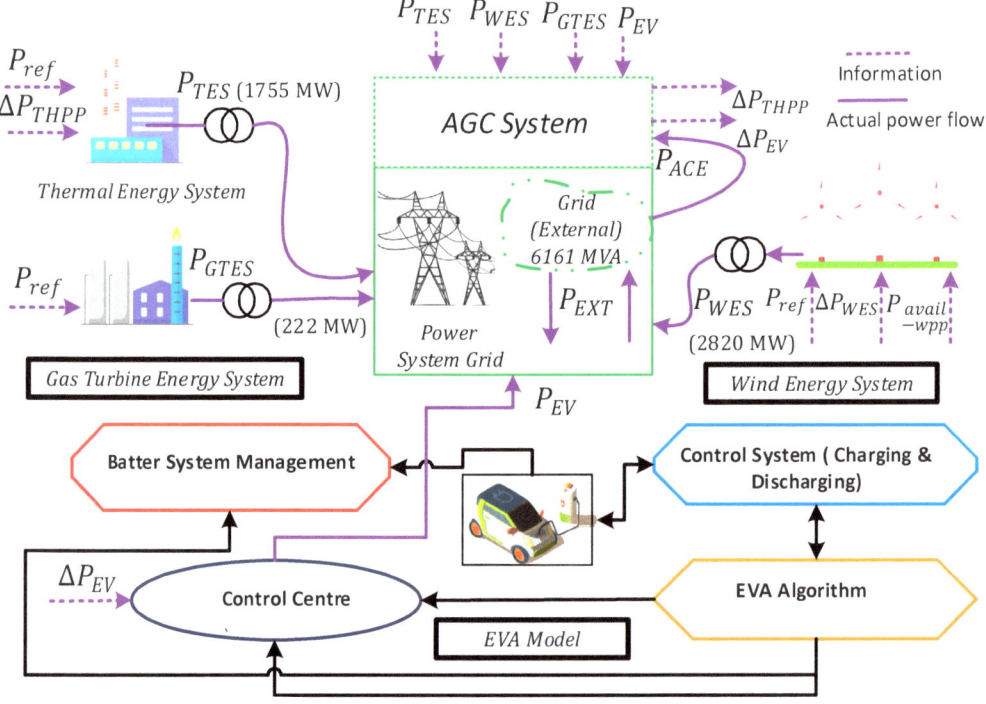

Figure 6. Optimized power system AGC network.

4. Performance Validation

The efficacy of the proposed AGC model was assessed on the established power grid model, comprising TES, GTES, WES, and an EVA model. An external grid with an interval of 16 s and a response rate of 6161 MW/Hz was connected to support the grid. The details regarding the parameters related to the power plant units and the EVA model are listed in Table 1, along with the maximum limits of the secondary operating reserves.

Table 1. Parameter data for power plant units and the EVA system.

Generating Units (MW)	TES	GTES	WES	EVA
Maximum Power (MW)	1755	222	2820	127.5
Operating Reserves (MW)	±100	0	−500	±75

Figure 7a illustrates the actual power generation from various power plant units, comprising TES, GTES, and WES, over 24 h of the Pakistan power system. For data acquisition, a winter day in 2023, was carefully chosen as input data for the TES and WES. However, the GTES power remained constant, maintaining a fixed value throughout the entire period. An essential aspect to consider is that the real inputs of the WES differ from the reference values (forecasted values) initially used to calculate the load-generation balance. Hence, variations between the actual and forecasted values of WES and changing load demands cause a power disparity between power demand and supply, significantly impacting the overall power system performance.

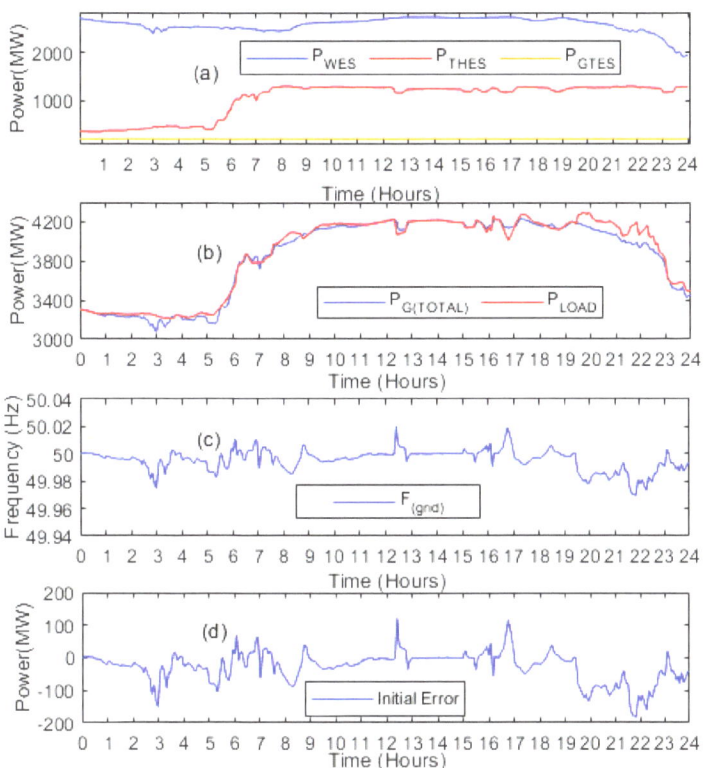

Figure 7. (a) Power generation, (b) net load and power demand, (c) frequency, (d) power demand error.

Figure 7b exhibits the disparateness amid the load exigency and cumulative generation from all three sources. The constant alterations in the frequency response of the developed power grid model are clearly illustrated in Figure 7c, revealing the dynamic principle engendered by the oscillating load and generation behaviors. To rectify these power incongruities within the network, this investigation advanced a control strategy for the AGC regulator to optimally use the operating reserves sourced from generating units and EVs. Figure 7d depicts the consequential power asymmetry within the load demand and overall power generation, represented for subsequent comparison. Active power balance management entails multifarious phases. At the outset, disparities in the power grid result in frequency oscillations identified by the governors on individual generating units. Consequently, the governors trigger the utilization of FCRs based on the features of a power plant and the synchronous power of the entire network. These intricate adaptations are crucial for reestablishing balance and stability within the power system. FCRs quickly stabilize the system frequency using governors' droop characteristics. Secondary reserves fine-tune frequency back to the nominal level. AGC dispatches balancing power to minimize ACE, however, lacks flexibility, increasing costs and risking system security. Hence, a more appropriate approach is required for efficient and secure power system operation.

Implementing an AGC system with a dynamic dispatch approach is crucial. This study proposes a smart power distribution approach for AGC, supporting grid integration with abundant renewable energy resources. The suggested system utilizes EVs and a thermal energy system for power regulation operation in wind energy-based power systems. The AGC system's dispatch strategy overcomes challenges by integrating EVs. This intelligent power system effectively balances the grid, reducing the need for conventional regulation sources. Hence, it reduces costs and operational stress, while providing an eco-friendly solution, curbing greenhouse effects. The case study demonstrates the efficacy of combining

EV storage capacities with TES reserve power for a secondary response, ensuring a stable and responsive power system.

Case Study: Power Balancing through EVA and TES

This case study explores the integration of EVs and TES for power-balancing operations. The study demonstrates how EVs contribute to AGC by providing regulatory power to handle intermittent wind power. The AGC effectively minimizes system frequency deviations by regulating reserves from TES and EVs. The intelligent allocation of operating reserves optimizes power balancing, enhancing overall power system stability and reliability. As shown in Figure 8, the AGC dispatch strategy was developed utilizing EVs' ability to provide positive and negative regulation strength. The process depicted in Figure 2 systematically determined these capacities. Initial measurements of various parameters led to calculating PRCs and NRCs for each dispatch interval. When PRC $\Delta P_s > 0$, the AGC commanded the EVA to employ all existing reserves before the TES responded. This involved either decreasing the load impact or supplying battery power to the grid. Conversely, during NRC, the battery's discharge power was increased to counteract the frequency deviation. Integrating EVs facilitated a more advanced and dynamic AGC dispatch, enabling better power balancing control and overall power system efficiency and stability.

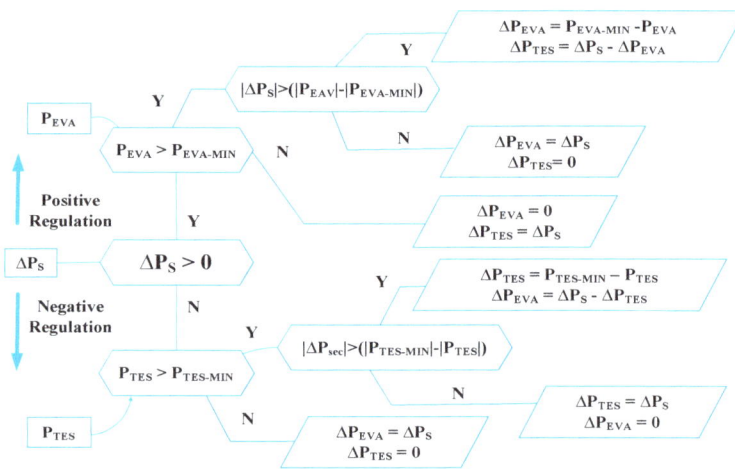

Figure 8. EVs and TES integration process.

The observed occurrence can be recognized as the inferior incremental cost of generating power from EVs. In the NRC, battery power loading is increased solely when the TES hits its lower limit ($P_{TES, min}$), fixed at 20% of its capacity, or when the AGC's secondary dispatch touches its lower limit ($\Delta P_{TES, min}$), which equals -100 MW. Figure 7d visually depicts the initial demand and generation imbalance, effectively compensated through the AGC system. The AGC achieves this by dispatching the operating reserves from EVs and the TES. This responsive and dynamic control mechanism efficiently mitigates power imbalances, ensuring grid stability and a dependable power supply.

The cumulative secondary dispatch (ΔP_{Sec}) from the power sources in the secondary response is visually depicted in Figure 9a, closely tracking the P_{ACE} error. The relatively sluggish response can be ascribed to the inherent delays linked to the AGC system and the power plant units. Figure 9b presents the dispatch (secondary) power producing units (ΔP_{EV} and ΔP_{WPP}). TES responded only after all the reserve power from EVs was utilized during the up-regulation process. This highlights the prioritization of using EV reserves before engaging the TES resources. In scenarios where power generation is exceeded, operating reserves from the TES are rendered before dispatching power from EVs. This approach effectively reduces the incremental cost associated with power generation and enhances overall system efficiency. The strategic deployment of secondary reserves and the coordination between generating units and EVs contribute to successfully managing power imbalances, ensuring optimal grid operation, and minimizing operational costs.

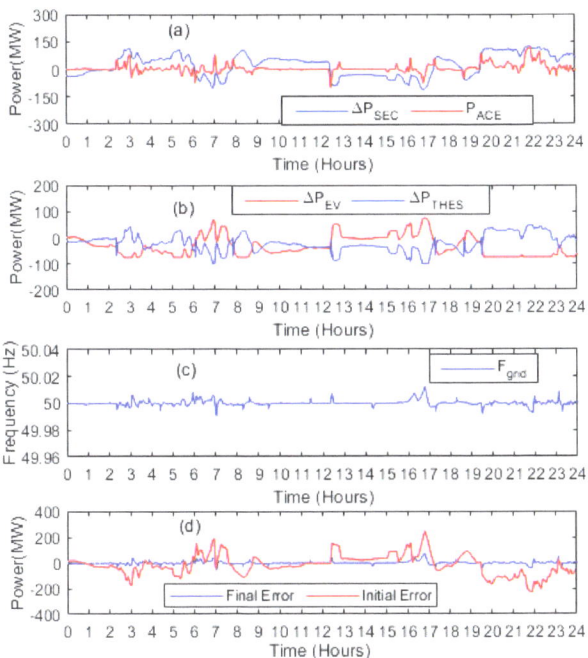

Figure 9. (**a**) ACE and power dispatch, (**b**) TES and EVs individual dispatch, (**c**) resulted system frequency, (**d**) power imbalances comparison.

Figure 9c presents the frequency variations observed in the system grid after the AGC's response. The AGC actions effectively mitigated the frequency deviations, leading to a more stable and controlled power system operation. Figure 9d compares real-time power disparities before and after the AGC response. The AGC intervention significantly reduced power imbalances in generation deficit and generation excess scenarios. This underscores the crucial role of AGC in maintaining grid stability and ensuring that power demand is met efficiently. To measure the EV response, the AGC activated 1.8 GWh of energy from EVs during generation shortages and surpluses. This indicates the substantial contribution of EVs in addressing active power imbalances in a power system characterized by large wind power integration. Without leveraging EVs for power balancing, many conventional power plants would be required to compensate for the imbalances, leading to higher operational costs and reduced sustainability. Thus, integrating EVs into the AGC process is a valuable solution for power system operators, providing dynamic and cost-effective reserve resources to enhance grid reliability and accommodate the intermittent nature of renewable energy sources.

Additionally, the study conducted a quantitative analysis to enhance the understanding, and compare the outcomes, of the AGC control system with the initial system error. This involved calculating the area under the positive and negative curves depicted in Figure 9d; the results have been presented in Table 2. These findings demonstrated a significant reduction in power error due to integrating large-scale wind energy systems into the network.

Table 2. Quantitative comparative analysis.

Case Studies	Up-Regulation Area (10^6)	Down-Regulation Area (10^6)	% Reduction in Positive Regulation Error	% Reduction in Negative Regulation Error
Initial Error	3.137	4.135	0.00%	0.00%
Case Study	0.3471	0.2145	90.0%	93.25%

5. Conclusions and Future Directions

This research conducted an extensive analysis of providing active power support to power systems heavily reliant on wind integration, utilizing the capacities of EVs in conjunction with TES. Wind-based power systems inherently possess an intermittent character, leading to forecasting errors that cause power imbalances between demand and generation. Additional operating reserves of traditional power plants often seek to meet the increased reserve requirements. However, such an approach is economically impractical and burdensome to the environment. Hence, in the current study, a real-time dynamic dispatch strategy was formulated for the AGC system to utilize EV capabilities in secondary power dispatch processes. A case study was conducted, integrating EVs into the proposed AGC system alongside the TES to offer regulation services. The performance analysis demonstrates that the integration of EVs with TES can substantially alleviate real-time power imbalances stemming from extensive wind power integration, elevating system operational security and reliability. Further, the quantitative comparison conducted in this study highlights the significant cost savings achieved through the reduced reliance on conventional power plants, underscoring the valuable role of EVs in the power system. Hence, this research provides valuable insights for power system operators to strategically leverage EVs as a dynamic and cost-effective solution to address power imbalances and improve power systems' overall sustainability and efficiency.

The study lays a strong foundation for future extensions, particularly in an artificial intelligence (AI)-based AGC system. The power system's operational parameters can be accurately forecasted using AI techniques like machine learning and predictive analytics. Moreover, while the current control system was tailored for a power system with substantial inertia, it holds promise for application in future micro-grid scenarios where system inertia is minimal due to the massive integration of renewable energy sources. This adaptability highlights the versatility of the suggested control system, making it well-suited for diverse power system configurations and ensuring its relevance in the face of evolving energy landscapes. Additionally, integrating building loads, particularly those utilized for heating or cooling purposes, represents a crucial avenue for future exploration. By incorporating building loads alongside EVs in the proposed control strategies, the potential for harnessing energy from diverse sources becomes more comprehensive. This integration could lead to a scenario where the reserve capacity obtained from traditional generation sources is fully replaced, resulting in a more sustainable and eco-friendly power system.

Author Contributions: Conceptualization, Z.U., K.U., C.D.-L., G.G. and A.B.; Methodology, Z.U., K.U. and C.D.-L.; Software, Z.U., K.U. and A.B.; Validation, Z.U., K.U. and G.G.; Formal analysis, C.D.-L. and G.G.; Investigation, Z.U., K.U. and A.B.; Resources, C.D.-L. and G.G.; Data curation, K.U. and A.B.; Writing—original draft, Z.U., K.U., A.B. and C.D.-L.; Writing—review & editing, K.U., C.D.-L., G.G. and A.B.; Visualization, K.U., C.D.-L. and A.B.; Supervision, G.G.; Project administration, C.D.-L. and G.G. All authors have read and agreed to the published version of the manuscript.

Funding: This research received no external funding.

Data Availability Statement: This study did not report any data.

Conflicts of Interest: The authors declare no conflict of interest.

Abbreviations

Acronym	Definition
BESs	Battery Energy Storage System
PA	Up-regulation area
CESs	Capacitive Energy Storage System
CEV	Environmental Burning Capacity
CFM	Baseload function
CIGRE	International Council on Large Electric Systems
CSEV	Sequential Environmental burner capacity
CVGV	Variable inlet guide vane position compressor capacity
FCR	Frequency Containment Reserve
FRR	Frequency Regulation Reserves
GTDB	Gas turbine dynamics block
GTES	Gas Turbine Energy System
NRC	Negative regulation capacity
NA	Down-regulation area
PDB	Power distribution block
PLB	Power limitation block
PJM	Regional Transmission Company
RPS	Reference Power Signal
SEV	Sequential environmental combustion
SMA	Smart Management Approach
STC	Steam Temperature Control
SEV	Sequential environmental combustion
TSO	Transmission system operator
TES	Thermal Energy System

References

1. Lee, T.-Y. Optimal Spinning Reserve for a Wind-Thermal Power System Using EIPSO. *IEEE Trans. Power Syst.* **2007**, *22*, 1612–1621. [CrossRef]
2. Nassar, I.A.; Abdella, M.M. Impact of replacing thermal power plants by renewable energy on the power system. *Therm. Sci. Eng. Prog.* **2018**, *5*, 506–515. [CrossRef]
3. Albadi, M.; El-Saadany, E. Comparative study on impacts of wind profiles on thermal units scheduling costs. *IET Renew. Power Gener.* **2011**, *5*, 26–35. [CrossRef]
4. Hashmi, M.H.; Ullah, Z.; Asghar, R.; Shaker, B.; Tariq, M.; Saleem, H. An Overview of the current challenges and Issues in Smart Grid Technologies. In Proceedings of the 2023 International Conference on Emerging Power Technologies (ICEPT), Topi, Pakistan, 6–7 May 2023; pp. 1–6. [CrossRef]
5. Hussain, S.; Lai, C.; Eicker, U. Flexibility: Literature review on concepts, modeling, and provision method in smart grid. *Sustain. Energy Grids Netw.* **2023**, *35*, 101113. [CrossRef]
6. Asghar, R.; Fulginei, F.R.; Wadood, H.; Saeed, S. A Review of Load Frequency Control Schemes Deployed for Wind-Integrated Power Systems. *Sustainability* **2023**, *15*, 8380. [CrossRef]
7. Ullah, K.; Ullah, Z.; Aslam, S.; Salam, M.S.; Salahuddin, M.A.; Umer, M.F.; Humayon, M.; Shaheer, H. Wind Farms and Flexible Loads Contribution in Automatic Generation Control: An Extensive Review and Simulation. *Energies* **2023**, *16*, 5498. [CrossRef]
8. Asghar, R.; Ullah, Z.; Azeem, B.; Aslam, S.; Hashmi, M.H.; Rasool, E.; Shaker, B.; Anwar, M.J.; Mustafa, K. Wind Energy Potential in Pakistan: A Feasibility Study in Sindh Province. *Energies* **2022**, *15*, 8333. [CrossRef]
9. Guille, C.; Gross, G. A conceptual framework for the vehicle-to-grid (V2G) implementation. *Energy Policy* **2009**, *37*, 4379–4390. [CrossRef]
10. Khooban, M.-H. Secondary Load Frequency Control of Time-Delay Stand-Alone Microgrids With Electric Vehicles. *IEEE Trans. Ind. Electron.* **2017**, *65*, 7416–7422. [CrossRef]
11. Li, Z.; Su, S.; Jin, X.; Chen, H.; Li, Y.; Zhang, R. A hierarchical scheduling method of active distribution network considering flexible loads in office buildings. *Int. J. Electr. Power Energy Syst.* **2021**, *131*, 106768. [CrossRef]
12. Mignoni, N.; Scarabaggio, P.; Carli, R.; Dotoli, M. Control frameworks for transactive energy storage services in energy communities. *Control Eng. Pract.* **2023**, *130*, 105364. [CrossRef]
13. Venkatesan, K.; Govindarajan, U. Optimal power flow control of hybrid renewable energy system with energy storage: A WOANN strategy. *J. Renew. Sustain. Energy* **2019**, *11*, 015501. [CrossRef]
14. Hernández, J.; Sanchez-Sutil, F.; Vidal, P.; Rus-Casas, C. Primary frequency control and dynamic grid support for vehicle-to-grid in transmission systems. *Int. J. Electr. Power Energy Syst.* **2018**, *100*, 152–166. [CrossRef]

15. Falahati, S.; Taher, S.A.; Shahidehpour, M. A new smart charging method for EVs for frequency control of smart grid. *Int. J. Electr. Power Energy Syst.* **2016**, *83*, 458–469. [CrossRef]
16. Giordano, F.; Arrigo, F.; Diaz-Londono, C.; Spertino, F.; Ruiz, F. Forecast-Based V2G Aggregation Model for Day-Ahead and Real-Time Operations. In Proceedings of the 2020 IEEE Power & Energy Society Innovative Smart Grid Technologies Conference (ISGT), Washington, DC, USA, 17–20 February 2020; pp. 1–5. [CrossRef]
17. Diaz-Londono, C.; Gruosso, G.; Maffezzoni, P.; Daniel, L. Coordination Strategies for Electric Vehicle Chargers Integration in Electrical Grids. In Proceedings of the 2022 IEEE Vehicle Power and Propulsion Conference (VPPC), Merced, CA, USA, 1–4 November 2022; IEEE: Piscataway, NJ, USA; pp. 1–6.
18. Diaz-Londono, C.; Vuelvas, J.; Gruosso, G.; Correa-Florez, C.A. Remuneration Sensitivity Analysis in Prosumer and Aggregator Strategies by Controlling Electric Vehicle Chargers. *Energies* **2022**, *15*, 6913. [CrossRef]
19. Mignoni, N.; Carli, R.; Dotoli, M. Distributed Noncooperative MPC for Energy Scheduling of Charging and Trading Electric Vehicles in Energy Communities. *IEEE Trans. Control Syst. Technol.* **2023**, *31*, 2159–2172. [CrossRef]
20. Hosseini, S.M.; Carli, R.; Parisio, A.; Dotoli, M. Robust Decentralized Charge Control of Electric Vehicles under Uncertainty on Inelastic Demand and Energy Pricing. In Proceedings of the 2020 IEEE International Conference on Systems, Man, and Cybernetics (SMC), Toronto, ON, Canada, 11–14 October 2020; pp. 1834–1839. [CrossRef]
21. Knezovic, K.; Martinenas, S.; Andersen, P.B.; Zecchino, A.; Marinelli, M. Enhancing the Role of Electric Vehicles in the Power Grid: Field Validation of Multiple Ancillary Services. *IEEE Trans. Transp. Electrif.* **2016**, *3*, 201–209. [CrossRef]
22. Falvo, M.C.; Sbordone, D.; Bayram, I.S.; Devetsikiotis, M. EV charging stations and modes: International standards. In Proceedings of the IEEE International Symposium on Power Electronics, Electrical Drives, Automation and Motion, Ischia, Italy, 18–20 June 2014; pp. 1134–1139.
23. Cui, Y.; Hu, Z.; Luo, H. Optimal Day-Ahead Charging and Frequency Reserve Scheduling of Electric Vehicles Considering the Regulation Signal Uncertainty. *IEEE Trans. Ind. Appl.* **2020**, *56*, 5824–5835. [CrossRef]
24. Tushar, M.H.K.; Zeineddine, A.W.; Assi, C.M. Demand-Side Management by Regulating Charging and Discharging of the EV, ESS, and Utilizing Renewable Energy. *IEEE Trans. Ind. Inform.* **2018**, *14*, 117–126. [CrossRef]
25. Liu, H.; Qi, J.; Wang, J.; Li, P.; Li, C.; Wei, H. EV Dispatch Control for Supplementary Frequency Regulation Considering the Expectation of EV Owners. *IEEE Trans. Smart Grid* **2016**, *9*, 3763–3772. [CrossRef]
26. Pham, T.N.; Trinh, H.; Van Hien, L. Load Frequency Control of Power Systems With Electric Vehicles and Diverse Transmission Links Using Distributed Functional Observers. *IEEE Trans. Smart Grid* **2015**, *7*, 238–252. [CrossRef]
27. Sanki, P.; Basu, M.; Pal, P.S.; Das, D. Application of a novel PIPDF controller in an improved plug-in electric vehicle integrated power system for AGC operation. *Int. J. Ambient. Energy* **2021**, *43*, 4767–4781. [CrossRef]
28. Khezri, R.; Oshnoei, A.; Hagh, M.T.; Muyeen, S. Coordination of Heat Pumps, Electric Vehicles and AGC for Efficient LFC in a Smart Hybrid Power System via SCA-Based Optimized FOPID Controllers. *Energies* **2018**, *11*, 420. [CrossRef]
29. Gruoup, I.W. Dynamic Models For Fossil FUELED Steam Units In Power System Studies. *IEEE Trans. Power Syst.* **1991**, *6*, 753–761.
30. Suwannarat, A. *Integration and Control of Wind Farms in the Danish Electricity System*; Institut for Energiteknik, Aalborg Universitet: Aalborg East, Denmark, 2008. Available online: https://vbn.aau.dk/ws/portalfiles/portal/7254073/Suwannarat.pdf (accessed on 13 September 2023).

Disclaimer/Publisher's Note: The statements, opinions and data contained in all publications are solely those of the individual author(s) and contributor(s) and not of MDPI and/or the editor(s). MDPI and/or the editor(s) disclaim responsibility for any injury to people or property resulting from any ideas, methods, instructions or products referred to in the content.

Article

Optimal Planning Strategy for Reconfigurable Electric Vehicle Chargers in Car Parks

Bingkun Song [1,*], Udaya K. Madawala [1,*] and Craig A. Baguley [2]

[1] Department of Electrical, Computer and Software Engineering, Faculty of Engineering, The University of Auckland, Auckland 1023, New Zealand

[2] Department of Electrical and Electronic Engineering, School of Engineering, Computer and Mathematical Sciences, Faculty of Design and Creative Technologies, Auckland University of Technology, Auckland 1142, New Zealand

* Correspondence: bson550@aucklanduni.ac.nz (B.S.); u.madawala@auckland.ac.nz (U.K.M.)

Abstract: A conventional electric vehicle charger (EVC) charges only one EV concurrently. This leads to underutilization whenever the charging power is less than the EVC-rated capacity. Consequently, the cost-effectiveness of conventional EVCs is limited. Reconfigurable EVCs (REVCs) are a new technology that overcomes underutilization by allowing multiple EVs to be charged concurrently. This brings a cost-effective charging solution, especially in large car parks requiring numerous chargers. Therefore, this paper proposes an optimal planning strategy for car parks deploying REVCs. The proposed planning strategy involves three stages. An optimization model is developed for each stage of the proposed planning strategy. The first stage determines the optimal power rating of power modules inside each REVC, and the second stage determines the optimal number and configuration of REVCs, followed by determining the optimal operation plan for EV car parks in the third stage. To demonstrate the effectiveness of the proposed optimal planning strategy, a comprehensive case study is undertaken using realistic car parking scenarios with 400 parking spaces, electricity tariffs, and grid infrastructure costs. Compared to deploying other conventional EVCs, the results convincingly indicate that the proposed optimal planning strategy significantly reduces the total cost of investment and operation while satisfying charging demands.

Keywords: electric vehicle (EV); reconfigurable electric vehicle chargers (REVCs); planning; EV car park; operation

1. Introduction

The speed of the global transition of the transport industry towards environmental friendliness is critically based on the rate of adoption of EVs [1–3]. Therefore, to increase the rate of adoption, more EV chargers (EVCs) are required. Further, with growing EV charging demands, more EVCs are expected to be publicly available [4–6]. Careful consideration must be given to various factors to properly plan EVC investments for public charging purposes. These include satisfying charging demands that vary in location and the charging time allowed [7–12], minimizing waiting time [13], and peak-load shaving and valley filling to mitigate grid loading impacts [14–20].

Satisfying public charging demands depends on properly planning the number and location of EVCs [7–13]. Meeting these demands can ensure the utilization rate of EVCs is high, which improves the cost-effectiveness of investment. However, the average utilization of public level-two and fast chargers is only 25% and 37%, respectively [21]. Numerous EVC planning issues for public charging have been addressed in the literature, including locations and installing appropriate numbers of fast and slow EVCs to meet charging demands with minimum investment costs [7–13,22]. However, existing works only considered the number of EVCs to be installed without determining the optimal power rating of each EVC. Nevertheless, satisfying charging demands also depends on determining the appropriate power ratings of EVCs. For example, EVCs that are undersized in terms of power rating may be highly utilized but unable to fully satisfy all charging demands. Not being able to satisfy charging demands introduces a risk that EV users may run out of charge, which lowers the attractiveness of owning and using an EV. If EVCs are oversized in terms of power rating, the level of utilization is lowered, and the cost-effectiveness of the investment is reduced. Currently, the average utilization of the rated capacity of EV chargers is only from 35% to 71% [21]. Therefore, EVC planning must be appropriate to ensure EVC investment is cost-effective without compromising on the ability to satisfy charging demands.

Less work exists on the planning of EVCs for large car parks. A planning approach for large car parks to determine the optimal number of fast EVCs and waiting queue size was presented in [23]. In [24], a planning method was developed to optimally plan the number of fast EVCs and EV charge waiting spaces, with the charging demands modeled using queuing theory and the Markov chain. An energy management strategy was presented in [25] to maximize the EV penetration level and revenue in a car park with a transformer capacity limit. In each of the studies reported in [23–25], different issues related to EV car parks were addressed. However, the EVCs used in those studies were conventional types that allow for charging only one EV at a time. This leads to underutilization of EVCs whenever the charging power demanded is less than the rated capacity of EVCs. This can occur for several reasons, for example, when the battery state-of-charge (SOC) is high and the charging power demanded from the EVC is low, or when the charging capacity allowed for a particular EV model is less than the power rating of the EVC. Thus, the cost-effectiveness of conventional EVC investments is limited. This can be addressed by raising the flexibility of EVCs, allowing one EVC to serve multiple EVs. An approach to improving the flexibility of EVCs was presented in [26] by using a four-way switch to adjust the connection of the EVC to only one of four connected EVs parked at four different charging spaces. However, this approach does not allow for concurrent charging of several EVs and gives limited improvement. In [27], the authors proposed a method to improve the flexibility of EVCs by allowing an EVC to connect multiple EVs. Thus, the charging schedules of the connecting EVs on an EVC can be controlled. By applying their method, fewer EVCs can be installed, increasing the utilization of EVCs. The authors in [28] designed a charging system to improve the flexibility of EVCs by connecting serval EVCs to serval charging spaces. In their method, the random parking behaviors of EV users at charging spaces can be solved by controlling the power flow from an available EVC to a charging space via the connection cables among the charging spaces and EVCs. However, the methods in [27,28] have some limitations. Firstly, the charging power of EVCs is not flexible, making EVCs less likely to adapt to the charging power requirements of different EV models. In addition, each EVC can only charge one EV at a time, so concurrent charging of multiple EVs is not available. Moreover, the fact that their methods are practically viable is unknown because they did not mention any EVC manufacturing companies that had produced similar products.

To address the shortcomings of the existing methods, conventional EVCs have been specifically adapted to overcome flexibility and utilization issues. EVCs of this type are referred to as reconfigurable EVCs (REVCs).

A REVC comprises internal PMs that can be dynamically allocated to realize different levels of output power at multiple outlets [29]. Several outlets can be coupled to a power cabinet housing the PMs, with a control unit used to connect and disconnect PMs in a power cabinet between outlets. This flexibility allows for better use of charging resources. For example, if an EV needs only a portion of the power rating of a REVC, idle PMs can be used for charging other EVs concurrently, which cannot be achieved by conventional EVCs. Further, since the PMs of a REVC are self-contained, they are easily installed and removed. Moreover, the wide power range makes a REVC suited to a wide range of charging powers. These are significant benefits, and numerous manufacturers have developed REVC products with similar functionality [30–38]. REVCs have been utilized in public areas, and some customers have provided positive feedback on using REVCs in terms of flexibility [39]. A comparison of REVCs and conventional slow and fast EVCs is shown in Table 1.

Table 1. Comparison of REVC and conventional types of EVCs.

EVC Type	Slow EVC	Conventional Fast EVC	REVC
AC or DC	AC	DC	DC
Number of EVs Charged Concurrently	1	1	>1
Flexibility	Low	Low	High
Degree of Resource Utilization	Low	Low	High

There are two types of REVCs on the market. One is the integrated-type REVCs, where each REVC has multiple identical PMs inside, as illustrated in Figure 1a. Generally, the number of EVs that can be charged concurrently by an integrated-type REVC depends on the number of PMs inside and the number of charging cables connected to a REVC. A disadvantage of the integrated-type REVCs is the need for long charging cables to reach charging spaces distant from the REVCs. For example, the lower-level charging spaces in Figure 1a are distant from the REVCs in the upper charging spaces. This need for long cables is an inconvenience and safety issue that practically limits the number of charging cables fitted to an integrated-type REVC to two or three. Consequently, the number of charging spaces served by an integrated-type REVC is limited.

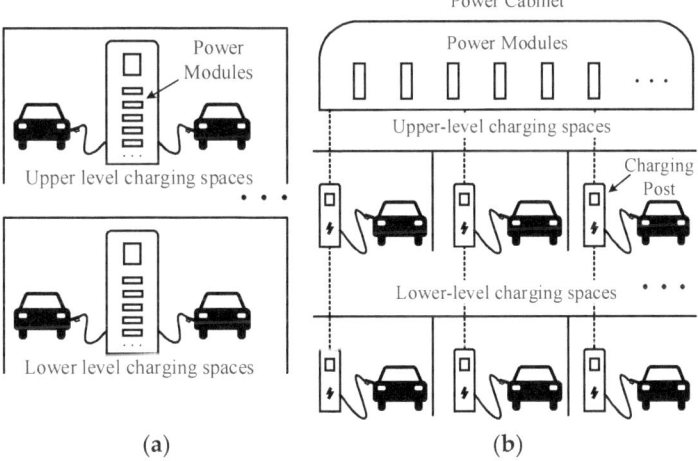

Figure 1. Schematic of (a) Integrated-type REVCs; (b) a split-type REVC.

The other type of REVC is the split-type, shown in Figure 1b, which has a power cabinet and separate charging posts (CPs). For split-type REVCs, power outlets are CPs connected to a REVC. The CPs are distributed at each charging space and connected to a power cabinet via unobtrusive underground cables (dashed lines in the figure). Charging cables of a convenient length connect to the CPs and are used to plug into the EVs. The use of underground cables and charging posts allows charging spaces to be served that are distant from the power cabinet, such as lower-level charging spaces. The PMs within the split and integrated-type REVCs are controlled and reconfigured as necessary to deliver the required power levels to different EVs.

The planning of REVCs has not previously been reported, despite key differences with conventional EVCs that mean higher levels of flexibility and utilization can potentially be realized. The key differences include the ability to charge multiple EVs concurrently and to change the power allocated to each connected EV dynamically. Especially for split-type REVCs, each one can connect many CPs, which introduces the numerous combinations of REVC numbers and corresponding CP numbers for planning in large car parks. Therefore, it is critical to have an optimal planning strategy to obtain the optimal plan for deploying REVCs in large car parks. An optimal planning strategy that incorporates these key differences and is suited to the planning of REVCs for small or large car parks is needed to realize the full potential of REVC technology. Accordingly, an optimal planning strategy for REVCs is presented in this paper. Compared with existing research, the key contributions of this paper are summarized as follows:

- The proposed optimal planning strategy can be applied as a planning tool for car parks deploying REVCs under various scenarios.
- The proposed optimal planning strategy can be applied as an operation controller for REVCs to allocate charging powers and spaces for individual EVs.
- The proposed optimal planning strategy can not only determine the size (number of REVCs) of EV car parks but also determine the optimal power rating for PMs and each REVC.
- A grouping method is proposed to aggregate EV charging demands for scenarios with a large number of EVs.

The remainder of the paper is organized as follows: The overall framework for the proposed optimal planning strategy is illustrated in Section 2. In Section 3, the modeling for determining the power rating of power modules is described. The optimization model for determining the number and configuration of REVCs is formulated in Section 4. In Section 5, the optimization model to determine the operation plan of REVCs is formulated. In Section 6, a comprehensive case study is undertaken to illustrate the effectiveness of the optimal planning strategy under various conditions and to show the benefits of using REVCs compared to conventional EVCs. Finally, conclusions are given in Section 7.

2. Overall Framework for Optimal Planning Strategy

In this section, the overall framework of the proposed three-stage optimal planning strategy is introduced. To provide an integrated plan, each stage of the proposed strategy should be closely related. The outputs of the previous stage are applied as parameters to the next stage. The first stage is to determine the optimal power rating of PMs. In this stage, the degree of utilization of PMs is modeled. Therefore, the objective is to determine the optimal power rating of PMs with a maximized degree of utilization. After the optimal power rating of PMs is determined, this optimal power rating is used as a parameter for the second stage. The second stage is to determine the optimal number and configuration of REVCs, including determining the number of REVCs, the number of PMs in each REVC, and the number of CPs connected to each REVC. Then, in the third stage, the optimal operation plan can be determined based on the EV charging demands, which include the charging profile for the whole car park, the additional transformer capacity to be upgraded, and allocations of charging space and charging power for individual EVs. The determined number and configuration of REVC in the second stage and the collected charging demands are applied as parameters and constraints for the third stage to allocate the optimal charging space and charging power for each EV. To obtain EV charging demands, government travel reports or data mining methods can be utilized [40,41]. Once the data on EV charging demand is obtained, advanced forecasting strategies can be applied to predict future EV charging behaviors [42,43]. The forecasting of the data is out of the scope of this paper, and please refer to the references for detailed explanations. Detailed explanations of the proposed strategy and modeling are demonstrated in the following sections. The overall framework of the proposed optimal planning strategy is displayed in Figure 2.

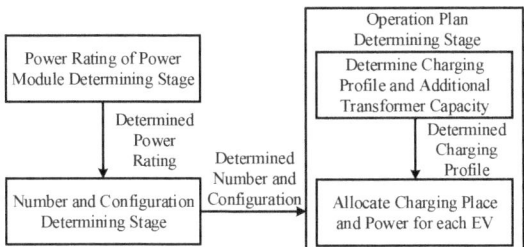

Figure 2. Framework of the proposed three-stage optimal planning strategy.

3. Model for Determining the Power Rating of Power Modules

Because a REVC is made up of multiple identical PMs, and PMs are key components of a REVC, determining the optimal power rating of PMs is essential. The objective is to select the optimal power rating of PMs to achieve the highest degree of utilization. For example, in Figure 3, an EV with a low battery SOC allows for charging at a 50 kW rate. If the EVCS comprises five 10 kW PMs, all PMs will be fully utilized. If the EVCS comprises four 15 kW PMs or three 20 kW PMs, underutilization will occur on the last PM. From this example, it is apparent that the highest degree of utilization is achieved with 10 kW PMs. In addition, the degree of utilization also depends on the charging power of an EV.

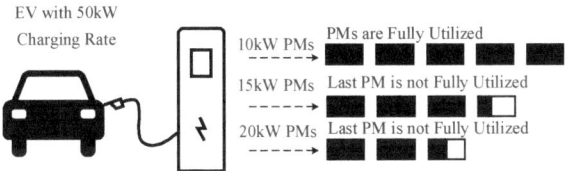

Figure 3. Block diagram of the degree of utilization of PMs.

When an individual EV is applying a REVC, mathematically, the number of working PMs is expressed as:

$$N_{individual} = ceil(\frac{P_x^{EV}}{P_u^{PM}}), \quad (1)$$

where the charging rate of this EV is at P_x^{EV}, and the power rating of PMs is P_u^{PM}. Consequently, for this EV, the degree of utilization for the last PM is:

$$UD_{individual} = 1 - [N_{individual} - \frac{P_x^{EV}}{P_u^{PM}}]. \quad (2)$$

The degree of utilization can be generalized to a scenario with several different EV types with diverse charging rates:

$$UD_{all} = \sum_{x=1}^{x=X} w_x \cdot UD_{individual} \quad (3)$$

where X is the number of EV types, and w_x is the weighting factor of each EV type.

Therefore, the objective is to choose the optimal power rating of PMs when the degree of utilization is maximized, which can be expressed as:

$$\dot{P}_u^{PM} = \underset{P_u^{PM}}{argmax} UD_{all}(P_u^{PM}). \quad (4)$$

4. Model for Determining the Number and Configuration of REVCs

After determining the optimal power rating of PMs, the optimal number and configuration of RVECs can be determined. The number and configuration of REVCs are determined by the parking behaviors of EV users. However, parking behavior is uncertain. Therefore, some typical scenarios with different parking behaviors are considered to represent uncertainty. Consequently, for determining the number and configuration under uncertainty, an optimization model is formed as a robust optimization problem. The purpose is to calculate an optimal solution when the worst-case scenario occurs within a number of scenarios "$\delta \in \Delta$". From the EV car park operator's perspective, at this stage, the objective includes minimizing the equivalent annual investment cost of REVCs and the degree of dissatisfaction due to losing some EV customers, as expressed in (5). The degree of dissatisfaction is described by the time at which EV users look for charging spaces at other EV car parks. The constraints account for the technical specifications of REVC and the number of parking spaces, as formulated in (9)–(16). Thus, the optimization model is formulated as follows:

$$\min_{a} \max_{\delta \in \Delta} CRF \cdot [cost_{inv}(a,\delta) + cost_{time}(a,\delta)], \tag{5}$$

where:

$$CRF = \frac{d(1+d)^m}{(1+d)^m - 1}, \tag{6}$$

$$cost_{inv}(a,\delta) = (N_{PM}(\delta) \cdot \dot{P}_u^{PM} \cdot cost_{PM}^{kW} + N_{CP}(\delta) \cdot cost_{CP} + N_{CP}(\delta) \cdot cost_{cable} + cost_{other}) \cdot N(\delta), \tag{7}$$

$$cost_{time}(a,\delta) = N_{loss}(\delta) \cdot cost_{time}^{perEV}, \tag{8}$$

subject to:

$$P_{REVC}^{min} \leq N_{PM}(\delta) \cdot \dot{P}_u^{PM} \leq P_{REVC}^{max}, \tag{9}$$

$$n_{CP}^{min} \leq N_{CP}(\delta) \leq n_{CP}^{max}, \tag{10}$$

$$N_{pt}^{EV}(\delta) = N_{CP}(\delta) \cdot N(\delta) + N_{loss}(\delta), \tag{11}$$

$$N_{CP}(\delta) \cdot N(\delta) + N_{loss}(\delta) \leq S, \tag{12}$$

$$N_{PM}(\delta) \in \{\mathbb{Z}^+\}, \tag{13}$$

$$N_{CP}(\delta) \in \{\mathbb{Z}^+\}, \tag{14}$$

$$N(\delta) \in \{\mathbb{Z}^+\}, \tag{15}$$

$$N_{loss}(\delta) \in \{\mathbb{Z}_0^+\}. \tag{16}$$

The objective function and constraints are further explained as follows.

4.1. Objective Function

The objective function is to minimize the equivalent annual investment cost of REVCs and the time cost (dissatisfaction degree) of some EV customers, which is expressed in (5). Decision variables are the number of power modules for a REVC ($N_{PM}(\delta)$), number of charging posts for a REVC ($N_{CP}(\delta)$), number of REVCs to be installed ($N(\delta)$), and number of lost EVs ($N_{loss}(\delta)$). In (6), d is the discount rate, m is the life cycle of the project, and CRF is the capital recovery factor of the costs. The capital recovery factor represents the conversion relationship from the present value to equivalent annual costs in years. Equation (7) defines the investment costs of deploying the split-type REVCs, including PMs, CPs, underground and charging cables, and other expenses. If the integrated-type REVCs are deployed, the second term can be removed from (7). The time cost (dissatisfaction degree) is defined in (8), which indicates the inconvenience of the user experience due to losing some EV customers. The time cost (dissatisfaction degree) is calculated by the product of the number of lost EVs and the cost of losing an EV. Because the number of charging EVs changes on different days of the year, it is necessary to determine the optimal number of REVCs to guarantee that most EVs have charging spaces. However, if too many REVCs are installed to satisfy all the charging EVs on peak days, the cost-effectiveness of the REVCs will be reduced. The reason is that many charging spaces will be idle on off-peak days due to too many REVCs. Therefore, the purpose of having the time cost (dissatisfaction degree) in the objective function is to save investment costs by not installing too many REVCs but accommodating most EVs.

4.2. Constraints

REVC must follow the technical specifications of the manufacturers. These technical specifications include the power rating of each REVC and the number of CPs allowed for each split-type REVC. These constraints are expressed in (9) and (10). For certain places, such as workplaces, commercial areas, and residential areas, EV users normally park their EVs in parking spaces for longer durations. EV users may set up their charging demands and connect charging cables on arrival. Therefore, to accommodate peak-time charging, the sum of the total number of CPs to be installed ($N_{CP}(\delta) \cdot N(\delta)$) with the number of lost EVs should be equal to the number of charging EVs during peak time ($N_{pt}^{EV}(\delta)$), defined in (11). In addition, the total number of CPs with the number of losing EVs should not be greater than the total number of parking spaces (S) in a car park, as defined in (12). The product of the number of CPs connected to each REVC ($N_{CP}(\delta)$) and the number of REVC ($N(\delta)$) defines the total number of parking spaces offering charging (charging spaces). All the variables in this optimization model are integers, as shown in (13)–(16).

5. Model for Determining the Operation Plan

After the optimal number and configuration of REVCs are determined, the operation plan can be determined. The operation plan is also an important factor to be determined when planning EV charging infrastructure. This is because the operation cost is a large proportion of the total cost. In most of the studies, the operation cost of EV charging is defined by the charging schedule of each EV [26–28]. To determine the charging schedule of each EV, a group of decision variables is applied to control how much power is delivered to each EV during each time slot. However, this is not suitable for planning scenarios with a large number of EVs because it is impractical and inefficient for the power distribution network (PDN) operator to control individual EVs directly. Therefore, to accommodate the large-scale EV charging scenarios, a hierarchical method is proposed to determine the operation plan, as displayed in Figure 4. Before EVs arrive at the EV car park, the EV users set up their charging demands via the mobile app. Some mobile apps have been commercialized for public EV charging [44–46]. Next, the charging demands are sent to the communication and control center (CCC) of the EV car park. Then, the CCC aggregates individual charging demands and sends the aggregated charging demand to the PDN operator on behalf of the EV car park. After the PDN operator receives the power request from the EV car park, the PDN operator dispatches power to the EV car park. The EV car park obtains the daily charging profile based on the dispatched power. Afterwards, based on the charging profile, the CCC calculates the optimal operation plan for each EV and sends the allocation information for charging spaces to each EV via the mobile app. Finally, each EV goes to the allocated charging space and gets charged. This method fits the large-scale EV charging infrastructure planning problems and day-ahead scheduling of EV charging problems since the EV charging demands are normally forecasted beforehand.

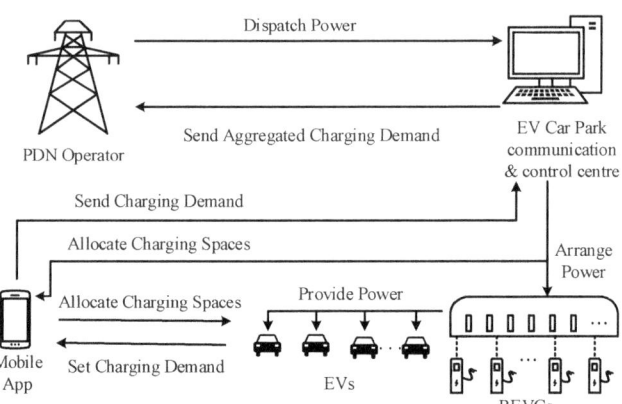

Figure 4. Framework for determining the operation plan.

To apply the proposed hierarchical method, there are two sub-stages to determine the optimal operation plan, as shown in Figure 2. In the first sub-stage, the individual EV charging demands are aggregated. Then, the aggregated charging demand is used to determine the optimal charging profile and additional transformer capacity to be upgraded. Afterwards, in the second sub-stage, the determined charging profile and additional transformer capacity are applied as constraints to determine the allocation of charging spaces and powers for individual EVs.

5.1. Determining the Charging Profile and Additional Transformer Capacity to Be Upgraded

In this sub-stage, individual EV charging demands are aggregated, and EVs are classified into different groups depending on their arrival and departure times, as demonstrated in Figure 5. Different colors are applied to represent different arrival and departure times of EVs. For example, in Figure 6, EV1, EV2, and EV5 arrive and depart during the same time slot, as highlighted in blue; therefore, they are grouped into the same set. EV3 and EV4 arrive and depart during the same time slot, as highlighted in orange; therefore, they are grouped into the same set.

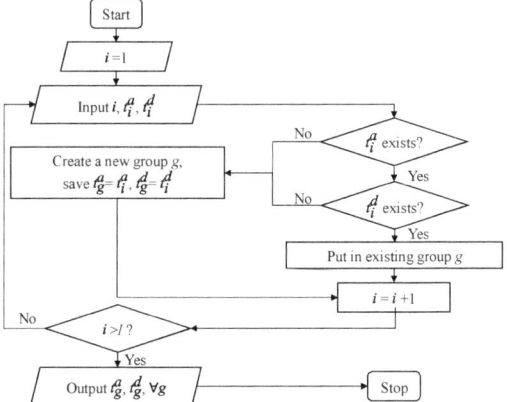

Figure 5. Flowchart for aggregating and grouping EVs.

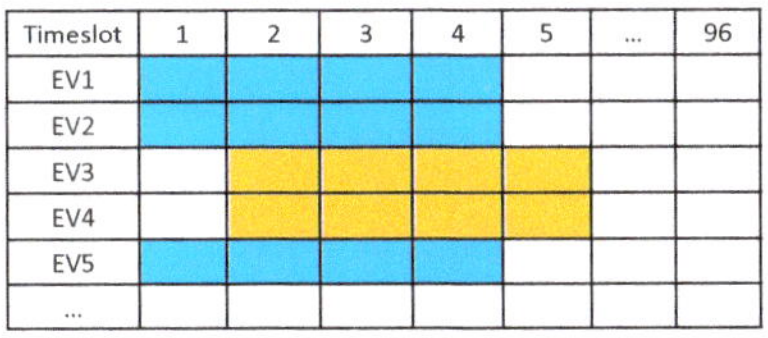

Timeslot	1	2	3	4	5	...	96
EV1							
EV2							
EV3							
EV4							
EV5							
...							

Figure 6. Example of parking duration for EVs.

For a group of EVs ($g \in G$) under scenario δ, to satisfy the energy demand, the total energy supplied from the grid should not be less than the total energy demands, as expressed in (17). To protect the EV batteries, (18) defines the sum of the total charged energy and energy at arrival for each group to be not more than the group battery capacity.

$$\sum_{t=t_g^a}^{t=t_g^d} P_{g,t}^{tot}(\delta) \cdot \Delta t \cdot \eta \geq \sum_{gi \in GI}(SOC_{gi}^d - SOC_{gi}^a) \cdot BC_{gi}, \forall g \in G, \quad (17)$$

$$\sum_{gi \in GI} SOC_{gi}^a \cdot BC_{gi} + \sum_{t=t_g^a}^{t=t_g^d} P_{g,t}^{tot}(\delta) \cdot \Delta t \cdot \eta \leq \sum_{gi \in GI} BC_{gi}, \forall g \in G, \quad (18)$$

where $P_{g,t}^{tot}(\delta)$ is the total charging power of a group in a time slot t; Δt is the duration of a time slot; η is the charging efficiency; SOC_{gi}^d is the desired SOC; SOC_{gi}^a is the SOC at arrival; and BC_i is the battery capacity for an EV. Apart from utilizing the SOCs of an EV, the energy level during a time slot of an EV can be utilized to represent the charging process as well. In this study, to simplify the representation of the model, SOCs of EVs are applied to the model.

The total charging power for each group should not be smaller than zero, as expressed by (19). During each time slot ($t \in T$), which is taken as 15 min in this paper, the sum of the total charging power in all the groups cannot exceed the sum of the maximum allowed charging power of all EVs, as expressed by (20).

$$P_{g,t}^{tot}(\delta) \geq 0, \forall t \in T, \forall g \in G, \quad (19)$$

$$\sum_{g \in G} P_{g,t}^{tot}(\delta) \leq \sum_{i \in I} P_i^{max}, \forall t \in T, \quad (20)$$

where P_i^{max} is the maximum allowed charging power for an EV i.

To protect the transformer of the PDN, during each time slot, the total charging power of all EV groups must not exceed the difference between the upgraded transformer capacity and the baseload (L_t^{base}). The upgraded transformer capacity is defined by adding the existing transformer capacity (C_{ex}) with additional transformer capacity to be upgraded ($C_{transformer}(\delta)$) to accommodate EV charging demands, as formulated in (21) and (22). Therefore, the peak demand ($P_{demand}^{peak}(\delta)$) is not larger than the upgraded transformer capacity but not smaller than the addition of charging power for all EV groups and the baseload, as expressed by (23).

$$\sum_{g \in G} P_{g,t}^{tot}(\delta) \leq (C_{ex} + C_{transformer}(\delta)) \cdot PF - L_t^{base}, \forall t \in T, \quad (21)$$

$$C_{transformer}(\delta) \geq 0, \quad (22)$$

$$C_{ex} + C_{transformer}(\delta) \geq P_{demand}^{peak}(\delta) \geq \sum_{g \in G} P_{g,t}^{tot}(\delta) + L_t^{base}, \forall t \in T, \quad (23)$$

where PF is the power factor of the transformer.

From the EV car park operators' perspective, the objective is to minimize the expected operational cost in all the scenarios. The corresponding optimization model is formulated as a linear programming problem below.

$$\min_b \mathbb{E}_\delta \{cost_{operation}(b, \delta)\} = \min_b \{\sum_{\delta \in \Delta} w_\delta \cdot cost_{operation}(b, \delta)\}, \quad (24)$$

where:

$$cost_{operation}(b,\delta) = CRF \cdot C_{transformer}(\delta) \cdot cost_{transformer}^{kVA},$$
$$+365 \cdot \sum_{g=1}^{G} \sum_{t=1}^{T} \left[P_{g,t}^{tot}(\delta) \cdot ET_t \cdot \Delta t \right] + 12 \cdot P_{demand}^{peak}(\delta) \cdot cost_{demand}^{kW}, \quad (25)$$

subject to (17)–(23).

The charging demands of EVs are uncertain. Therefore, some typical scenarios are taken to represent the uncertainty of the charging demands. These typical scenarios have a different probability (w_δ). Consequently, to comprehensively consider these scenarios, the expected value of the annual operation cost of all the scenarios with different probabilities is calculated. Thus, the objective function in this sub-stage is to minimize the expected value of the annual operation cost, as expressed by (24). The annual operation cost incorporates three parts: the first term is the upgrade cost of the transformer; the second term is the EV charging cost; and the third term is the cost of demand charge, as expressed by (25). The decision variables are the transformer capacity to be upgraded ($C_{transformer}(\delta)$), total charging power of each group in each time slot ($P_{g,t}^{tot}(\delta)$), and the peak demand ($P_{demand}^{peak}(\delta)$). The upgrade cost of the transformer is the cost that the EV car park operator spends to upgrade the additional capacity of the transformer to accommodate EV charging demands. The EV charging cost is based on the EV car park operator purchasing electricity from the utility company. The electricity tariff ET_t is set up by the utility company. The demand charge is a monthly fee to maintain the amount of grid capacity needed to deliver enough power to a large energy user. It depends on the peak power that a large energy user puts on the grid.

5.2. Allocation of Charging Spaces and Powers for EVs

In this sub-stage, the charging spaces and powers can be allocated to individual EVs based on the EV charging demands, the determined number and configuration of REVCs, and the determined charging profile.

To satisfy the energy demand, the total energy supplied from the grid for each EV ($i \in I$) should not be less than the energy required by each EV, as expressed by (26). To protect the battery of each EV, (27) defines the sum of the charged energy and energy at arrival for each EV as not more than the battery capacity of each EV.

$$\sum_{t=t_i^a}^{t=t_i^d} \sum_{n \in N} P_{i,n,t}(\delta) \cdot \Delta t \cdot \eta \geq (SOC_i^d - SOC_i^a) \cdot BC_i, \forall i \in I, \quad (26)$$

$$\sum_{t=t_i^a}^{t=t_i^d} \sum_{n \in N} P_{i,n,t}(\delta) \cdot \Delta t \cdot \eta + SOC_i^a \cdot BC_i \leq BC_i, \forall i \in I, \quad (27)$$

where t_i^a and t_i^d are the arrival time and departure time of an EV, respectively; SOC_i^a and SOC_i^d are the arrival and departure SOC of an EV, respectively; BC_i is the battery capacity of an EV.

The charging power of each EV ($P_{i,n,t}(\delta)$), during each time slot, should not be smaller than zero or greater than the maximum charging power allowed by each EV, as expressed by (28).

$$0 \leq P_{i,n,t}(\delta) \leq S_{i,n,t}(\delta) \cdot P_i^{max}(\delta), \forall i \in I, \forall n \in N, \forall t \in T, \quad (28)$$

where $S_{i,n,t}(\delta)$ is a set of binary variables defining the connection between an EV i and a REVC $n \in N$, as shown in (29).

$$S_{i,n,t}(\delta) \in \{0,1\}. \quad (29)$$

The summation of the individual EV charging power during each time slot should not be greater than the total charging power for the whole EV car park, as expressed in (30).

$$\sum_{i \in I} \sum_{n \in N} \sum_{t \in T} P_{i,n,t}(\delta) \leq P_t^{tot}(\delta), \forall t \in T, \quad (30)$$

To avoid situations where the sum of the charging power of EVs exceeds the determined charging profile, a new set of variables $P_t^{add}(\delta)$ is applied to indicate that the CCC makes additional power requests to the PDN operator. In (31), the additional power request during each time slot should be greater or equal to zero because the power is dispatched from the PDN.

$$P_t^{add}(\delta) \geq 0, \forall t \in T, \tag{31}$$

Thus, the total charging power for the whole EV car park is the sum of the determined charging profile and the additional power requests, as expressed in (32).

$$P_t^{tot}(\delta) = \sum_{g \in G} P_{g,t}^{tot}(\delta) + P_t^{add}(\delta), \forall t \in T, \tag{32}$$

In (33), for each REVC, during each time slot, the total charging power provided by a REVC cannot exceed its power rating ($N_{PM}(\delta) \cdot \dot{P}_u^{PM}$).

$$\sum_{i \in I} P_{i,n,t}(\delta) \leq N_{PM}(\delta) \cdot \dot{P}_u^{PM}, \forall n \in N, \forall t \in T, \tag{33}$$

where $\dot{P}_u^{PM}$ is determined in Section 3, and $N_{PM}(\delta)$ is determined in Section 4.

Since the configuration of the REVCs is determined in Section 4, for each REVC, the number of connected EVs during each time slot cannot be more than the number of CPs linked to each REVC, as expressed by (34).

$$\sum_{i \in I} S_{i,n,t}(\delta) \leq N_{CP}, \forall n \in N, \forall t \in T, \tag{34}$$

An EV is only allowed to connect one REVC during the time it stays in the car park, as formulated in (35). In (36), an EV is not connected to any REVCs before it arrives or after it leaves the car park.

$$\sum_{n \in N} S_{i,n,t}(\delta) = 1, \forall i \in I, \forall t \in \left[t_i^a, t_i^d\right], \tag{35}$$

$$\sum_{n \in N} S_{i,n,t}(\delta) = 0, \forall i \in I, \forall t \in \left[1, t_i^{a-1}\right] \cup \left[t_i^{d+1}, T\right], \tag{36}$$

The primary purpose of this sub-stage is to ensure that the sum of the power allocated to EVs is close to the charging profile determined in the previous sub-stage. Therefore, the objective is to minimize the difference between the total charging power for the whole EV car park and the summation of the charging power of individual EVs in each time slot. Thus, the optimization model for allocating charging spaces and powers is formulated below as a mixed-integer linear programming problem.

$$\min_{P_{i,n,t}(\delta)} \left[P_t^{tot}(\delta) - \sum_{i \in I} \sum_{n \in N} \sum_{t \in T} P_{i,n,t}(\delta) \right], \tag{37}$$

subject to (26)–(36).

6. Case Study

6.1. Case Overview and Parameter Settings

The proposed optimal planning strategy is applied to an existing car park with 400 parking spaces on the south campus of Auckland University of Technology, New Zealand. The existing transformer capacity, C_{ex}, is 750 kVA. In this case study (University), there are four typical types of days based on the weather conditions in Auckland city. According to the annual calendar of the University, the probabilities of the occurrence of these types of days are 0.26 (University days in summer), 0.31 (University days in winter), 0.33 (holiday days in summer), and 0.10 (holiday days in winter) [47]. The existing baseload of the campus is illustrated in Figure 7. There are a total of 461 vehicles on a university day, and there are a total of 182 vehicles on a holiday day. Arrival times and parking durations of these vehicles were recorded during working hours from 9:00 to 17:00, as shown in Figures 8 and 9, respectively. The data were collected in the University car park from 9 am to 5 pm on University days and holiday days, respectively. Arrival and departure times and parking durations of vehicles were recorded manually. Then, the recorded data were summarized according to the University days and holiday days. All the vehicles parked were internal combustion engine vehicles. For the case study, the vehicles are assumed to be EVs. Popular EV models in New Zealand, including the Nissan Leaf, BMW i3, Hyundai Kona, and Renault Zoe, are applied in this case study. The proportions of these EV models are assumed as 40%, 30%, 20%, and 10%, respectively. Rated battery capacities and maximum charging powers for each EV type are shown in Table 2. In the summer season, the usable battery capacities were assumed to be rated capacities. However, in the winter season, due to the influence of temperature, the usable battery capacities were assumed to be 90% of the rated capacities [48]. The arrival SOC and departure SOC for each EV are taken as 40% and 80%, respectively [26,49–51]. Transformer and REVC parameters are displayed in Table 3 [52]. The electricity tariffs are listed in [53], used as electricity purchase prices paid by REVC operators, as shown in Table 4. The demand charge per kW ($cost^{kW}_{demand}$) is taken as 8.46 $/kW [54]. The discount rate (d) is assumed to be 6% [55]. The life cycle of the project (m) is assumed to be 10 years. The time cost (dissatisfaction degree) of losing an EV is set as the time cost of an EV looking for an available charging space in another EV car park. The time cost per hour is taken as the median hourly wage, which is NZD 29.66 [56]. Therefore, in 10 years, the time cost (dissatisfaction degree) of losing an EV is the multiplication of time cost per hour (NZD 29.66/hour), time spent every year (17 h) [57], and the project life cycle (10 years).

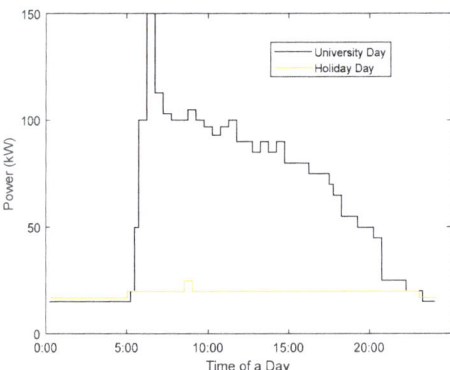

Figure 7. Baseload of the transformer.

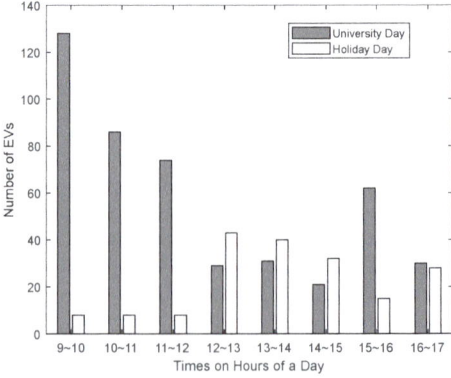

Figure 8. Arrival times of EVs.

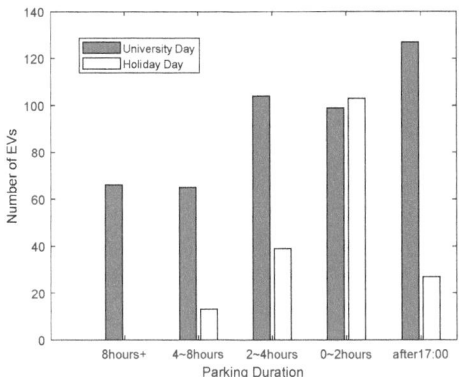

Figure 9. Parking duration of EVs.

Table 2. Rated battery capacity and maximum charging power.

EV Type	Nissan Leaf	BMW i3	Hyundai Kona	Renault Zoe
Battery Capacity	62 kWh	42.2 kWh	67.5 kWh	54.7 kWh
Maximum Charging Power	100 kW	49 kW	77 kW	46 kW

Table 3. Transformer and REVC parameters.

Parameter	Value		Parameter	Value
P_{REVC}^{max}	480 kW (split-type)	160 kW (integrated-type)	$cost_{cable}$	NZD 1500
P_{REVC}^{min}	120 kW (split-type)	100 kW (integrated-type)	$cost_{other}$	NZD 1500
n_{CP}^{max}	12 (split-type)	3 (integrated-type)	PF	0.99
n_{CP}^{min}	2 (either type)		C_{ex}	750 kVA
$cost_{PM}^{kW}$	NZD 100/kW		$cost_{transformer}^{kVA}$	NZD 150/kVA
$cost_{CP}$	NZD 1500		η	95%

Table 4. Electricity tariffs.

Period Type	Tariff (NZD/kWh)	Times (Hour)
Peak	0.2167	(12:00~19:00]
Shoulder	0.1116	(10:00~12:00] and (19:00~21:00]
Off-peak	0.0837	(21:00~10:00]

In Section 6.3, the comparison primarily focuses on split-type REVCs and conventional EVCs, as split-type REVCs are considered to have the highest flexibility. After that, Section 6.5 compares split-type REVCs and integrated-type REVCs.

6.2. Optimal Power Rating of PMs

According to REVC manufacturers, five commonly available power ratings of PMs are used. These power ratings include 10 kW, 15 kW, 16 kW, 20 kW, and 30 kW [58,59].

Based on (3) and (4), calculated degrees of utilization of PMs are displayed in Table 5. In this scenario, PMs with a 10 kW power rating have the highest degree of utilization. Consequently, 10 kW PMs are selected and will be utilized as a parameter in the number and configuration determining stage.

Table 5. Degrees of utilization of PMs.

Power Rating of PMs (kW)	Degree of Utilization
10	87%
15	38.14%
16	36.875%
20	73.5%
30	48.74%

6.3. Planning Results

The optimization in the number and configuration determining stage is conducted according to the selected power rating of PMs and other parameters. The numerical simulation results of planning are shown in Table 6. The outputs of the number and configuration determining stages are utilized as parameters in the operation plan determining stage. The planning results of REVC are compared with slow and fast EVCs under coordinated and uncoordinated charging. Coordinated charging is an operation to control the charging schedules of EVs to realize some objectives, such as peak shaving or reducing charging costs. Uncoordinated charging is an ordinary operation in which EVs get charged immediately after connecting EVCs. The optimization problems were solved by CPLEX using the branch-and-bound method [60]. Planning results show that the number of REVCs to be deployed is much fewer than the number of slow EVCs or fast EVCs, and the total cost of planning REVCs is less than the other two types of EVCs. This is because each REVC can serve several charging spaces concurrently and dynamically allocate charging power to different charging spaces. This contrasts with conventional slow or fast EVCs, which can serve only one charging space at a time.

Table 6. Planning results.

EVC Type	REVC (Split-Type)	Slow EVC	Fast EVC	Slow EVC	Fast EVC
Operation Manner	Coordinated	Coordinated	Coordinated	Uncoordinated	Uncoordinated
EVC Number	30	362	360	362	360
Charging Post/cable per EVC	12	1	1	1	1
EVC Power Rating (kW)	120	20	100	20	100
Upgraded Transformer Capacity (kVA)	750 + 771	750 + 771	750 + 771	750 + 1937	750 + 5762
Equivalent Annual Investment Cost (k NZD)	201.764	245.921	2445.623	245.921	2445.623
Equivalent Annual Time Cost (Dissatisfaction Degree) (k NZD)	1.370	0	1.370	0	1.370
Annual Operation Cost (k NZD)	102.674	102.674	102.674	319.620	986.262
Annual Cost of Demand Charge (k NZD)	154.412	154.412	154.412	281.515	661.098
Equivalent Annual Transformer Upgrade Cost (k NZD)	6.914	6.914	6.914	39.476	117.431
Total Annual Cost (k NZD)	428.789	471.576	2672.648	886.532	4211.784

According to the outputs of the number and configuration determining stage and EV charging demands, optimization in the operation plan determining stage is conducted. The coordinated charging strategy can be achieved by applying the proposed charging strategy, where the operation cost is reduced significantly, as indicated in Table 6. Even though the operation cost, the cost of demand charge, and the transformer upgrade cost of three different types of EVCs are the same with coordinated charging, the total annual cost of deploying REVCs is 9% and 84% less than slow and conventional fast EVCs, respectively. This is due to the fact that a smaller number of REVCs can be deployed than conventional EVCs, and a larger number of EVs can be charged concurrently by each REVC. The charging profiles of REVCs under different scenarios by applying the proposed optimal planning strategy are illustrated in Figure 10. Comparing Figure 10a,b, the peak load is larger in the summer than in the winter. This is because the driving range of an EV is short in winter due to the weather impact on the usable battery capacity [39]. Consequently, under the same arrival and departure SOCs, the energy delivered to an EV is less in winter. For charging profiles on a holiday day of the summer season or winter season, most of the charging sessions happen during the peak-tariff period. This is because most of the EVs arrive after 12:00 and depart before 18:00. Therefore, due to the constraints of arrival and departure times, their charging sessions are arranged during the peak-tariff period. Comparing the charging profiles on a university day and on a holiday day, the peak load is much smaller on a holiday day than on a university day. This is due to the fact that the number of charging EVs on a holiday day is much smaller than on a university day. The charging profiles of slow and fast EVCs with uncoordinated charging are shown in Figures 11 and 12, respectively. In Figure 10, many charging sessions are scheduled during off-peak and shoulder periods, compared to Figures 11 and 12. This is because by applying the optimization model in the third stage of the proposed strategy, the function of coordinated charging can be achieved, and charging sessions can be shifted to relatively lower tariff periods to minimize the operation cost. For EVs stayed overnight, charging sessions can be arranged during night periods with off-peak tariffs. For EVs parked for a short time, charging power can be designated to a proper level without compromising on charging demands.

In addition, the cost of the demand charge and the upgraded transformer capacity by applying coordinated charging are relatively smaller. However, for uncoordinated charging, charging sessions occur immediately after the arrival of EVs. Therefore, at the peak time of the arrival of numerous EVs, many charging sessions are carried out at the same time, resulting in a high peak load. Consequently, high peak loads result in a large demand charge and upgraded transformer capacity.

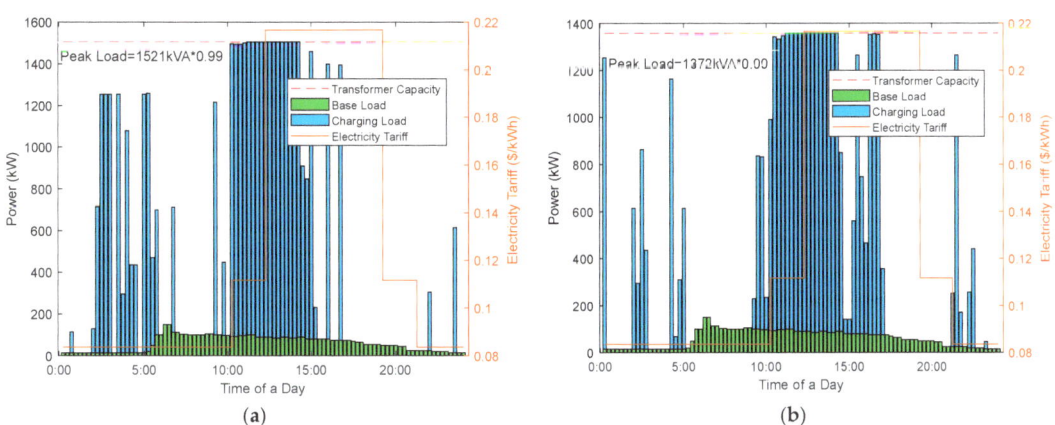

Figure 10. *Cont.*

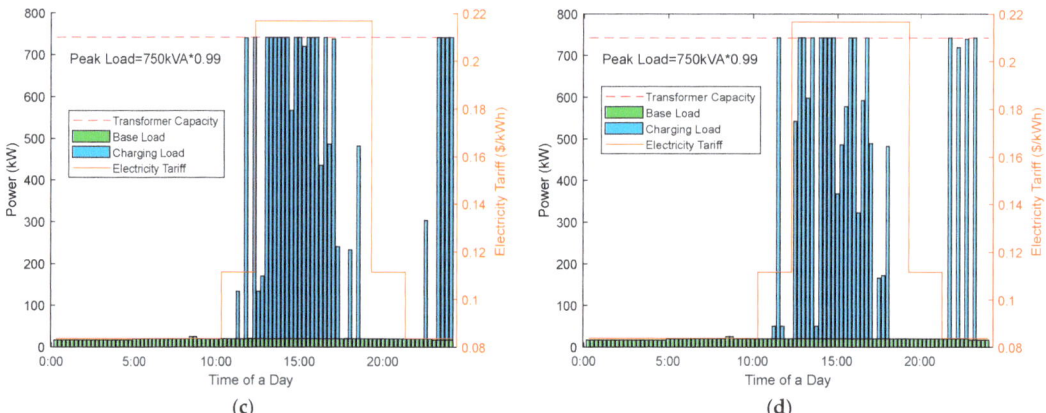

Figure 10. Charging profiles with the proposed strategy (**a**) a university day of the summer season; (**b**) a university day of the winter season; (**c**) a holiday day of the summer season; (**d**) a holiday day of the winter season.

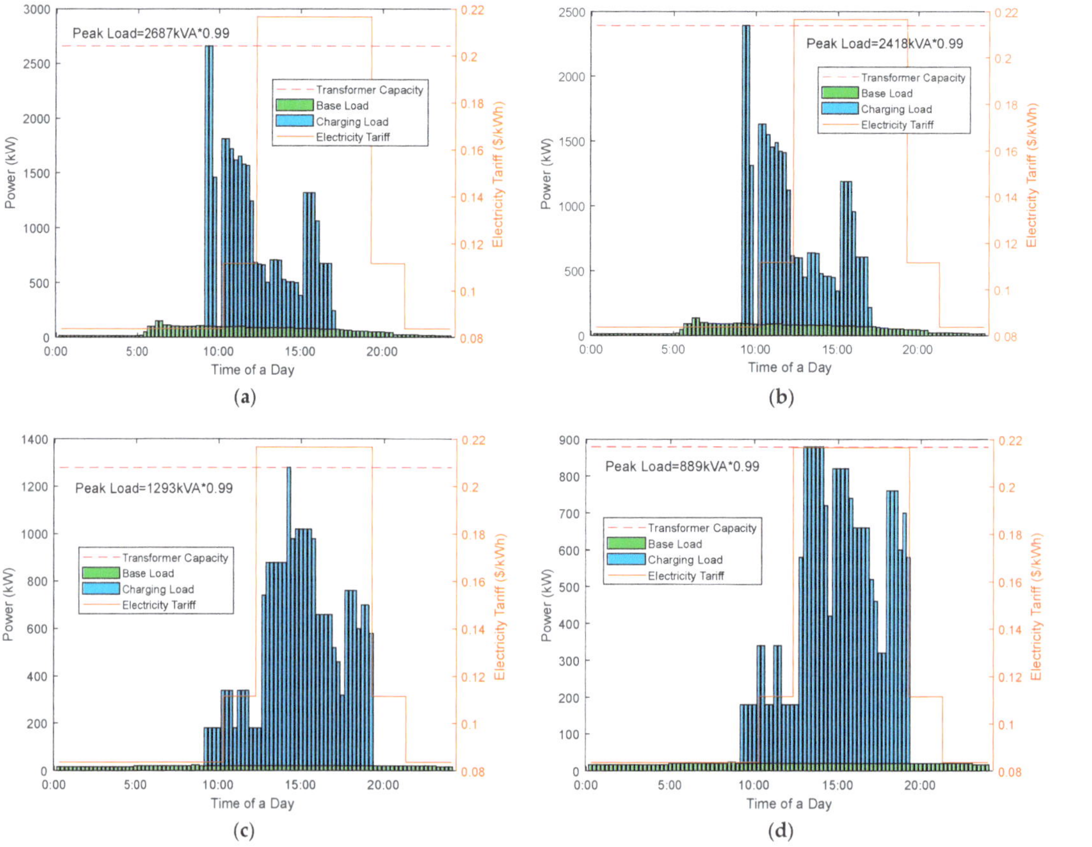

Figure 11. Charging profiles by applying slow EVCs with uncoordinated charging (**a**) a university day of the summer season; (**b**) a university day of the winter season; (**c**) a holiday day of the summer season; (**d**) a holiday day of the winter season.

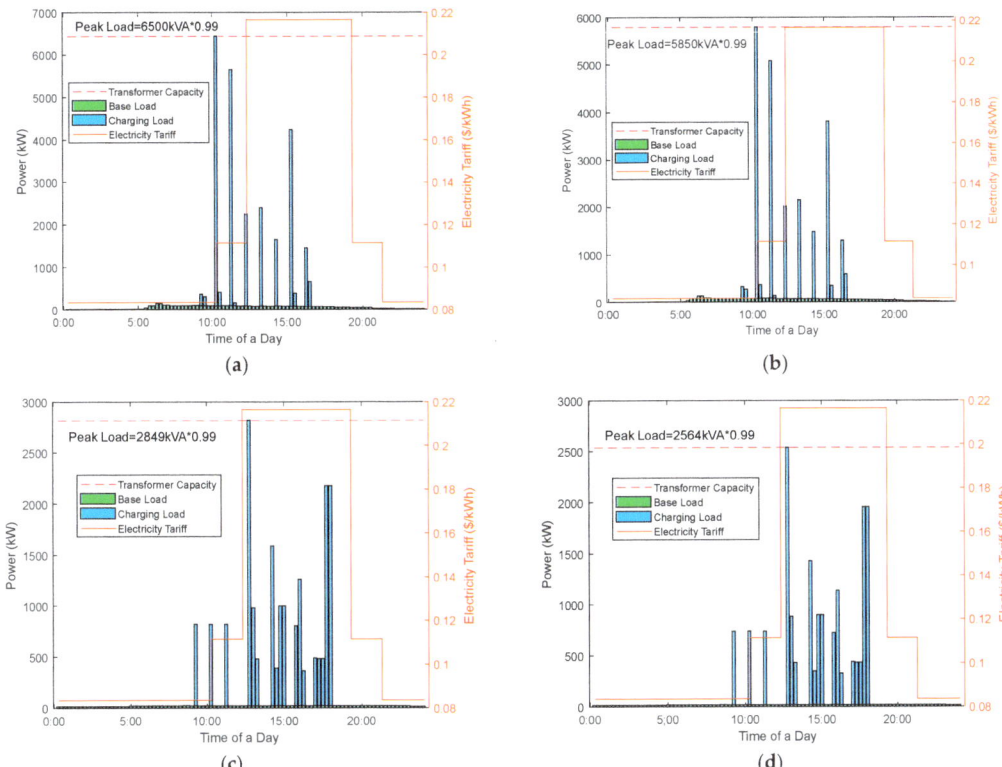

Figure 12. Charging profiles by applying conventional fast EVCs with uncoordinated charging (**a**) a university day of the summer season; (**b**) a university day of the winter season; (**c**) a holiday day of the summer season; (**d**) a holiday day of the winter season.

6.4. Influence of State-of-Charge

The arrival SOC and departure SOC depend on the battery energy left when each EV arrives at a car park and the energy required before each EV leaves, respectively. Thus, different arrival and departure SOCs would influence the charging demands of EVs and further influence the total annual cost. The total annual costs under different arrival SOCs and departure SOCs are given in Figure 13. The lower the arrival SOC and the higher the departure SOC, the higher the total annual cost. For the battery health of EVs, it is suggested that battery SOC be maintained between 20% and 80% [36]. Consequently, applying a 20% arrival SOC to 80% departure SOC is appropriate for a worst-case planning scenario.

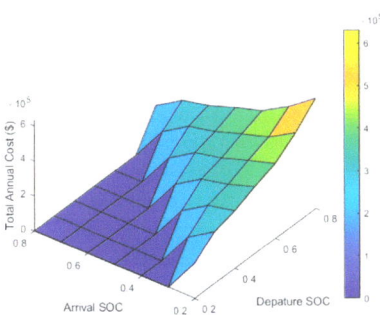

Figure 13. Total annual cost under different arrival and departure SOCs.

6.5. Comparison of Split-Type and Integrated-Type REVCs

The comparison of two types of REVCs is undertaken in this section under different charging spaces. Under a small number of charging spaces, the equivalent annual investment cost of planning integrated-type REVCs is less than that of planning split-type REVCs, as shown in Figure 14a. However, under a large number of charging spaces, the equivalent annual investment cost of planning split-type REVCs is less than that of planning integrated-type REVCs, as illustrated in Figure 14b.

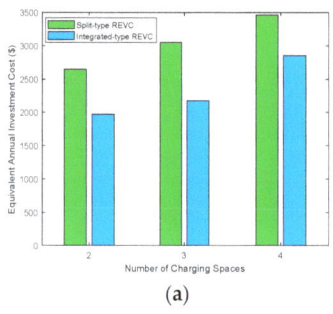
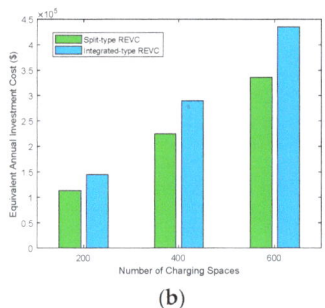

(a)   (b)

Figure 14. Equivalent annual investment cost comparison (**a**) with a small number of charging spaces; (**b**) with a large number of charging spaces.

A car park with a small number of charging spaces only requires a few EVCs. Due to the technical specifications, the power rating of each split-type REVC is larger than that of each integrated-type REVC. Consequently, the number of PMs in each split-type REVC is greater than that of each integrated-type REVC. Therefore, the cost of each split-type REVC is more expensive than each integrated-type REVC. Thus, the annual investment cost of deploying split-type REVC is more costly than deploying integrated-type REVC. However, because each split-type REVC can connect more EVs, the charging resources of a split-type REVC can be shared with more EVs compared to an integrated-type REVC. Consequently, for a car park with a large number of charging spaces, relatively fewer split-type REVCs are deployed compared to integrated-type REVCs. Thus, the annual investment cost of deploying split-type REVC is less expensive than deploying integrated-type REVC.

6.6. Sensitivity Analysis for REVC Component Costs

The component costs of REVCs and conventional EVCs vary depending on the manufacturers. Therefore, to evaluate the influence of the component costs on the equivalent annual investment cost of the planning, a sensitivity analysis is undertaken in this section. The main components of REVCs are PMs, CPs, and cables. The cost range for these components is presumed according to [44], shown in Table 7. The cost range for a 20 kW slow EVC is from NZD 5000 to NZD 8000. The cost range for a 100 kW fast EVC is from NZD 50,000 to NZD 75,000. To compare the equivalent annual investment costs with deploying REVCs, slow EVCs, and fast EVCs, the same scenario in Section 6.1 is applied. With the variations in the cost range, the equivalent annual investment cost of deploying REVCs is smaller than that of deploying slow or fast EVCs, as shown in Figure 15. This is because the number of REVCs to be deployed is much smaller than the other two types of EVCs, as each REVC can charge multiple EVs concurrently.

Table 7. Cost range for REVC components.

Component	Cost Range
Power Module	100~150 (NZD/kW)
Charging Post	1500~2500 (NZD)
Cable	1500~3500 (NZD)

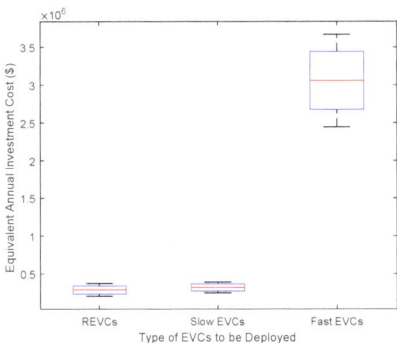

Figure 15. Comparison of equivalent annual investment costs with deploying different types of EVCs.

7. Conclusions

An optimal planning strategy for REVCs has been proposed in this paper. The proposed optimal planning strategy involves three stages, and an optimization model has been developed for each stage. The optimization model of the first stage determines the optimal power rating of internal PMs inside each REVC, and the optimization model of the second stage determines the optimal number and configuration of REVCs, followed by determining the optimal operation plan of REVCs from the optimization model of the third stage. To demonstrate the effectiveness of the proposed optimal planning strategy based on the optimization models, a comprehensive case study has been undertaken using realistic car parking scenarios with 400 parking spaces, electricity tariffs, and grid infrastructure costs. The results of the case study have shown that deploying REVCs by applying the proposed optimal planning strategy can reduce the total annual cost by 9% and 84%, respectively, compared to deploying other conventional types of EVCs.

Author Contributions: Conceptualization, B.S., U.K.M. and C.A.B.; methodology, B.S.; software, B.S.; validation, B.S.; formal analysis, B.S.; investigation, B.S., U.K.M. and C.A.B.; resources, B.S., U.K.M. and C.A.B.; data curation, B.S.; writing—original draft preparation, B.S.; writing—review and editing, B.S., U.K.M. and C.A.B.; visualization, B.S.; supervision, U.K.M. and C.A.B.; project administration, U.K.M.; funding acquisition, no funding. All authors have read and agreed to the published version of the manuscript.

Funding: This research received no external funding.

Data Availability Statement: The data presented in this study are available on request from the first author.

Conflicts of Interest: The authors declare no conflict of interest.

References

1. Singh, P.P.; Wen, F.; Palu, I.; Sachan, S.; Deb, S. Electric Vehicles Charging Infrastructure Demand and Deployment: Challenges and Solutions. *Energies* **2023**, *16*, 7. [CrossRef]
2. Aduama, P.; Al-Sumaiti, A.S.; Al-Hosani, K.H. Electric Vehicle Charging Infrastructure and Energy Resources: A Review. *Energies* **2023**, *16*, 1965. [CrossRef]
3. Town, G.; Taghizadeh, S.; Deilami, S. Review of Fast Charging for Electrified Transport: Demand, Technology, Systems, and Planning. *Energies* **2022**, *15*, 1276. [CrossRef]
4. Mathieu, L. *Recharge EU: How Many Charge Points Will Europe and Its Member States Need in the 2020s*; Transport & Environment: Brussels, Belgium, 2020.
5. Palmer, R.; Vipond, S.; Fisher, E.; Roling, S. *Progress and Insights Report 2021*; Climate Group EV 100: New York, NY, USA, 2021.
6. Song, B.; Madawala, U.; Baguley, C. A Review of Grid Impacts, Demand Side Issues and Planning Related to Electric Vehicle Charging. In Proceedings of the 2021 IEEE Southern Power Electronics Conference (SPEC), Kigali, Rwanda, 6–9 December 2021.
7. Schettini, T.; dell'Amico, M.; Fumero, F.; Jabali, O.; Malucelli, F. Locating and Sizing Electric Vehicle Chargers Considering Multiple Technologies. *Energies* **2023**, *16*, 4186. [CrossRef]

8. Campaña, M.; Inga, E.; Cárdenas, J. Optimal Sizing of Electric Vehicle Charging Stations Considering Urban Traffic Flow for Smart Cities. *Energies* **2021**, *14*, 4933. [CrossRef]
9. Liu, Z.; Wen, F.; Ledwich, G. Optimal planning of electric vehicle charging stations in distribution systems. *IEEE Trans. Power Del.* **2013**, *28*, 102–110. [CrossRef]
10. Guo, C.; Yang, J.; Yang, L. Planning of Electric Vehicle Charging Infrastructure for Urban Areas with Tight Land Supply. *Energies* **2018**, *11*, 2314. [CrossRef]
11. Wang, X.; Shahidehpour, M.; Jiang, C.; Li, Z. Coordinated Planning Strategy for Electric Vehicle Charging Stations and Coupled Traffic-Electric Networks. *IEEE Trans. Power Syst.* **2019**, *34*, 268–279. [CrossRef]
12. Micari, S.; Polimeni, A.; Napoli, G.; Andaloro, L.; Antonucci, V. Electric vehicle charging infrastructure planning in a road network. *Renew. Sustain. Energy Rev.* **2017**, *80*, 98–108. [CrossRef]
13. Dong, X.; Mu, Y.; Jia, H.; Wu, J.; Yu, X. Planning of fast EV charging stations on a round freeway. *IEEE Trans. Sustain. Energy* **2016**, *7*, 1452–1461. [CrossRef]
14. Wang, M.; Ismail, M.; Shen, X.; Serpedin, E.; Qaraqe, K. Spatial and temporal online charging/discharging coordination for mobile PEVs. *IEEE Wirel. Commun.* **2015**, *22*, 112–121. [CrossRef]
15. Maigha; Crow, M.L. Cost-constrained dynamic optimal electric vehicle charging. *IEEE Trans. Sustain. Energy* **2017**, *8*, 716–724. [CrossRef]
16. Gan, L.; Topcu, U.; Low, S.H. Optimal decentralized protocol for electric vehicle charging. *IEEE Trans. Power Syst.* **2013**, *28*, 940–951. [CrossRef]
17. Mou, Y.; Xing, H.; Lin, Z.; Fu, M. Decentralized optimal demand side management for PHEV charging in a smart grid. *IEEE Trans. Smart Grid* **2015**, *6*, 726–736. [CrossRef]
18. Kisacikoglu, M.C.; Erden, F.; Erdogan, N. Distributed control of PEV charging based on energy demand forecast. *IEEE Trans. Ind. Inform.* **2018**, *14*, 332–341. [CrossRef]
19. Xu, Z.; Su, W.; Hu, Z.; Song, Y.; Zhang, H. A Hierarchical Framework for Coordinated Charging of Plug-In Electric Vehicles in China. *IEEE Trans. Smart Grid* **2016**, *7*, 428–438. [CrossRef]
20. Jian, L.; Xue, H.; Xu, G.; Zhu, X.; Zhao, D.; Shao, Z.Y. Regulated Charging of Plug-in Hybrid Electric Vehicles for Minimizing Load Variance in Household Smart Microgrid. *IEEE Trans. Ind. Electron.* **2013**, *60*, 3218–3226. [CrossRef]
21. Borlaug, B.; Yang, F.; Pritchard, E.; Wood, E.; Gonder, J. Public Electric Vehicle Charging Station Utilization in the United States. *Transp. Res. Part D Transp. Environ.* **2023**, *114*, 103564. [CrossRef]
22. Zhang, H.; Hu, Z.; Xu, Z.; Song, Y. An Integrated Planning Framework for Different Types of PEV Charging Facilities in Urban Area. *IEEE Trans. Smart Grid* **2016**, *7*, 2273–2284. [CrossRef]
23. Ismail, M.; Bayram, I.S.; Abdallah, M.; Serpedin, E.; Qaraqe, K. Optimal planning of fast PEV charging facilities. In Proceedings of the 2015 First Workshop on Smart Grid and Renewable Energy (SGRE), Doha, Qatar, 22–23 March 2015.
24. Yang, Q.; Sun, S.; Deng, S.; Zhao, Q.; Zhou, M. Optimal Sizing of PEV Fast Charging Stations with Markovian Demand Characterization. *IEEE Trans. Smart Grid* **2019**, *10*, 4457–4466. [CrossRef]
25. Yang, Z.; Huang, X.; Gao, T.; Liu, Y.; Gao, S. Real-Time Energy Management Strategy for Parking Lot Considering Maximum Penetration of Electric Vehicles. *IEEE Access* **2022**, *10*, 5281–5291. [CrossRef]
26. Mokgonyana, L.; Smith, K.; Galloway, S. Reconfigurable Low Voltage Direct Current Charging Networks for Plug-In Electric Vehicles. *IEEE Trans. Smart Grid* **2019**, *10*, 5458–5467. [CrossRef]
27. Zhang, H.; Hu, Z.; Xu, Z.; Song, Y. Optimal Planning of PEV Charging Station with Single Output Multiple Cables Charging Spots. *IEEE Trans. Smart Grid* **2019**, *8*, 2119–2128. [CrossRef]
28. Chen, H.; Hu, Z.; Luo, H.; Qin, J.; Rajagopal, R.; Zhang, H. Design and Planning of a Multiple-Charger Multiple-Port Charging System for PEV Charging Station. *IEEE Trans. Smart Grid* **2019**, *10*, 173–183. [CrossRef]
29. Vaughan, P.; Baxter, D.; Hagenmaier, C.F.; Tran, J.P.K.; Matsuno, C.T.; Eldridge, G.A.; Romano, P. Dynamic Allocation of Power Modules for Charging Electric Vehicles. U.S. Patent 010150380B2, 11 December 2018.
30. ChargePoint Express 250 Specifications and Ordering Information. Available online: https://chargepoint.ent.box.com/v/CPE250-DS-EN-US (accessed on 27 September 2023).
31. INFY POWER Split Type HP Fast DC Charger. Available online: https://www.infypower.com/split-type-hp-fast-dc-charger.html (accessed on 27 September 2023).
32. ABB Electric Vehicle Infrastructure Terra High Power—GEN III. Available online: https://search.abb.com/library/Download.aspx?DocumentID=9AKK107991A9632&LanguageCode=en&DocumentPartId=&Action=Launch (accessed on 27 September 2023).
33. SIEMENS CPC150—The 150 kW Compact Power Charge. Available online: https://assets.new.siemens.com/siemens/assets/api/uuid:c9041d4a-79fe-4e80-8b2a-439fbea5946a/broschuere-sicharge-cpc150-print.pdf (accessed on 27 September 2023).
34. ChargePoint Express Plus. Available online: https://www.chargepoint.com/businesses/dc-stations/express-plus (accessed on 27 September 2023).
35. SENKU 150V-1000VDC 60KW—160KW DC EV Charger. Available online: http://www.senkumachinery.com/product/-ev-car-charging-station.html (accessed on 27 September 2023).
36. SCU EV Charging Stack EVMS EV Charging Stack (EV Power Unit + EV Charger Post). Available online: https://www.scupower.com/ev-charger/evse-ev-charging-stack/ (accessed on 27 September 2023).

37. EVTEC Sspresso&Charge 6in1. Available online: https://www.evtec.ch/application/files/9716/1329/3826/factsheet_espressocharge_6in1_en.pdf (accessed on 27 September 2023).
38. EVBox Troniq Modular. Available online: https://evbox.com/en/ev-chargers/troniq-modular (accessed on 27 September 2023).
39. ChargePoint Express 250 Review: New Charger on the Block. Available online: https://www.youtube.com/watch?v=xgKo-zxGw9k (accessed on 9 October 2023).
40. Annual Reports. Available online: https://at.govt.nz/about-us/reports-publications/annual-reports (accessed on 10 October 2023).
41. Wang, R.; Xing, Q.; Chen, Z.; Zhang, Z.; Liu, B. Modeling and Analysis of Electric Vehicle User Behavior Based on Full Data Chain Driven. *Sustainability* **2022**, *14*, 8300. [CrossRef]
42. Jonas, T.; Daniels, N.; Macht, G. Electric Vehicle User Behavior: An Analysis of Charging Station Utilization in Canada. *Energies* **2023**, *16*, 1592. [CrossRef]
43. Khan, S.; Brandherm, B.; Swamy, A. Electric Vehicle User Behavior Prediction using Learning-based Approaches. In Proceedings of the 2020 IEEE Electric Power and Energy Conference (EPEC), Edmonton, AB, Canada, 9–10 November 2020.
44. The New ChargePoint Mobile App Is Here. Available online: https://www.chargepoint.com/about/news/new-chargepoint-mobile-app-here/ (accessed on 10 October 2023).
45. EVgo Mobile App. Available online: https://www.evgo.com/download-app/ (accessed on 10 October 2023).
46. Mobile App. Available online: https://www.tesla.com/ownersmanual/model3/en_us/GUID-F6E2CD5E-F226-4167-AC48-BD021D1FFDAB.html (accessed on 10 October 2023).
47. AUT Semester Dates. Available online: https://www.aut.ac.nz/study/semester-dates (accessed on 27 September 2023).
48. Winter & Cold Weather EV Range Loss in 7000 Cars. Available online: https://www.recurrentauto.com/research/winter-ev-range-loss (accessed on 27 September 2023).
49. Kostopoulos, E.; Spyropoulos, G.; Kaldellis, J. Real-world study for the optimal charging of electric vehicles. *Energy Rep.* **2020**, *6*, 418–426. [CrossRef]
50. Common DC Fast Charging Curves and How to Find Yours. Available online: https://www.chargepoint.com/blog/common-dc-fast-charging-curves-and-how-find-yours (accessed on 9 October 2023).
51. How DC Fast Charging Really Works and an Intro to Charging Curves. Available online: https://www.chargepoint.com/blog/how-dc-fast-charging-really-works-and-intro-charging-curves (accessed on 9 October 2023).
52. Reducing EV Charging Infrastructure Costs. Available online: https://rmi.org/wp-content/uploads/2020/01/RMI-EV-Charging-Infrastructure-Costs.pdf (accessed on 27 September 2023).
53. Rate BEVT Business Electric Vehicle Time-of-Use. Available online: https://www.alabamapower.com/content/dam/alabama-power/pdfs-docs/Rates/BEVT.pdf (accessed on 27 September 2023).
54. Schedule GT (SC) General Service Time-of-Use. Available online: https://dms.psc.sc.gov/Attachments/Matter/0c8ffb07-d245-f293-036bd9115ebae6ad (accessed on 27 September 2023).
55. Schroeder, A.; Traber, T. The economics of fast charging infrastructure for electric vehicles. *Energy Policy* **2012**, *43*, 136–144. [CrossRef]
56. New Median Wage Will Apply from 27 February 2023. Available online: https://www.immigration.govt.nz/about-us/media-centre/news-notifications/new-median-wage-will-apply-from-27-february-2023 (accessed on 27 September 2023).
57. Drivers Spend an Average of 17 Hours a Year Searching for Parking Spots. Available online: https://www.usatoday.com/story/money/2017/07/12/parking-pain-causes-financial-and-personal-strain/467637001/ (accessed on 27 September 2023).
58. SCU EV Charger Module. Available online: https://www.scupower.com/ev-charger/ev-charger-module/ (accessed on 27 September 2023).
59. INFY POWER Module. Available online: https://www.infypower.com/module (accessed on 27 September 2023).
60. IBM ILOG CPLEX Optimization Studio. Available online: https://www.ibm.com/products/ilog-cplex-optimization-studio (accessed on 27 September 2023).

Disclaimer/Publisher's Note: The statements, opinions and data contained in all publications are solely those of the individual author(s) and contributor(s) and not of MDPI and/or the editor(s). MDPI and/or the editor(s) disclaim responsibility for any injury to people or property resulting from any ideas, methods, instructions or products referred to in the content.

Article

A Case Study of the Use of Smart EV Charging for Peak Shaving in Local Area Grids

Josef Meiers and Georg Frey *

Automation and Energy Systems, Saarland University, D-66123 Saarbrücken, Germany; josef.meiers@aut.uni-saarland.de
* Correspondence: georg.frey@aut.uni-saarland.de

Abstract: Electricity storage systems, whether electric vehicles or stationary battery storage systems, stabilize the electricity supply grid with their flexibility and thus drive the energy transition forward. Grid peak power demand has a high impact on the energy bill for commercial electricity consumers. Using battery storage capacities (EVs or stationary battery systems) can help to reduce these peaks, applying peak shaving. This study aims to address the potential of peak shaving using a PV plant and smart unidirectional and bidirectional charging technology for two fleets of electric vehicles and two comparable configurations of stationary battery storage systems on the university campus of Saarland University in Saarbrücken as a case study. Based on an annual measurement of the grid demand power of all consumers on the campus, a simulation study was carried out to compare the peak shaving potential of seven scenarios. For the sake of simplicity, it was assumed that the vehicles are connected to the charging station during working hours and can be charged and discharged within a user-defined range of state of charge. Furthermore, only the electricity costs were included in the profitability analysis; investment and operating costs were not taken into account. Compared to a reference system without battery storage capacities and a PV plant, the overall result is that the peak-shaving potential and the associated reduction in total electricity costs increases with the exclusive use of a PV system (3.2%) via the inclusion of the EV fleet (up to 3.0% for unidirectional smart charging and 8.1% for bidirectional charging) up to a stationary battery storage system (13.3%).

Keywords: bidirectional charging; electric vehicle; smart charging; peak-shaving

Citation: Meiers, J.; Frey, G. A Case Study of the Use of Smart EV Charging for Peak Shaving in Local Area Grids. *Energies* **2024**, *17*, 47. https://doi.org/10.3390/en17010047

Academic Editors: Cesar Diaz-Londono and Yang Li

Received: 17 November 2023
Revised: 12 December 2023
Accepted: 18 December 2023
Published: 21 December 2023

Copyright: © 2023 by the authors. Licensee MDPI, Basel, Switzerland. This article is an open access article distributed under the terms and conditions of the Creative Commons Attribution (CC BY) license (https:// creativecommons.org/licenses/by/ 4.0/).

1. Introduction

The amended Federal Climate Protection Act (KSG) passed by the German Bundestag in 2021 raises Germany's greenhouse gas reduction target from 55% to 65% compared to 1990. A reduction rate of 88% is to be achieved by 2040 and greenhouse gas neutrality by 2045. These climate protection targets will set in motion an extensive and far-reaching transformation process in Germany that will affect all sectors. With the increased use of decentralized, fluctuating generation systems (e.g., PV systems) and the penetration of e-mobility and other controllable loads such as heat pumps, the demands on the public and non-public grids (local area grids (LAG)) and their operators are growing. Cost-efficient measures and concepts for grid operation are becoming a key factor for an economical energy supply that meets the requirements of the customer and the regulatory framework. The backbone of future smart-grids is the infrastructure of information and communication technology (ICT) and automation technology (AT). Without communication connections, the use of information and the resulting targeted control of actuators in the network will not be possible. Wired (e.g., fiber optics) or wireless communication technologies (e.g., GSM, LTE, and LoRaWAN) can be used as transmission media. The distribution grid operator (DGO) or LAG operator (LAGO) can access measuring devices (MD) and controllable loads (CL) in their network structure with their own communication infrastructure.

The EU also classifies battery technology as an Important Project of Common European Interest (IPCEI) across the entire value chain from raw material extraction to recycling in a circular economy. The EU's efforts in this context also involve developing innovative battery systems including battery management systems. Over the next few years, the EU will invest 2.9 billion euros in research and development projects for renewable energies and energy storage [1]. As efficiency increases and supply expands, demand will rise rapidly and the price of energy storage systems will fall as a result. According to their data for the years 2010 to 2019 and a forecast up to 2025, the Statista Research Department sees a downward trend in the global price trend for lithium-ion batteries [2].

In 2022, the Renewable Energy Research Association (FVEE) formulated recommendations for the implementation of system integration that are aimed at industry, society, research, and politics. This refers to the technical and digital linking of energy system components and the development of various flexibility options for the use of high proportions of volatile renewable energies [3]. One requirement here is the rapid implementation of the anchoring of energy storage as an independent pillar of the energy system, as stipulated in the coalition agreement of the current federal government. In the area of digitization of an integrated energy system, the intensification of the standardization of digital interfaces and data formats is called for as well as the implementation of grid-supportive behavior of energy market participants. In particular, it highlights the need for research into system integration with joint research and development work between research institutes, energy suppliers, and municipal players.

The provision of flexibility is therefore indispensable in the future electricity grid, which will be characterized by a high proportion of fluctuating electricity generation from wind power and photovoltaic systems, and battery storage systems are absolutely essential. As a rule, the installation of battery storage systems is initially dependent on economic considerations. The revenue opportunities, and thus, the question of whether the storage system is worthwhile, are heavily dependent on the local conditions (renewable generation and consumption capacity). With this knowledge, it is then necessary to investigate combinations of several applications, so-called Multi-Use approaches, which, by providing flexibility, enable both profit maximization for the operators and economic optimization of grid expansion via their system-beneficial behavior.

The transmission system operators' 2022 draft of the grid development plan up to 2037 [4] lists forecasts in which battery storage in particular will become significantly more relevant in the future. This applies to both large-scale battery storage systems and decentralized PV home storage systems. The increase in PV home storage systems is based on the expansion of building PV systems. While around half of all new rooftop PV systems with storage systems have been installed in recent years, forecasts assume that the proportion will increase to 100% by 2035. The expansion of stationary large-scale battery storage systems is based on the expansion of ground-mounted PV systems, which is assumed to increase linearly. The forecast expansion rate is 30% by 2030, up to 70% in 2035, and up to 100% in 2040.

In the Ariadne Report [5], the authors also assume a necessary expansion of electrical storage capacities—stationary battery storage, but also mobile batteries in the Vehicle-2-Grid (V2G) network. Suitable market integration must be created for these storage systems in order to reduce any disadvantage compared to grid expansion and communication technologies must be established to ensure meaningful operation in the entire electricity system. The study recommends a review of the extent to which battery storage systems can contribute even more flexibility to the electricity grid. Flexibility is defined by the Federal Network Agency as a change in feed-in or withdrawal in response to an external signal (price signal or activation) with the aim of providing a service in the energy system [6].

In the Prognos study [7], the authors also assume a future electricity grid with a high degree of flexibility by 2045. This will be characterized by the rapid expansion of battery storage, load management, and intensive electricity trading with other countries.

Due to the immense ramp-up of battery storage technologies, the aspect of sustainability must also be given greater consideration in future product development [8]. Also, the European Parliament and the Council accounts for that in the new Batteries Regulation, adopted in 2023, which should minimize the environmental impact and strengthen the circular economy concerning battery storage appliances.

Grid peak power demand has a high impact on the energy bill for commercial electricity consumers. Using battery storage capacities (EVs or stationary battery systems) can help to reduce these peaks, applying peak shavings.

This study aims to address the potential of peak shaving using smart unidirectional and bidirectional charging technology for an EV fleet and a stationary battery storage system (BSS) in combination with a PV plant on the university campus of Saarland University (UdS) in Saarbrücken as a case study.

The major contribution of this paper is to answer the following questions:

- How much peak load and electricity cost can be reduced with peak shaving using an EV fleet with bi-directional charging technology?
- What is the impact on different sizes of the EV fleet?
- What is the impact of bi-directional charging technology compared to smart unidirectional charging?
- Is there more of less potential on using a stationary BSS of the same performance?

Based on an annual measurement of the grid demand power of all consumers on the campus, a simulation study was carried out to compare the peak shaving potential of a fleet of EVs (30 EVs/50 EVs) with, on the one hand both, smart unidirectional and bidirectional charging and, on the other hand, a stationary BSS (due to the comparability with the same capacity and performance as EV fleets).

For the sake of simplicity, it was assumed that the vehicles are connected to the charging station during working hours and can be charged and discharged within the user-defined charging limits.

In addition, we assume that the smart EV charging technology uses perfect prediction on the future grid power demand to control the EV charging process in an optimal way to minimize the grid demand power peaks for the whole day.

Furthermore, only the annual costs of electricity (per kWh and kW) that the end consumer has to pay were included in the profitability analysis; investment and operating costs were not taken into account. The scenarios with EV fleets and a stationary BSS were also combined with a PV system with a peak power of 1 MW. Additionally, the case with a 1 MW PV system without storage capacity (EV fleets/stationary BSS) was also considered.

The structure of this contribution is as follows: After describing the importance of the flexibility of battery storage capacities in the context of the German energy transition in this section, the next section presents marketing options for the flexibility of battery storage, in particular the peak-shaving functionality. Furthermore, the next section analyzes the measured consumption data and presents the basis for calculating the electricity procurement costs for the evaluation of the simulation studies based on the price sheet of the local distribution grid operator. The vehicle-to-everything (V2X) concepts are presented and the scenarios under consideration are assigned to them. In addition, the models for the PV system and the electric vehicles are presented and the optimization problem is formulated. Section 3 presents and discusses the results of the simulation study for the seven use cases considered. Section 4 concludes this article with a summary and an outlook.

2. Materials and Methods

2.1. Review Marketing of Battery Storage Flexibility

In order to operate battery storage systems (BSS) for both stationary and EV in an economically viable manner and to develop business models, a regulatory framework is required. In many cases, the flexibility provided is currently only used for a single application (SINGLE-USE). However, there is additional potential in using several applications at the same time and thus utilizing different sources of income. This is referred to as a MULTI-USE approach. However, the application service does not necessarily have to be provided simultaneously (in parallel). For example, it can also be provided sequentially at fixed times or dynamized [9].

In the following, a distinction is made between four options for marketing the flexibility provided by storage capacities such as EVs or BSS [10,11]:

- System owner-friendly operation
  End consumer-related applications lie in the self-consumption optimization of the emergency power supply and are used in conjunction with e-mobility or the local Virtual-Power-Line (VPL). In the VPL concept, battery storage serves as a buffer after energy sources and before energy sinks in order to limit power peaks on the intermediate supply line and guarantee a constant energy flow.
- Market-serving operation
  This includes participation in arbitrage trading on the electricity markets and the day-ahead and intraday markets. The storage capacities charges at times when prices are low and discharges when electricity is scarce and prices are therefore high. Power-to-X models should also be mentioned.
- System-serving operation
  The storage capacities participate in the balancing power market and maintains capacity to stabilize the electricity grid. Depending on the dimensions and response time of the system and within the European Network of Transmission System Operators for Electricity (ENTSO-E) grid, three frequency regulation products are offered: Frequency Containment Reserve (FCR), Frequency Restoration Reserve (FRR), and Replacement Reserve (RR), whereas, depending on the duration of the imbalance in the grid, FRR replaces FCR and RR replaces FRR after a fixed time period. Battery storage systems are particularly suitable for participation in the FCR market due to their short response time. Other applications include black start capability and voltage stabilization of the power grid by providing reactive power.
- Grid-serving operation
  While the system-serving operation is aimed at stabilizing the electricity grid at the national and European level, the grid-serving operation of the battery has a different focus: The focus here is on the local grid and local congestion management. As the expansion of decentralized renewable energy production plants progresses and the number of electric cars increases, this is becoming increasingly important, as line bottlenecks will occur more frequently due to a delay in the expansion of electricity grid capacities. The provision of battery flexibility represents an alternative to the expensive grid expansion.

However, the terms system-serving operation and grid-serving operation are not used consistently.

Furthermore, battery storage applications can be divided into Front-of-The-Meter (FTM) and Behind-The-Meter (BTM) applications. FTM applications take place on the side of the public grid and BTM applications take place on the side of the end consumer. Figure 1 summarizes the described flexibility options once again.

The electrification of the transport sector is a key pillar of the German energy transition. The German government's goal is to have six million electric vehicles on German roads by 2030. The resulting ramp-up of electric vehicles and their charging facilities and, in the future, battery storage systems, will pose major challenges for public distribution grids and non-public LAG, such as the campus of Saarland University as a result of the university's own vehicle fleet and of the EVs of university staff due to considerably higher low-voltage power consumption and significantly higher simultaneity. In most cases, however, the EVs' charging and, in case of bidirectional charging, discharging behavior can be controlled. However, there are limits to the extent to which this leads to a noticeable loss of comfort for consumers or the aging of battery cells due to cyclic stress. Fraine et al. [12] interviewed 89 persons from groups of young people (18–25 years), parents, and non-parents (29–56 years), among others, concerning their driving behavior. The average driving time was between 9.3 and 10.6 h per week, which means 1.4 h in average per day. This means that the car stands around unused for an average of 22.6 h (94%) a day and therefore EVs, in conjunction with the currently developed technology of bidirectional charging to control the power flow in the connected public or non-public electricity grid, offer great potential for flexibility in the electricity grid, which will give e-mobility a new boost in the future. By using the storage capacity of EVs to support the grid, grid expansion costs can be reduced, e.g., by reducing grid bottlenecks or applying peak shaving.

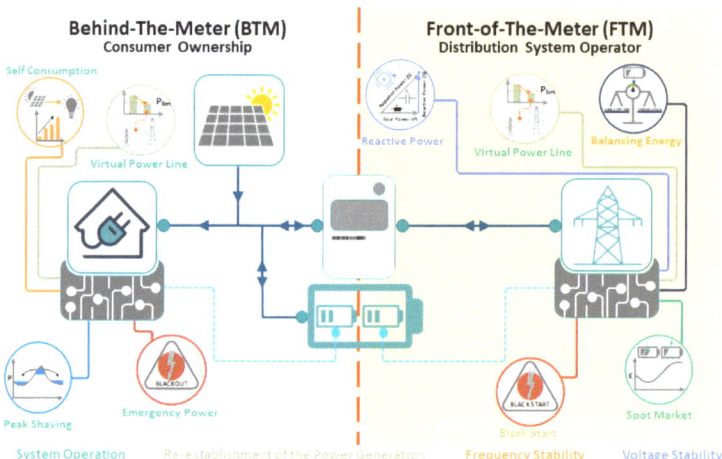

Figure 1. Behind-The-Meter and Front-of-The-Meter applications for battery storage systems.

A lot of effort has already been invested in scientific research into the potential of applications that consider EV and/or BSS capacities in the energy system.

In some studies, only smart unidirectional charging was examined [13]; in others, the bidirectional charging of EVs [14–19]. Some focused on a building energy system ([18]), while others examined the impact on local grids (commercial, industrial, and parking lots) [20,21].

Either renewable energy sources (e.g., PV systems) are considered in the system [17] or not [13,14,16,22]. Other studies look at stationary battery storage for peak shaving during the charging process of electric vehicles [22]. And still, other researchers are investigating the positive effect of exclusively stationary battery storage systems on local power grids [23,24].

Ioakimidis et al. [20] examined the Vehicle-to-Building (V2B) functionality on a parking lot for a maximum of 65 vehicles. The real parking lot occupancy was measured and used as the basis for the simulation-based investigation of three scenarios with 8, 35, and 65 randomly selected parking spaces. The results show that the power peaks could be reduced to between 3% and 20% depending on the scenario.

Minhas et al. [19] presented a multi-timescale, cost-effective scheduling and control strategy of energy distribution in a model predictive intelligent home energy management system comprising EVs and PV. In their study, the authors found that electricity energy costs from the grid supply could be reduced by 13%.

Mahmud et al. [21] have investigated the peak shaving of a commercial building using six EVs with bidirectional charging functionality in a parking lot. In their results, the industrial peak loads can be reduced by 50% and the energy cost can be reduced by 27.3%.

In their study, Fenner et al. [25] investigated the potential of peak shaving in parking areas in the Helsinki region. Based on real measurement data on the charging behavior of 25,000 charging cycles, a peak load reduction of 55% was achieved using optimization algorithms.

Van Kriekinge et al. [26] investigated the effect of smart unidirectional and bidirectional EV charging on electricity costs and peak load reduction for a commercial building with a connected PV system near Brussels based on measured energy consumption and production data. According to their results, all MPC-based charging strategies were able to achieve a peak load reduction between 14.6% and 33.7% and total electricity cost reductions between 6.71% and 7.67% compared to uncoordinated charging, with bi-directional charging delivering the best results.

Peak shaving serves to stabilize the electricity grid, which is heavily stressed by short-term, heavy loads and must also be permanently available for these load cases. Short-term, particularly high electricity loads from large industrial or commercial consumers, drive up their electricity prices enormously; as such, electricity customers not only have to pay the energy price, but also a demand power charge. With peak shaving, the costs for high electricity loads can be reduced by means of Demand Response (DR) measures or BSS.

BSS are ideal for smoothing out dynamic load peaks within the scope of their performance characteristics. Assuming an appropriate charging/discharging strategy, battery storage capacities in EVs and a stationary BSS connected to the power grid are capable of realizing fast and reliable load peak compensation.

DR is understood as a short-term, deliberate change in consumer load in response to price signals in the market. DR is achieved either via load shifting or flexibilization of the load profile or a load reduction. Electricity consumption is brought forward, delayed, or avoided altogether. Therefore, peak shaving can be performed in three ways:

- On the consumer side
  A consumer reduces its electricity consumption quickly and at short notice (load shedding), so as not to cause a peak load. This can be achieved by throttling production.
- On the self-generation system side
  By switching on self-generation plants based on renewable energy sources (e.g., PV or wind power plants) or conventional energy sources (e.g., diesel generators), the electricity demand from the supply grid is reduced on balance depending on the ratio of generation and consumer output. In this way, self-generated electricity is used to balance out the impending peak load.
- On the electricity storage side
  Similar to the generation systems, battery storage systems can also smooth out the grid demand peak power by discharging. Due to their technology, battery storage systems and electrical storage systems can quickly provide high current densities and are therefore particularly suitable for compensating for short-term load peaks.

2.2. Load Demand Analysis

The measured data on the electricity demand of all consumers on the university campus is available with a time resolution of 15 min. The meter readings were recorded at these intervals. These values were then converted into average power values in an initial processing step. According to this dataset, the annual consumption for 2022 is 25,003.810 MWh. Analysis of the available electrical consumption power for the year 2022 shows a maximum peak power of 4.38 MW (30 June 2022 11:45 a.m., day 181) and a minimum power of 1.61 MW (4 June 2022 5:30 a.m., day 155) (Figure 2).

Including the information about the day type [27], Figure 3 shows that the daily peak loads of consumers on weekdays are almost twice as high as on Sundays and public holidays. On Saturdays, they are somewhere in between. A seasonal course of the daily peak loads can also be seen. These are higher on summer and winter days than in spring and fall. This is presumably caused by the cooling loads in summer and the heating demand in winter. Figures 2 and 3 clearly show the reduced demand on weekends. It can also be seen that consumer demand is lower between June 4 and 6 (days 155 to 157) and November 12 and 13 (days 316 and 317) than on weekdays and weekends. Both periods are weekends.

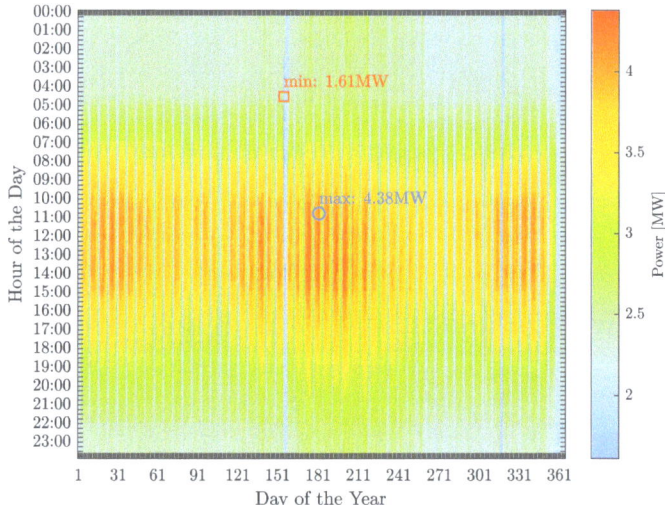

Figure 2. Electrical load profile for the UdS campus in 2022.

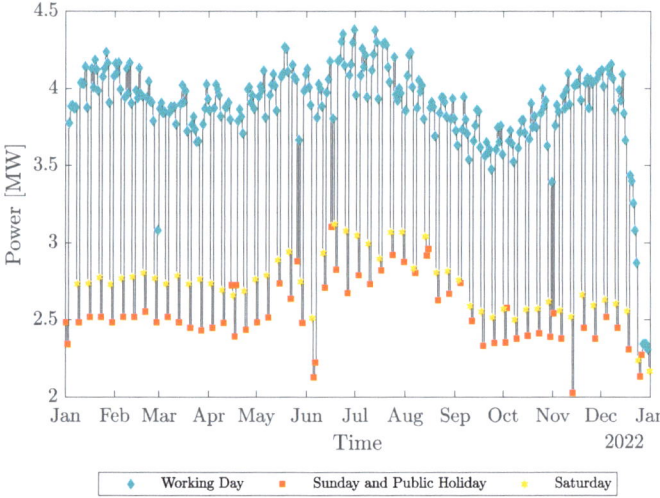

Figure 3. Daily consumption peaks, differentiated by type of day (Working Day/Sunday and Public Holiday/Saturday).

2.3. Peak Loads and Grid Usage Fees

Grid operators do not like load peaks as the electricity grid is planned and designed on the basis of the maximum power in the grid.

Nevertheless, many industrial companies that are connected to the different grid levels (high-, medium-, or low-voltage grid) cause fluctuating loads in everyday operation; for example, when starting up production facilities, heating up, or pumping processes. The source of the sudden increase in load, i.e., the commercial or industrial electricity customer, is reliably detected via consumption metering.

In Germany, electricity consumers are divided into two groups according to the type of consumption metering [28]:

- Customers with Recording Power Metering (RPM customers);

- Customers without recording power metering (SLP customers). Their consumption will be estimated based on Standard Load Profiles (SLP).

In the energy industry, peak shaving refers to the smoothing of load peaks and the associated grid consumption peaks for industrial and commercial electricity consumers (RPM customers). These peaks in electricity consumption are not only relevant for grid stability, but also, as explained above, for electricity costs. As the grid usage fees, which make up a large proportion of the total costs, are based on the highest power consumption in the billing period.

RPM customers are large commercial and industrial consumers whose annual electricity consumption exceeds 100,000 kWh and whose power requirement is at least 30 kW (see Section 12 of the Electricity Grid Access Ordinance (StromNZV)).

In Germany, distribution grids are managed with two voltage levels: Medium voltage and low voltage. Medium-voltage distribution grids generally have a voltage of 10 kV or 30 kV. In the medium-voltage grid, electricity is distributed between grid stations within the individual urban or rural districts. The low-voltage grid is the grid that transports electricity to the end consumer. The low-voltage grid is operated at a voltage of 0.4 kV and is connected to the medium-voltage grid via grid stations.

The total annual electricity costs for an industrial or commercial consumer (RPM customer) are made up of three components:

- the energy costs;
- the capacity costs;
- the basic annual costs for the metering equipment.

The costs for power measurement in the medium-voltage grid are slightly higher than measurement in the low-voltage grid due to higher technical requirements in terms of dielectric strength.

The total annual charge $C_{total,a}$ is the sum of three price components:

- fixed basic annual charge for the metering equipment $C_{bc,a}$;
- costs for grid capacity, that is, the product of the grid demand power peak price C_{kW} and the maximum annual grid power consumption $P_{max,a}$;
- costs for energy from the grid, that is, the product of the electricity energy price C_{kWh} in EUR per kWh and the annual energy demand from the grid E_a in kWh

Therefore, the following equation generally applies to the calculation of grid usage fees for RPM customers:

$$C_{total,a} = C_{bc,a} + C_{kW} P_{max,a} + C_{kWh} E_a \qquad (1)$$

The annual usage period is an important key figure in the energy industry. It is the quotient of the annual energy and the maximum output of a system. It indicates how many hours of electricity would have been drawn in a year if the maximum output had been constantly drawn. In the ideal case, with absolutely constant consumption without interruption, the annual usage period is 365×24 h $= 8760$ h. The annual usage period also has an impact on the costs for grid usage. The higher the annual usage period, the higher the capacity price for grid usage, but conversely, the lower the energy price for grid usage. Therefore, constant consumption, which results in a low maximum output and a high annual usage period, is economically advantageous. When pricing the use of electricity grid infrastructure, a distinction is often made between two or three ranges of annual usage periods. In this case, the local DGO distinguishes between two ranges: less than or equal to 2500 h and greater than 2500 h.

The annual usage period t_a is calculated from the annual energy E_a and the annual maximum grid demand power $P_{max,a}$ according to the following equation:

$$t_a = \frac{E_a}{P_{max,a}} \qquad (2)$$

According to the fee table of the local DGO for 2023 [29], Stadtwerke Saarbrücken GmbH charges the annual basic fee for metering point operation with consumer/feed-in power metering from a medium voltage of 485.01 EUR for power metering, a capacity price of 26.07 EUR/kW, and a energy price of 0.0649 EUR/kWh for an annual usage period of up to 2500 h. For an annual usage period of more than 2500 h, the price is 159.28 EUR/kW and 0.0116 EUR/kWh (Table 1).

Table 1. Charges of the local DGO for consumption from medium voltage and annual power system.

Charges	Value
Annual Usage Period $t_a \leq 2500$	
Capacity Price C_{kW}	26.07 EUR/kW
Energy Price C_{kWh}	0.0649 EUR/kWh
Annual Usage Period $t_a > 2500$	
Capacity Price C_{kW}	159.28 EUR/kW
Energy Price C_{kWh}	0.0116 EUR/kWh
Basic Fee for Metering per Year $C_{bc,a}$	485.01 EUR

According to the measured load profile for 2022, the annual maximum consumer capacity as a 15 min average is 4.38239 MW. With an annual electricity demand of 25,003.810 MWh, this results in an annual utilization period according to Equation (2) of 5705 h, and thus, more than 2500 h. A peak load of 4.38239 MW in the corresponding tariff at a capacity price of 159.28 EUR/kW leads to an annual capacity price of 698,027 EUR per year. The energy price of 0.0116 EUR/kWh results in annual costs of around 290,044 EUR and, with the basic price of 485.01 EUR, results in total electricity costs of around 988,556 EUR in 2022 according to Equation (1). This value serves as a reference for the simulation studies with a PV system and storage capacities of EVs or a stationary BSS.

2.4. Smart Unidirectional Charging and Bidirectional Charging for Electric Vehicles

In addition to uncoordinated unidirectional charging, where the battery will be charged with maximum power given by the charging characteristics, unidirectional smart charging (V1G, smart charging) for electric vehicles offers the possibility of using dynamic charging tariffs and times via adapted charging in order to save costs or to make optimum use of a supply of renewable energy. According to Hildermeier et al. [30], charging technology can be considered smart or intelligent if it meets the following minimum requirements:

- It can measure consumer energy consumption in real-time or near real-time;
- It can transmit this data to the consumer and to other authorized parties;
- It has the ability to automatically control consumption and is also below the maximum charging power.

Therefore, smart charging technology enables customers to apply DR, due to reacting on control signals (e.g., price signals) quickly with the help of ICT.

With smart charging, the charging infrastructure can be optimized by distributing the available power efficiently and flexibly. This means that even charging stations with limited power capacity can be used optimally at all times.

The technology of bidirectional charging comprises several applications that are generally referred to as V2X ("Vehicle-to-Everything") and, as explained briefly below, can be divided into several categories (see Figure 4):

- Vehicle-to-Load (V2L)

  V2L is a bidirectional function that enables an electric vehicle to use its built-in high-voltage battery to charge or supply low-voltage devices. It is sometimes also referred to as vehicle-to-device (V2D). Depending on the type of device to be charged or powered, V2L can be used while the electric vehicle is driving or parked. This conversion from a direct current to an alternating current is integrated into the vehicle. Electric vehicles usually offer one or both of two options for V2L charging: an AC socket (in the vehicle)

and a V2L adapter (vehicle-to-charging plug) that is used with the electric vehicle's charging port.
- Vehicle-to-Home (V2H)
  With V2H, the battery is used as a power backup to feed a local building or local grid downstream of the grid connection point. The electricity temporarily stored in the battery, for example, from renewable energy sources, can be used to optimize your own electricity requirements. However, no electricity is fed back into the public grid. With V2H, it is important that not all of the battery capacity is available as electricity storage, so that you always have sufficient range when you set off.
- Vehicle-to-Building (V2B)
  V2B works in a similar way to V2H, but on a larger scale. By bundling several electric vehicles or entire fleets, the energy requirements of buildings in an area network are optimized. Typical areas of application are properties or industrial plants. With your significantly larger battery capacity and total output, line losses and imbalances can be corrected, particularly in industrial plants with high inductive loads, and effective measures can be taken to smooth out grid power peaks.
- Vehicle-to-Vehicle (V2V)
  This concept provides for the connection of two electric vehicles via a cable; for example, to charge a broken-down vehicle or to use parked, provided vehicles as charging stations.
- Vehicle-to-Grid (V2G)
  – Self-Consumption Optimization
    Electricity from the vehicle battery is provided for direct consumption on site behind the grid connection point in the respective property as part of comprehensive in-house optimization via a (local) energy management system;
  – Grid-serving Charging
    This means that the grid operator influences the charging behavior of the EV against the background of its load monitoring in order to reduce/avoid the grid consumption of the existing consumption devices for a limited period of time;
  – Electricity Trading
    Electricity is fed into the distribution grid on the basis of a contract with a supplier/dealer or made available to them. The supply/feed-in takes place in accordance with the specifications or a control signal from the supplier/dealer and in coordination with the local grid operator;
  – System-serving Charging
    Electricity is fed into or supplied to the grid on the basis of a contract with the transmission system operator (TSO). The supply/feed-in takes place according to the specifications or via a control signal from the TSO and in coordination with the local grid operator.

V2G technology is covered by the international ISO 15118 [31] standard. ISO 15118 defines the basic standards that apply to bidirectional communication between vehicles and charging stations and also regulates plug-and-charge and payment at charging stations.

The scenario considered here of integrating electric vehicles into the Saarbrücken campus grid is assigned to the V2B topology.

The technical requirements for bidirectional charging must be taken into account in all components involved and the communication between them. The charging process is controlled either by the EVs' integrated Battery Management System (BMS) in AC charging mode (On-board-charger) or via communication to the DC charging station (Off-board charger) that controls the power flows via given control signals from the EVs' BMS [32].

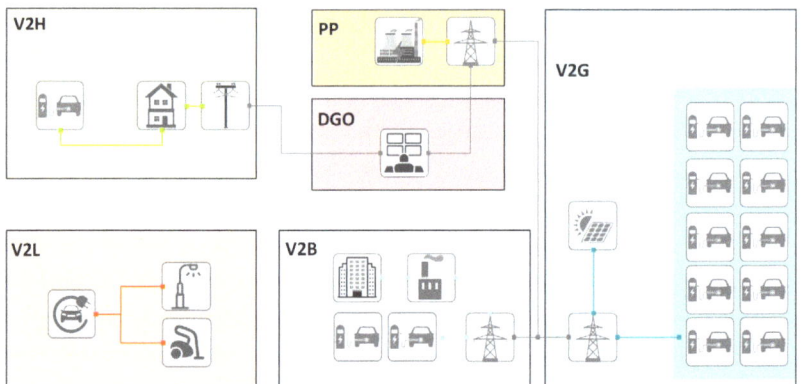

Figure 4. V2X Topology: V2G = Vehicle-to-Grid; V2L = Vehicle-to-Load; V2H = Vehicle-to-Home; V2B = Vehicle-to-Building; DGO = Distributed Grid Operator; PP = Power Plant.

In order to benefit from the advantages of bidirectional charging, three key conditions must be met:

- The wallbox must support bidirectional charging
- The vehicle must support bidirectional charging
- The vehicle and wallbox must have compatible DC connections (e.g., CCS, CHAdeMO)

When it comes to charging infrastructure, a distinction must be made between alternating current (AC) charging stations and direct current (DC) charging stations. Bidirectional charging makes sense where vehicles are parked for long periods and remain connected to a charging station, i.e., particularly at home or at work. AC charging stations with outputs of up to 22 kW are currently mainly used there. DC charging stations in this power range are currently only available from individual providers. On the vehicle side, there are also two approaches to implementing bidirectionality, which differ according to where the electricity is converted from DC to AC voltage. This can take place either in the vehicle or in the charging station. This means that, depending on the charging technology, modifications to the vehicle or the charging station are necessary in order to use bidirectionality. These changes are associated with additional costs for the charging infrastructure or the vehicle side.

The ISO 15118-20 [33] communication standard, which will be used by European and American vehicle manufacturers together with the Combined Charging System (CCS), enables bidirectional charging via both three-phase AC (maximum 44 kW) and DC fast charging (maximum 50 kW). Depending on the car manufacturer, both directions are currently being pursued.

Vehicles that use the CHAdeMO standard (DC) have already supported bidirectional charging for several years. CHAdeMO is an acronym for "CHArge de MOve" (charging to move). The first regenerative vehicle models (AC and DC) based on the Chinese GBT standard also already exist. The implementation of ISO 15118-20 together with CCS as the predominant standard for communication between the vehicle and the charging station will enable intelligent and grid-friendly charging in the future and create a basis for V2X as a way of integrating e-mobility.

The availability of the EEBUS communication standards will also support its use. EEBUS is a communication interface to support interoperability and data exchange between the components of an energy management system (e.g., PV, storage, and e-mobility). The OCPP protocol (Open Charge Point Protocol), which has been transferred to the international standard IEC 63110 [34], has become established for controlling the charging infrastructure (communication between charging station and charging station management system) in public charging. Communication between electric vehicles and charging stations, on the other hand, is described in the international standard ISO 15118. As things stand at present, the application of bidirectional charging is still in the early stages. Although there are already vehicle manufacturers offering this technology (see Table 2), the appropriate infrastructure does not yet exist. Suitable wallboxes are not expected until the second half of 2023. In addition, various manufacturers are still limiting the use of the function. This is due to the warranty conditions regarding the service life or mileage of the battery. Volkswagen limits the discharge energy of the ID models to 10,000 kWh and 4000 h [35].

Table 2. Overview of some electric vehicles that support bidirectional charging [36].

Model	Plug Type	AC/DC	V2X Functionality
Hyundai Ioniq 6	Schuko plug	AC (single-phase)	V2L
Ford F-150 Lightning	CCS	DC	V2H/V2G
Honda	CCS	DC	V2H/V2G
Nissan eNV200	CHAdeMO	DC	V2H/V2G
Nissan Leaf	CHAdeMO	DC	V2H/V2G
VW ID.3,4,5	CCS	DC	V2H/V2G
Volvo EX90	Schuko plug/Typ 2/CCS	AC (single phase)/DC	V2H/V2G

At present, the range of bidirectional charging stations is still limited. Some of them have been listed in Table 3 and the price for a bidirectional charging station is significantly higher than for a normal unidirectional wallbox. Depending on the model, the cost of a bidirectional wallbox can amount to several thousand euros. As the supply of V2H, V2B, and V2G charging stations is likely to increase in the future, lower prices can be expected. The manufacturer data in Table 3 shows that the current maximum output of bidirectional wallboxes is 22 kW.

Table 3. Overview of some wallboxes that support bidirectional charging [37].

Manufacturer	Model	Plug Type	Max Power [kW]
Wallbox Chargers	Quasar 1	ChAdeMO	7.4
Wallbox Chargers	Quasar 2	CCS	12.8
Kostal	BDL Wallbox	CCS	11
Eaton	BDL Wallbox	CCS	22
Ambibox	ambiCHARGE	CCS	22
Silla	Duke 44	CCS	22 (2x)

On the basis of the market situation described above, a VW ID.4 with a battery capacity of 77 kWh was selected for the electric vehicle fleet in this case study. The charging curve can be seen in Figure 5, but this is limited in both the charging and discharging directions by the wallbox's maximum output of 22 kW, which, according to the charging curve, corresponds to the charging power at full charge (SOC = 100%). The charging curve was linearly interpolated using the five interpolation points from the data collection provided in [?]. The maximum charging power of the EV is 125 kW at a SOC between 0% and 30%, has a constant power of 65 kW between 70 and 80%, and decreases linearly to 22 kW at 0%.

The basic behavior in the use case with bidirectional charging follows the sequence shown in Figure 6. The figure shows the time of day of the charging process and the resulting potential for flexibility services (e.g., peak shaving). The electric vehicle arrives at the charging point on the campus at the arrival time $t_{a,w}$ at 8:00 a.m., with a certain state of charge (SOC) which is assumed to be 50% ($SOC_{EV,a}$) for all scenarios and for the entire EV fleet. During the idle time up to the departure time $t_{d,w}$ at 18:00, the battery capacity can be used freely within the lower discharge limit $SOC_{EV,min}$, which is 30%, and the upper discharge limit $SOC_{EV,max}$. However, the state of charge at departure ($SOC_{EV,d}$) is chosen to be 90% and must be reached again at the departure time $t_{d,w}$. The lower state of discharge is defined as a buffer for spontaneous mobility. The upper discharge limit $SOC_{EV,max}$ corresponds to the desired state of charge $SOC_{EV,d}$ for all vehicles at the time of departure. As hard boundary conditions in the optimization algorithm, these limit values cannot be exceeded or undercut. The flexibility range is limited by the maximum charging capacity towards the departure time.

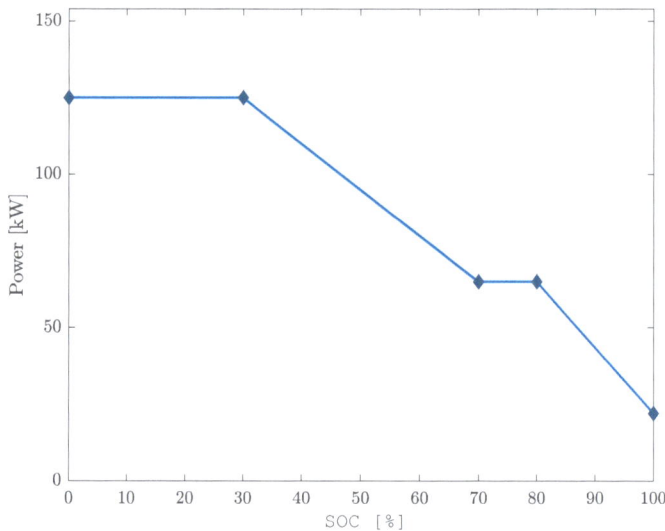

Figure 5. Charging Curve of Volkswagen ID.4 with 77 kWh.

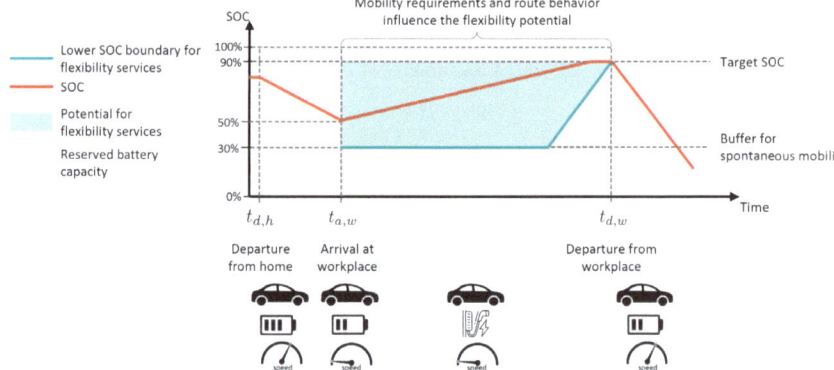

Figure 6. Schematic representation of the flexibility potential of an EV within the arrival and departure time at the workplace.

An entity E_{EV} was created for the simulation, which has the properties described above and listed in the following Table 4. At the current state and in this use case, a fleet of a single entity is considered. In future model development, several entities with different properties will be considered.

A constant power conversion efficiency of 90% was assumed for charging ($\eta_{batt,ch}$) and discharging ($\eta_{batt,disch}$) for both the EV fleet and the stationary BSS [39,40]. Round trip efficiency has not been considered as well as temperature-depending efficiency or dependencies on the SOC. The stationary BSS was modeled with the same capacity, performance, and SOC limits as the EV fleet.

Table 4. EV entity properties considered in this use case.

EV Property	Value
State of charge on arrival at the charging station on campus $SOC_{EV,a}$	50%
Total battery capacity of the EV (W_{EV})	77 kWh
Nominal charging and discharging power of the wallbox ($P_{nom,wallbox}$)	22 kWh
Daily arrival time $t_{a,w}$	8:00 a.m.
Daily departure time $t_{d,w}$	18:00 a.m.
Destination SOC at departure time $SOC_{EV,d}$	90%
Minimum discharge depth during charging time $SOC_{EV,min}$	30%
Maximum discharge depth during charging time $SOC_{EV,max}$	90%

2.5. PV System

The MATLAB library PVlib [41] was used to model the PV system. The data for the solar irradiation of a Typical Meteorological Year (TMY) was retrieved from the PVGIS platform [42] for the Saarbrücken location. In addition, the NREL (National Renewable Energy Laboratory) sun position algorithm (SPA) [43] was used to calculate the position of the sun as this provides very accurate sun positions.

The individual steps and parameters are listed below in Table 5.

Table 5. Steps and parameters of the PV system model.

Step	Description	PVlib Function/Parameter
1	Set location (Saarbrücken)	latitude = 49.233°; longitude = 7°; elevation = 193 m
2	Retrieve the solar radiation data for a TMY from PVGIS	https://re.jrc.ec.europa.eu/api/v5_2/tmy?lat=49.233&lon=7&outputformat=json (accessed on 2 November 2023)
3	Set PV array parameters	Tilt Angle = 30°; Azimut Angle = 180° (South); 12 PV modules in series; 12 parallel strings
4	Calculate the sun position with SPA algorithm	location (step 1); time; air pressure/dry bulb temperature (step 2)
5	Define the PV module	pvl_sapmmoduledb(); BP Solar SX150 (No. 100)
6	Define the PV Inverter	SNLInverterDB(); Agepower AP 20000 TL3-US 277V 20.4 kW (No. 80)
7	Calculate Relative Air Mass	pvl_relativeairmass(); sun elevation position (step 4)
8	Calculate Absolute Air Mass	pvl_absoluteairmass(); relative air mass (step 7); air pressure (step 2)
9	Determine Angle of Incidence	pvl_getaoi(); PV array orientation (step 1); sun position (step 4)
10	Calculate Beam Radiation Component on Array	Direct Normal Irradiance (step 2); Angle of Incidence (step 9)
11	Determine extraterrestrial radiation from day of year	pvl_extraradiation(); Day of the Year
12	Calculate Sky Diffuse Radiation Component on Array using Perez model and france1988 coefficients	pvl_perez(); PV array orientation (step 3); sun position (step 4); Horizontal, Direct Beam (step 2) and Horizontal Extraterrestrial Irradiation (step 11)
13	Determine Ground Reflected Radiation Component on Array	pvl_grounddiffuse(); PV array orientation (step 3); Global Horizontal Irradiation (step 2), albedo = 0.2
14	Calculate Total Diffuse Radiation Component on Array	Sky Diffuse Radiation (step 12) + Ground Reflected Radiation (step 3)
15	Calculate Total Radiation Component on Array	Sky Diffuse Radiation (step 12) + Ground Reflected Radiation (step 3) + Beam Radiation (step 10)
16	Determine PV Module Cell Temperature	pvl_sapmcelltemp(); total incident irradiance (step 15); wind speed/dry bulb temperature (step 2); reference irradiance = 1000 W/m²; PV module parameters (step 5)

Table 5. *Cont.*

Step	Description	PVlib Function/Parameter
17	Calculates the SAPM effective irradiance using the SAPM spectral loss and SAPM angle of incidence loss functions	PV module parameters (step 5); absolute air mass (step 8); angle of incidence (step 9); beam radiation component on array (step 10); diffuse radiation on array (step 14); soiling factor = 0.98
18	Determine Module/Array I-V Performance (DC power, voltage, current output) using Sandia PV Array Performance Model (SAPM)	pvl_sapm(); PV module parameters (step 5); cell temperature (step 16); SAPM effective irradiance (step 17)
19	DC Power to AC Power Conversion	pvl_snlinverter(); Inverter parameters (step 6); PV array I-V performance parameters (step 18)
20	Scaling PV array to 1 MWp PV plant	

Applying the PV system model described in Table 5 results in the electricity production of the PV system shown in Figure 7 over the course of a year.

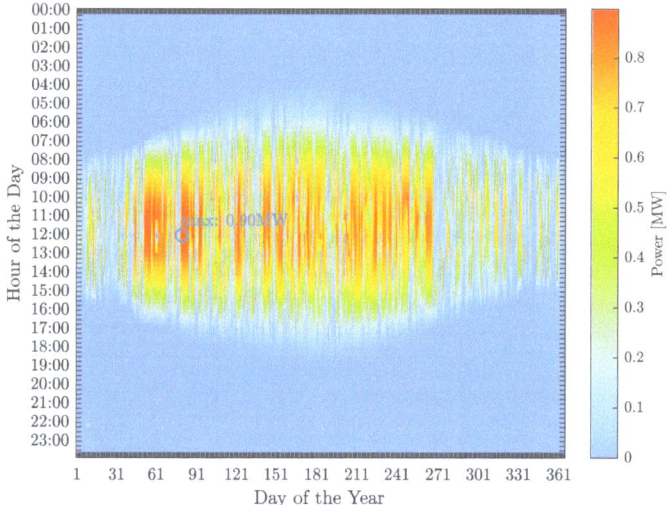

Figure 7. AC power production of the modeled PV plant using solar irradiance data from PVGIS for a TMY and PVlib.

2.6. Optimization Problem Formulation

The optimization problem was formulated as a mixed integer problem in MATLAB R2021b [44] using the YALMIP toolbox, R20210331 [45] and Gurobi 9.5 [46] as a solver. The simulation was carried out in 15 min steps. All results are based on the 15-min based average time base. The optimization cycle takes place once to create an optimized schedule for the entire next day. This assumes a perfect forecast of consumption and PV production. The behavior of the EV fleet can also be seen as a perfect prediction as there is no spontaneous mobility and the assumed availability of the vehicles with the assumed parameters is fixed (see Table 4).

The energy balance equation is:

$$P_{pv}(t) + P_{grid,dem}(t) = P_{grid,feedIn}(t) + P_{batt}(t) + P_{pv,loss}(t) + P_{load}(t) \tag{3}$$

where P_{pv} is the PV production, $P_{grid,dem}$ is the energy demand from the grid, $P_{grid,feedIn}$ is the energy that is fed into the grid, $P_{pv,loss}$ is the loss of unused PV power, and P_{load} is the load power consumption. Due to the design of the PV system, there is effectively no feed-in of PV electricity; the minimum consumption power of 1.61 MW exceeds the peak power of the PV system (see Figure 2). Nevertheless, $P_{pv,loss}$ can also be considered as a slack variable to ensure a solution to the optimization problem.

The PV production P_{pv} and the consumer power demand P_{load} serve as input and the other power values in Equation (3) are optimization variables.

When formulating the optimization problem, several binary variables were introduced to cover all use cases. These can be divided into switches, parameters, and optimization variables. A description of the state values is given in Table 6. The switches are essentially used to set user-defined properties of the system configuration and are selected before the simulation. They include the variables s_{EV}, s_{stat} and s_{bidi}. Parameters, on the other hand, change their values at the simulation runtime. They include the variables s_{work} and $s_{EV,avail}$. The values of the optimization variables are determined by the solver at the simulation runtime in order to minimize the objective function. They include $s_{batt,ch}$ and $s_{batt,disch}$.

Table 6. Binary variables in the Optimization Problem.

Type	Description
s_{bidi}	switch to choose whether EV fleet has bidirectional charging capability (1: yes, 0: no)
s_{batt}	switch to choose whether there is battery capacity in the model (scenarios 2–7) or not (scenario 1) (1: yes, 0: no)
s_{stat}	switch to choose whether battery is EV or stationary BSS (1: BSS, 0: EV)
s_{work}	parameter showing if current day is working day (1: yes, 0: no)
$s_{EV,avail}$	parameter showing if EV is available, i.e., day time is between $t_{a,w}$ and $t_{d,w}$ (1: yes, 0: no)
$s_{batt,disch}$	optimization variable indicating battery storage is in discharge state (1: yes, 0: no)
$s_{batt,ch}$	optimization variable indicating battery storage is in discharge state (1: yes, 0: no)

The state of charge of the battery (EV fleet or BSS) at time t ($SOC(t)$) results from the state of charge at time $t-1$ and the relative amount of energy supplied or dissipated in the time step Δt (15 min), which results from the battery power $P_{batt}(t)$ and the nominal total battery capacity. This is the product of the number of EVs (n_{EV}) and the storage capacity of the individual EV (W_{batt}).

$$SOC(t) = SOC(t-1) + \frac{P_{batt}(t)\Delta t}{W_{batt} n_{EV}} \quad (4)$$

The battery power $P_{batt}(t)$ is the sum of the discharging power $P_{batt,disch}(t)$ and the charging power $P_{batt,ch}(t)$, taking into account the corresponding efficiencies $\eta_{batt,disch}$ and $\eta_{batt,disch}$. This should only be the case at times when the battery is available, i.e., $s_{EV,avail} = 1$.

$$P_{batt}(t) = (-\frac{1}{\eta_{batt,disch}} P_{batt,disch}(t) + \eta_{batt,ch} P_{batt,ch}(t)) s_{EV,avail} \quad (5)$$

A distinction must be made between two cases of availability. If the EV fleet is considered, this depends on the time of day and the type of day (working day or no working day). If, on the other hand, a stationary BSS is considered s_{stat}, the day type and time of day are irrelevant as the following equation shows.

$$s_{EV,avail} = \begin{cases} 1: & (t_{a,w} \leq t \leq t_{d,w} \text{ and } s_{work} = 1) \text{ or } s_{stat} = 1 \\ 0: & \text{else} \end{cases} \quad (6)$$

where s_{work} is a binary parameter that is 1 if the current day is a working day, otherwise it is 0.

The maximum charging power of the battery $P_{batt,ch,max}$ is controlled by the EVs' BMS. In case of charging, the upper power limit is given by the charging curve (Figure 5), where the minimum value is 22 kW at 100% SOC. Therefore, in both cases, charging and discharging, the power is limited by the nominal power of the wallbox $P_{nom,wallbox}$ that is 22 kW. For the maximum power of the fleet, the number of EVs in the fleet n_{EV} (30 or 50) must be taken into account. In addition, the presence of storage capacities is realized with s_{batt} as a switch in order to be able to select the corresponding scenarios.

$$P_{batt,ch,max} = P_{nom,wallbox} n_{EV} s_{batt} \quad (7)$$

Two additional conditions must be taken into account, that of the bidirectional charging function and the presence of battery capacities as given in Equation (8).

$$P_{batt,disch,max} = P_{nom,wallbox} n_{EV} s_{bidi} s_{batt} \quad (8)$$

where s_{bidi} is the binary switch indicating that the EV fleet has a bidirectional charging capability or not and s_{batt} is a user-defined binary switch to choose whether the model has a battery capacity (scenarios 2–7) or not (scenario 1).

The charging and discharging power of the battery is limited in each case by the maximum values $P_{batt,ch,max}$ and $P_{batt,disch,max}$ explained above and the respective binary optimization variables $s_{batt,ch}$ and $s_{batt,disch}$ as formulated in Equations (9) and (10).

$$0 \leq P_{batt,ch}(t) \leq P_{batt,ch,max} s_{batt,ch} \quad (9)$$

$$0 \leq P_{batt,disch}(t) \leq P_{batt,disch,max} s_{batt,disch} \quad (10)$$

As already explained, the state of charge $SOC(t)$ for both the EV fleet and the stationary BSS is limited by the two limits $SOC_{EV,min}$ and $SOC_{EV,max}$.

$$SOC_{EV,min} \leq SOC(t) \leq SOC_{EV,max} \quad (11)$$

In the use cases with the EV fleet, the initial SOC on each day is the defined starting value $SOC_{EV,a}$ (50%), while in the scenarios with a stationary BSS, this only applies at the start of the simulation on the first day.

The grid reference power is limited upwards by the maximum value $P_{grid,dem,max}$.

$$0 \leq P_{grid,dem}(t) \leq P_{grid,dem,max} \quad (12)$$

where $P_{grid,dem,max}$ has been chosen as 20 MW, which is high enough to give no constraint on the grid demand power.

The grid feed-in power, on the other hand, is limited upwards by $P_{pv,nom}$, that is, the peak power of the PV plant.

$$0 \leq P_{grid,feedIn}(t) \leq P_{pv,nom} \quad (13)$$

The binary variables $s_{batt,ch}$ and $s_{batt,disch}$ were introduced, turning the optimization problem into a mixed integer problem. They indicate whether the battery storage is in a discharge or charge state, but not both at the same time.

$$s_{batt,ch}(t) + s_{batt,disch}(t) \leq 1 \quad (14)$$

The objective function is a weighted sum of three terms for the EV fleet (J_{EV}, Equation (15)) and two terms for the use cases with stationary battery storage (J_{BSS}, Equation (16)). Weighting factor w is chosen be 0.9. For both cases, the maximum grid consumption (peak) (Equation (18)) and the power loss of PV production (Equation (19)) should be minimized. In the cases with an EV fleet, the state of charge at the end of the working time $t_{d,w}$ should correspond to the target value $SOC_{EV,d}$, so the difference between these two values must be minimized (Equation (17)).

$$J_{EV} = min(w(J_0 + J_2) + (1-w)J_1) \quad (15)$$

$$J_{BSS} = min(wJ_2 + (1-w)J_1) \quad (16)$$

$$J_0 = \|SOC_{EV}(t) - SOC_{EV,d}\|_2 \quad (17)$$

$$J_1 = max(P_{grid,dem}) \quad (18)$$

$$J_2 = \|P_{pv,loss}\|_2 \quad (19)$$

During optimization, a schedule for the next day is generated using a perfect prediction of the consumption profile and PV production.

3. Results and Discussion

As aforementioned, to increase comparability, the stationary BSS was modeled with the same capacity and performance as the EV fleet. The same limits were also assumed for the SOC ($SOC_{EV,min}$, $SOC_{EV,max}$). In contrast to the use of electric vehicles, however, restrictions such as the target SOC at departure time ($SOC_{EV,d}$) or of time-limited charging and discharging (availability only on working days and between arrival and departure) were omitted for the scenarios with a stationary BSS.

The following scenarios in Table 7 are considered as use cases.

Table 7. Considered scenarios.

Scenario	PV	EV/BSS	uni/bi [1]	Number of EVs	Accumulated Storage Capacity [MWh]/ Maximum Peak Power [MW]
1	yes	-	-		
2	yes	EV	uni	30	2.31/0.66
3	yes	EV	uni	50	3.85/1.1
4	yes	EV	bi	30	2.31/0.66
5	yes	EV	bi	50	3.85/1.1
6	yes	BSS	-	-	2.31/0.66
7	yes	BSS	-	-	3.85/1.1

[1] uni = uni-directional smart charging; bi = bi-directional charging.

The current system state without a PV system and storage capacities (EV fleet/stationary BSS) serves as the reference scenario.

This study assumes that the charging of EVs is free of charge for the participants. In the case of bidirectional charging, this is understood as an incentive and compensation for providing the battery storage capacity of the electric vehicle.

The results of a 1-year simulation of all the scenarios and the reference system are shown in Figures 8–10.

In the following figure, Figure 8, energy flows and the SOC of the reference scenario (Figure 8a) and scenarios 1, 4, 5, and 7 are shown (Figure 8b–e). When looking at the system with a PV plant (scenario 1; without storage capacities), in Figure 8b, in comparison to the reference case (Figure 8a), it is noticeable that there is a good overlap between PV production and the consumption profile on the campus. The consumption peaks are also at midday. With a suitable design of the PV system, there could be a high potential for peak shaving here alone, at least in the summer months.

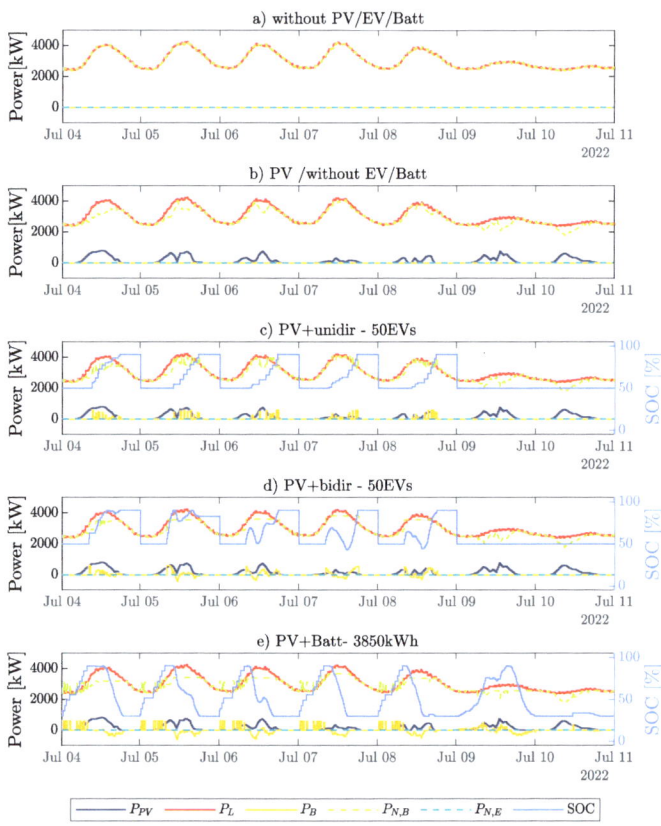

Figure 8. Energy flows and battery SOC shown as an example for one week (Monday to Sunday, 4–11 July 2023) for the reference case without storage capacities and PV plant in (**a**) and scenarios 1, 4, 5, and 7 (**b**–**e**).

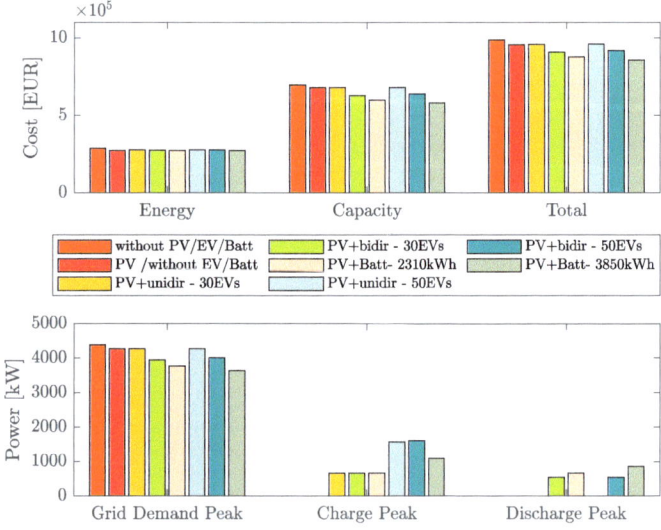

Figure 9. Comparison of costs and maximum annual peak power for the scenarios under consideration.

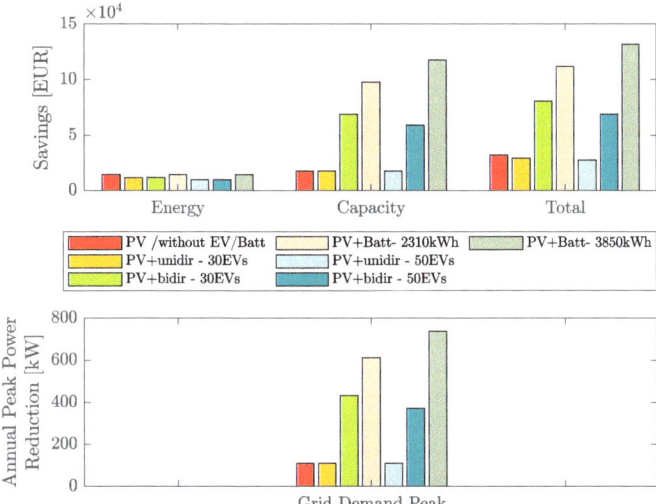

Figure 10. Cost savings and peak power reduction compared to the scenarios under consideration.

Figure 8c shows use case 4. The behavior of the EV before arrival at the workplace ($t < t_{a,w}$) and after the end of working hours ($t > t_{d,w}$) was not modeled, whereby the SOC on each working day corresponds to the start value $SOC_{(EV,a)} = 50\%$ for $t < t_{a,w}$ and the target value $SOC_{(EV,d)} = 90\%$ for $t > t_{d,w}$. It can be seen that the smart charging optimization algorithm charges the EV fleet particularly during periods of high PV production and on days with low PV production (see July 7 and 8), where charging takes place in the afternoon, i.e., after the midday consumption peaks.

Use case 5 (Figure 8d) shows the EV fleet feeding back into the local area grid. As expected, this occurs at midday, when the load is at its highest. Furthermore, it can be seen that the EVs are already pre-charged in the morning in order to be able to reduce the power peaks at midday by discharging.

If a stationary BSS is used instead of the EV fleet (Figure 8e, scenario 7), the time restrictions on storage use no longer apply. As the PV system is not sufficient to cover consumption, the stationary storage system is also pre-charged from the electricity grid like the EV fleet in use case 5 in order to achieve peak shaving. This already takes place at night.

The annual electricity costs of grid supply energy and the maximum grid supply power peaks that occurred during the year (c.f. Section 2.3, Equation (1)) and the savings and reductions of the same via utilization of storage capacities (EVs and BSS) are discussed below.

For the reference system, the total annual costs amount to 988,556 EUR, with energy costs of 290,044 EUR and capacity costs of 698,027 EUR.

With a 1 MWp PV system (scenario 1), the total costs can be reduced by 31,874 EUR, of which 14,259 EUR is attributable to energy costs and 17,615 EUR to capacity costs. The maximum annual grid power consumption was reduced by 110.59 kW.

As expected, the additional use of 30 electric vehicles in scenario 2 with a maximum charging capacity of 22 kW and the given boundary conditions (Table 4) reduces the savings again. The total savings compared to the reference case then only amount to 29,184 EUR, whereby the grid capacity costs do not increase due to the use of Smart Charging technology in this case. For the energy costs, the savings amount to only 11,569 EUR. Smart Charging optimization can effectively reduce an increase in the peak grid consumption and distribute the grid consumption together with the additional PV power more evenly.

The same applies to the scenario with a PV system and 50 vehicles in unidirectional charging mode (scenario 3). The total savings are further reduced to 27,390 EUR, which is also only due to the reduced savings in energy costs of 9775 EUR.

In contrast, the use of electric vehicles with bidirectional charging technology and V2B integration by means of optimization-based charging and discharging management can increase the savings in electricity costs to a far greater extent. In scenario 4, the total savings amount to 80,705 EUR, which corresponds to a share of 8.1%. The energy costs account for 11,715 EUR and increase by 2544 EUR due to the additional energy required for charging compared to the scenario with a PV system but without EVs (scenario 1). Consequently, the high savings are due to a reduction in capacity costs of 68,989 EUR. The maximum grid power peak was reduced by 433 kW (9.9%) compared to the reference scenario.

However, a savings in capacity costs does not follow an increase in the number of electric vehicles with bidirectional charging technology—on the contrary, as the scenario with 50 EVs in scenario 5 shows. The savings per year then only amount to 69,039 EUR (7%), with a capacity cost reduction of 59,177 EUR and an energy cost reduction of 9862 EUR compared to the reference scenario. In this scenario, the maximum grid power peak consumption can only be reduced by 371 kW. This is due to the increased energy requirement in order to achieve the charging target of a 90% state of charge at departure time.

It can be concluded from this that even in the bidirectional case, there is a certain optimum number of EVs participating in the campus area network on a case-specific basis.

The evaluation of the additional energy for charging the EVs in unidirectional smart charging mode results in a demand of 257,693 kWh (30 EVs) or 429,488 kWh (50 EVs). At the stated energy price of 0.0116 EUR/kWh (Table 1), this results in additional energy costs of 2989 EUR (30 EVs) or 4982 EUR (50 EVs). In this tariff, the additional costs of charging the electric vehicles are therefore marginal compared to the previous costs.

In addition to the scenarios with electric vehicles and a PV system, two scenarios with a stationary battery storage system with comparable values for storage capacity (2310 kWh corresponds to 30 EVs/3850 kWh corresponds to 50 EVs) and nominal power were selected for comparison with the two cases with 30 and 50 vehicles and bidirectional charging technology.

The savings could be further increased in both cases. In scenario 6 with a 2310 kWh battery capacity, 112,035 EUR were saved, which corresponds to 11.3% of the total annual costs of the reference scenario. The savings were made both in terms of energy costs (14,284 EUR) and capacity costs (97,751 EUR). In scenario 7 with a storage capacity of 3850 kWh, the savings even increased to a total of 131,864 EUR (energy cost savings of 14,270 EUR, capacity cost savings of 117,593 EUR). At the same time, it was possible to reduce the grid power demand peak by 613 kW (scenario 6) and 738 kW (scenario 7), respectively.

Contrary to the results from the scenarios with EVs, the electricity costs can be further reduced with the increasing storage capacity and performance of the stationary BSS. This can be explained by the elimination of the boundary conditions regarding availability (presence on campus due to working hours) and driver comfort (reaching a target SOC at departure time).

Given the summarized results in Table 8, one can answer the research question from the introduction chapter:

- Peak load can be reduced with peak shaving technology between 8.5% and 9.9% and the total electricity cost between 7% and 8.1% for an EV fleet with a size of 30 or 50, respectively, with bi-directional charging technology.
- Peak load reduction and cost savings do not increase with growing size of the EV fleet. There exists an optimal number of EVs that is beneficial for the operator of the local grid.
- Bi-directional charging has a significant positive impact on peak load and electricity cost reduction. Peak loads can be reduced up to 7.4% and the total electricity costs can be further reduced by 5.1% compared to smart uni-directional charging.

- Using a stationary BSS of the same storage capacity and performance as the two considered EV fleets has a further significant positive impact. Peak load can be reduced by 6.9% and total electricity costs by 5.2% compared to the EV fleet with bidirectional charging.

Table 8. Summarized results.

Scenario	Peak Reduction [kW]	Peak Reduction [%]	Total Cost Savings [EUR]	Total Cost Savings [%]
1	110.59	2.5	31,874	3.2
2	110.59	2.5	29,184	3.0
3	110.59	2.5	27,390	2.8
4	433.13	9.9	80,705	8.1
5	371.53	8.5	69,039	7.0
6	613.7	14	112,035	11.3
7	738.28	16.8	131,864	13.3

4. Conclusions

Electricity storage systems, whether electric vehicles or stationary battery storage systems, stabilize the electricity supply grid with their flexibility and thus drive the energy transition forward. This study aims to address the potential of peak shaving using a PV plant and smart unidirectional and bidirectional charging technology for two fleets of electric vehicles and two comparable configurations of stationary battery storage systems on the university campus of Saarland University in Saarbrücken as a case study. Based on an annual measurement of the grid demand power of all consumers on the campus, a simulation study was carried out to compare the peak shaving potential of seven scenarios with a fleet of electric vehicles with, on the one hand, both smart unidirectional and bidirectional charging, and on the other hand, stationary battery storage systems. For the sake of simplicity, it was assumed that the vehicles are connected to the charging station during working hours and can be charged and discharged within a user-defined charging status. Furthermore, only the electricity costs were included in the profitability analysis; investment and operating costs were not taken into account.

Overall, the simulation results show that

1. An optimization-based unidirectional charging technology (Smart Charging) in combination with a PV system increases the potential for peak load smoothing. The scenarios with the PV system and electric vehicle in unidirectional charging mode show that the grid capacity peak is at the same level as the scenario with a PV system only.
2. The bidirectional charging technology enables a further reduction in the maximum grid supply power, but there is an optimum in the number of participating EVs.
3. The limiting boundary conditions of bidirectional charging (time-limited storage use, target charging status at departure time) are circumvented by using a comparable stationary BSS, thus enabling a further significant reduction in total grid supply costs. In addition, this solution offers a controlled risk reduction in power shaving, as the number of EVs effectively connected to the grid cannot be predicted with certainty and, therefore, the decisive load peak cannot be covered with certainty.

Therefore, the peak-shaving potential and the associated reduction in capacity costs from the grid increases with the exclusive use of a PV system via the inclusion of the EV fleet up to a stationary battery storage system when considering only the capacity costs from the grid.

The model described here was created with some simplifying assumptions. For example, an ideal prediction for the load profile and the PV power generation is used in the optimization process. Furthermore, no investment costs either for the installation of the charging infrastructure and the ICT required for the optimization algorithm used here nor investment or operating cost of the stationary BSS were taken into account. In the market model applied here, which provides for the free provision of EV storage capacity in exchange for free charging, these investment costs are eliminated. When using a stationary BSS, these not inconsiderable costs are incurred in addition to other operating costs. It remains to be examined whether the degree of increased flexibility takes account of the higher procurement costs. Future work will also focus on the resulting cost for the EV owner due to battery aging and further financial compensation models for the provision of the EV's battery capacity. Furthermore, the real availability of EVs (arrival and departure times and number of vehicles, initial state of charge, and the charging preferences of car owners) and their performance spectrum in terms of battery capacity and power must be measured and statistically evaluated in order to create a behavior model. In addition, more realistic forecasts of the load profile (based on historical time series) and PV production (e.g., with the help of weather forecasts) must be created. As already mentioned, some vehicle manufacturers have limited the possibilities of using bidirectional charging in order to avoid premature aging of the battery, among other things. It is therefore important to add an aging model to the battery model for a complete evaluation.

Author Contributions: Conceptualization, J.M. and G.F.; methodology, J.M.; software, J.M.; validation, J.M. and G.F.; formal analysis, J.M.; investigation, J.M.; resources, G.F.; data curation, J.M.; writing—original draft, J.M.; Writing—review & editing, G.F. All authors have read and agreed to the published version of the manuscript.

Funding: This research received no external funding

Data Availability Statement: Data are contained within the article.

Acknowledgments: Thanks go to the university's Facility Management for providing the measurement data.

Conflicts of Interest: The authors declare no conflict of interest.

References

1. IPCEI European Battery Innovation (EuBatIn). Available online: https://www.ipcei-batteries.eu/about-ipcei (accessed on 13 November 2023).
2. Statista Research Department. Lithium-Ionen-Batterien—Kosten pro kWh bis 2025. Available online: https://de.statista.com/statistik/daten/studie/534429/umfrage/weltweite-preise-fuer-lithium-ionen-akkus/ (accessed on 13 November 2023).
3. ForschungsVerbund Erneuerbare Energien (FVEE) . Handlungsempfehlungen für die nächste Phase der Energiewende. Available online: https://www.fvee.de/wp-content/uploads/2022/03/FVEESystemintegration.pdf (accessed on 13 November 2023).
4. Drees, T.; Dederichs, T.; Meinecke, M.; Dolak, A. Szenariorahmen zum Netzentwicklungsplan Strom 2037 mit Ausblick 2045 (Version 2023). Available online: https://www.netzentwicklungsplan.de/nep-aktuell/netzentwicklungsplan-20372045-2023 (accessed on 13 November 2023).
5. Luderer, G.; Günther, C.; Sörgel, D.; Kost, C.; Benke, F.; Auer, C.; Koller, F.; Herbst, A.; Reder, K.; Böttger, D.; et al. Deutschland auf dem Weg zur Klimaneutralität 2045-Szenarien und Pfade im Modellvergleich (Zusammenfassung). Available online: https://ariadneprojekt.de/publikation/deutschland-auf-dem-weg-zur-klimaneutralitat-2045-szenarienreport/ (accessed on 13 November 2023).
6. Bundesnetzagentur für Elektrizität, Gas, Telekommunikation, Post und Eisenbahnen. Flexibilität im Stromversorgungssystem: Bestandsaufnahme, Hemmnisse und Ansätze zur verbesserten Erschließung von Flexibilität. Available online: https://www.bundesnetzagentur.de/SharedDocs/Downloads/DE/Sachgebiete/Energie/Unternehmen_Institutionen/NetzentwicklungUndSmartGrid/BNetzA_Flexibilitaetspapier.pdf?__blob=publicationFile&v=1 (accessed on 13 November 2023).
7. Dambeck, H.; Ess, F.; Falkenberg, H.; Kemmler, A.; Kirchner, A.; Koepp, M.; Kreidelmeyer, S.; Lübbers, S.; Piégsa, A.; Scheffer, S.; et al. Towards a Climate-Neutral Germany by 2045. Available online: https://www.agora-energiewende.org/fileadmin/Projekte/2021/2021_04_KNDE45/A-EW_213_KNDE2045_Summary_EN_WEB.pdf (accessed on 13 November 2023).
8. Zhang, Y.; Chen, J.; Teng, S.; Zhang, H.; Wang, F.Y. Sustainable Lifecycle Management for Automotive Development via Multi-Dimensional Circular Design Framework. *IEEE Trans. Intell. Veh.* **2023**, *8*, 4151–4154. [CrossRef]
9. Englberger, S.; Jossen, A.; Hesse, H. Unlocking the Potential of Battery Storage with the Dynamic Stacking of Multiple Applications. *Cell Rep. Phys. Sci.* **2020**, *1*, 100238. [CrossRef]
10. Fridgen, G.; Haupt, L. Batterien als Schlüsseltechnologie Interdisziplinäre Batterieforschung verhilft der Energiewende zum Erfolg Nachhaltigkeit. *Spektrum* **2019**, *15*, 64–67.
11. Zander, W.; Lemkens, S.; Macharey, U.; Langrock, T.; Nilis, D.; Zdrallek, M.; Friedrich Schäfer, K.; Steffens, P.; Kornrumpf, T.; Hummel, K.; et al. Dena-NETZFLEXSTUDIE: Optimierter Einsatz von Speichern für Netz-und Marktanwendungen in der

Stromversorgung. Available online: https://www.dena.de/fileadmin/dena/Dokumente/Pdf/9191_dena_Netzflexstudie.pdf (accessed on 13 November 2023).
12. Fraine, G.; Smith, S.; Zinkiewicz, L.; Chapman, R.; Sheehan, M. At home on the road? Can drivers' relationships with their cars be associated with territoriality? *J. Environ. Psychol.* **2007**, *27*, 204–214. [CrossRef]
13. Debnath, B.; Biswas, S.; Uddin, M.F. Optimization of Electric Vehicle Charging to Shave Peak Load for Integration in Smart Grid. In Proceedings of the 2020 IEEE Region 10 Symposium (TENSYMP), Dhaka, Bangladesh, 5–7 June 2020; pp. 483–488. [CrossRef]
14. Khan, S.U.; Mehmood, K.K.; Haider, Z.M.; Rafique, M.K.; Khan, M.O.; Kim, C.H. Coordination of Multiple Electric Vehicle Aggregators for Peak Shaving and Valley Filling in Distribution Feeders. *Energies* **2021**, *14*, 352. [CrossRef]
15. Sami, I.; Ullah, Z.; Salman, K.; Hussain, I.; Ali, S.M.; Khan, B.; Mehmood, C.A.; Farid, U. A Bidirectional Interactive Electric Vehicles Operation Modes: Vehicle-to-Grid (V2G) and Grid-to-Vehicle (G2V) Variations Within Smart Grid. In Proceedings of the 2019 International Conference on Engineering and Emerging Technologies (ICEET), Lahore, Pakistan, 21–22 February 2019; pp. 1–6. [CrossRef]
16. Schlund, J.; German, R.; Pruckner, M. Synergy of Unidirectional and Bidirectional Smart Charging of Electric Vehicles for Frequency Containment Reserve Power Provision. *World Electr. Veh. J.* **2022**, *13*, 168. [CrossRef]
17. Müller, M.; Blume, Y.; Reinhard, J. Impact of behind-the-meter optimised bidirectional electric vehicles on the distribution grid load. *Energy* **2022**, *255*, 124537. [CrossRef]
18. Seo, M.; Kim, C.; Han, S. Peak shaving of an EV Aggregator Using Quadratic Programming. In Proceedings of the 2019 IEEE Innovative Smart Grid Technologies—Asia (ISGT Asia), Chengdu, China, 21–24 May 2019; pp. 2794–2798. [CrossRef]
19. Minhas, D.M.; Meiers, J.; Frey, G. Electric Vehicle Battery Storage Concentric Intelligent Home Energy Management System Using Real Life Data Sets. *Energies* **2022**, *15*, 1619. [CrossRef]
20. Ioakimidis, C.S.; Thomas, D.; Rycerski, P.; Genikomsakis, K.N. Peak shaving and valley filling of power consumption profile in non-residential buildings using an electric vehicle parking lot. *Energy* **2018**, *148*, 148–158. [CrossRef]
21. Mahmud, K.; Hossain, M.J.; Ravishankar, J. Peak-Load Management in Commercial Systems With Electric Vehicles. *IEEE Syst. J.* **2019**, *13*, 1872–1882. [CrossRef]
22. Koivuniemi, E.; Lepistö, J.; Heine, P.; Takala, S.; Repo, S. Smart EV charging in office buildings. In Proceedings of the CIRED 2020 Berlin Workshop (CIRED 2020), online, 22–23 September 2020; Volume 2020, pp. 403–406. [CrossRef]
23. Barchi, G.; Pierro, M.; Moser, D. Predictive Energy Control Strategy for Peak Shaving and Shifting Using BESS and PV Generation Applied to the Retail Sector. *Electronics* **2019**, *8*, 526. [CrossRef]
24. Bereczki, B.; Hartmann, B.; Kertész, S. Industrial Application of Battery Energy Storage Systems: Peak shaving. In Proceedings of the 2019 7th International Youth Conference on Energy (IYCE), Bled, Slovenia, 3–6 July 2019; pp. 1–5. [CrossRef]
25. Fenner, P.; Rauma, K.; Rautiainen, A.; Supponen, A.; Rehtanz, C.; Järventausta, P. Quantification of peak shaving capacity in electric vehicle charging—Findings from case studies in Helsinki Region. *IET Smart Grid* **2020**, *3*, 777–785. [CrossRef]
26. Van Kriekinge, G.; De Cauwer, C.; Sapountzoglou, N.; Coosemans, T.; Messagie, M. Peak shaving and cost minimization using model predictive control for uni- and bi-directional charging of electric vehicles. *Energy Rep.* **2021**, *7*, 8760–8771. [CrossRef]
27. Richter, J. Feiertage API. Available online: https://github.com/bundesAPI/feiertage-api (accessed on 14 November 2023).
28. Bundeskartellamt Bundesnetzagentur. Monitoringbericht 2020. 2020. Available online: https://www.bundeskartellamt.de/SharedDocs/Publikation/DE/Berichte/Energie-Monitoring-2020.html (accessed on 16 November 2023).
29. GmbH, S.S. Entgelte für die Netznutzung sowie den Messstellenbetrieb (einschließlich Messung) Strom—Gültig ab 01.01.2023. Available online: https://www.saarbruecker-stadtwerke.de/media/download-63a548f43a8de (accessed on 14 November 2023).
30. Hildermeier, J.; Kolokathis, C.; Rosenow, J.; Hogan, M.; Wiese, C.; Jahn, A. Smart EV Charging: A Global Review of Promising Practices. *World Electr. Veh. J.* **2019**, *10*, 80. [CrossRef]
31. *ISO 15118*; Road Vehicles—Vehicle to Grid Communication Interface. ISO: Geneva, Switzerland, 2019.
32. Elma, O.; Cali, U.; Kuzlu, M. An overview of bidirectional electric vehicles charging system as a Vehicle to Anything (V2X) under Cyber–Physical Power System (CPPS). *Energy Rep.* **2022**, *8*, 25–32. [CrossRef]
33. *ISO 15118-20*; Road Vehicles. Vehicle to Grid Communication Interface. Part 20: 2nd Generation Network Layer and Application Layer Requirements. ISO: Geneva, Switzerland, 2022.
34. *IEC 63110*; Protocol for Management of Electric Vehicles Charging and Discharging Infrastructures. IEC: Geneva, Switzerland, 2022.
35. Reitberger, S. E-auto von VW Kann Nun Strom in Netz Speisen: Doch Die Technik stößt an Grenzen. 2022. Available online: https://efahrer.chip.de/news/e-auto-von-vw-kann-nun-strom-in-netz-speisen-doch-die-technik-stoesst-an-grenzen_108599 (accessed on 14 November 2023).
36. Bidirektionales Laden 2023. Available online: https://www.elektroauto-news.net/wissen/bidirektionales-laden (accessed on 14 November 2023).
37. Babcicky, P. Marktübersicht Bidirektionaler Ladestationen. Available online: https://bidirektionale-wallboxen.de/marktuebersicht/ (accessed on 14 November 2023).
38. Hösl, N.; Teulings, T. Chargeprice/Open-EV-Data. 2022. Available online: https://github.com/chargeprice/open-ev-data (accessed on 14 November 2023).
39. Trentadue, G.; Lucas, A.; Otura, M.; Pliakostathis, K.; Zanni, M.; Scholz, H. Evaluation of Fast Charging Efficiency under Extreme Temperatures. *Energies* **2018**, *11*, 1937. [CrossRef]

40. Rezaeimozafar, M.; Eskandari, M.; Savkin, A.V. A Self-Optimizing Scheduling Model for Large-Scale EV Fleets in Microgrids. *IEEE Trans. Ind. Informatics* **2021**, *17*, 8177–8188. [CrossRef]
41. Stein, J.S.; Holmgren, W.F.; Forbess, J.; Hansen, C.W. PVLIB: Open source photovoltaic performance modeling functions for Matlab and Python. In Proceedings of the 2016 IEEE 43rd Photovoltaic Specialists Conference (PVSC), Portland, OR, USA, 5–10 June 2016; pp. 3425–3430. [CrossRef]
42. Huld, T.; Müller, R.; Gambardella, A. A new solar radiation database for estimating PV performance in Europe and Africa. *Sol. Energy* **2012**, *86*, 1803–1815. [CrossRef]
43. Mahooti, M. NREL's Solar Position Algorithm (SPA). Available online: https://www.mathworks.com/matlabcentral/fileexchange/59903-nrel-s-solar-position-algorithm-spa (accessed on 15 November 2023).
44. MATLAB. The MathWorks Inc.: Natick, MA, USA, 2021. Available online: https://www.mathworks.com (accessed on 15 November 2023).
45. Löfberg, J. YALMIP: A Toolbox for Modeling and Optimization in MATLAB. In Proceedings of the CACSD Conference, Taipei, Taiwan, 2–4 September 2004.
46. Gurobi Optimization, LLC. Gurobi Optimizer Reference Manual. 2023. Available online: https://www.gurobi.com (accessed on 15 November 2023).

Disclaimer/Publisher's Note: The statements, opinions and data contained in all publications are solely those of the individual author(s) and contributor(s) and not of MDPI and/or the editor(s). MDPI and/or the editor(s) disclaim responsibility for any injury to people or property resulting from any ideas, methods, instructions or products referred to in the content.

Article

Categorization of Attributes and Features for the Location of Electric Vehicle Charging Stations

Andrea Mazza [1], Angela Russo [1], Gianfranco Chicco [1,*], Andrea Di Martino [2], Cristian Giovanni Colombo [2], Michela Longo [2], Paolo Ciliento [3], Marco De Donno [3], Francesca Mapelli [3] and Francesco Lamberti [3]

[1] Dipartimento Energia "Galileo Ferraris", Politecnico di Torino, Corso Duca degli Abruzzi 24, 10129 Torino, Italy; andrea.mazza@polito.it (A.M.); angela.russo@polito.it (A.R.)
[2] Department of Energy, Politecnico di Milano, 20156 Milan, Italy; andrea.dimartino@polimi.it (A.D.M.); cristiangiovanni.colombo@polimi.it (C.G.C.); michela.longo@polimi.it (M.L.)
[3] Atlante Srl, Piazzale Lodi 3, 20137 Milan, Italy; paolo.ciliento@atlante.energy (P.C.); marco.dedonno@atlante.energy (M.D.D.); francesca.mapelli@atlante.energy (F.M.); francesco.lamberti@atlante.energy (F.L.)
* Correspondence: gianfranco.chicco@polito.it

Abstract: The location of Electric Vehicle Charging Stations (EVCSs) is gaining significant importance as part of the conversion to a full-electric vehicle fleet. Positive or negative impacts can be generated mainly based on the quality of service offered to customers and operational efficiency, also potentially involving the electrical grid to which the EVCSs are connected. The EVCS location problem requires an in-depth and comprehensive analysis of geographical, market, urban planning, and operational aspects that can lead to several potential alternatives to be evaluated with respect to a defined number of features. This paper discusses the possible use of a multi-criteria decision-making approach, considering the differences between multi-objective decision making (MODM) and multi-attribute decision-making (MADM), to address the EVCS location problem. The conceptual evaluation leads to the conclusion that the MADM approach is more suitable than MODM for the specific problem. The identification of suitable attributes and related features is then carried out based on a systematic literature review. For each attribute, the relative importance of the features is obtained by considering the occurrence and the dedicated weights. The results provide the identification of the most used attributes and the categorization of the selected features to shape the proposed MADM framework for the location of the electric vehicle charging infrastructure.

Keywords: multi-attribute decision-making; criteria; attributes; features; electric vehicle; charging station; siting

1. Introduction

The environmental commitment is leading towards the progressive reconversion of the vehicle fleet from conventional Internal Combustion Engines (ICEs) to Electric Vehicles (EVs). The need to reduce vehicle-related CO_2 emissions is forcing the acceleration of the electrification process of circulating vehicles. The benefits can be observed by reducing the global emissions due to transportation from 19% to 33% [1]. The electrification process relating to the vehicle fleet used for public transport is more incentivized and supported, and charging solutions can be designed on purpose and easily installed inside hubs and maintenance depots. However, the situation is different for private vehicles based on their unique concerns. The need for a widespread charging infrastructure strongly influences the EV diffusion level. To ensure e-mobility to take place and develop, the charging needs must be satisfied in every condition, assisting an increase in the EV market. For advancing EV diffusion, the charging infrastructure must be available in terms of capillary diffusion. An acceptable EV penetration is generally accompanied by a satisfactory penetration of EV charging infrastructure, thus enabling charging operations for EV owners and drivers [2].

Moreover, more widespread EV charging infrastructure helps reduce the anxiety drivers experience with respect to the EV driving distance range. Therefore, the location of Electric Vehicle Charging Stations (EVCSs) nowadays represents a constraint for the diffusion of EVs, but also an opportunity for market share and competitors. EVCS location may be considered as a technical problem, searching for the most appropriate solutions in grids with photovoltaic and battery storage systems, from distribution systems [3] to micro-grids [4]. Considering the spatial target of EV charging, specific solutions can be studied for different cases of residential communities [5], cities and urban areas [6,7], highways [8], and regional areas of a country [9]. The present work aims to consider the EVCS location problem as the main topic of interest and is based on an overview of the scientific literature. The location of EVCS infrastructures is addressed as a preliminary step to create and design a geo-localization tool that aims to select the eligible location among an initial set of alternatives for the installation of EVCSs for a Charging Point Operator (CPO). The methodological approach followed herein addresses the EVCS location problem from the multi-criteria decision-making (MCDM) point of view. In particular, this paper presents the following:

- The first novelty of this work consists of the conceptual categorization of the criteria, tailored to the EVCS location problem considered, based on the results actually available in the scientific literature regarding this subject. Starting with the categorization results and from the attributes that refer to each criterion, a numerical assessment regarding the importance of the attributes is performed, with the aim of identifying the most relevant attributes that can be considered to assist the decision maker in the choice of the most suitable EVCS location. The numerical assessment is shaped through the computation of appearances and weights for each study contribution considered, whose results are aggregated into two separate matrices. This method is exploited not only to quantify and evaluate the distribution of criteria in the literature, but also to extrapolate which attributes are predominantly considered, thus establishing a hierarchical order.
- The second novelty of this paper is the release of the ranking of the most relevant attributes, which can be considered as a basis for the implementation of EVCS location tools to guide the decision makers towards consistent attribute-driven choices. Moreover, this aims to constitute a standard framework of criteria to be implemented in the future for further research projects regarding the EVCS location problem. Therefore, a common basis can ensure direct comparisons between different solutions and approaches.
- The greatest challenge in the application of this approach is the need to deal with the highly fragmented and non-homogeneous background that is fundamentally related to the scopes and achievements expressed by each study contribution, together with the different focus points set by different authors on the types of attributes and the assignment of weights. Different points of view addressed in the literature, i.e., the cases seen from the perspective of stakeholders or policy-makers that are not always clearly stated, and a variable framework of criteria among the papers considered make the research context uneven. If not handled and addressed correctly, all these aspects can lead to meaningless final judgements.

The first aspect is the identification of the most suitable MCDM method to apply. The related discussion is presented in Section 2, underlining advantages and drawbacks of the different options and indicating the preferred solution. Then, based on a systematic literature review, the most recurring criteria are identified and rearranged to create the novel framework of the proposed criteria illustrated in Section 3. The new categorization is provided in Section 4 with a focus on the relative relationships and importance assigned within the literature references examined. Then main achievements of this study are summarized with the final conclusions. Figure 1 clarifies the process followed in the implementation of this paper.

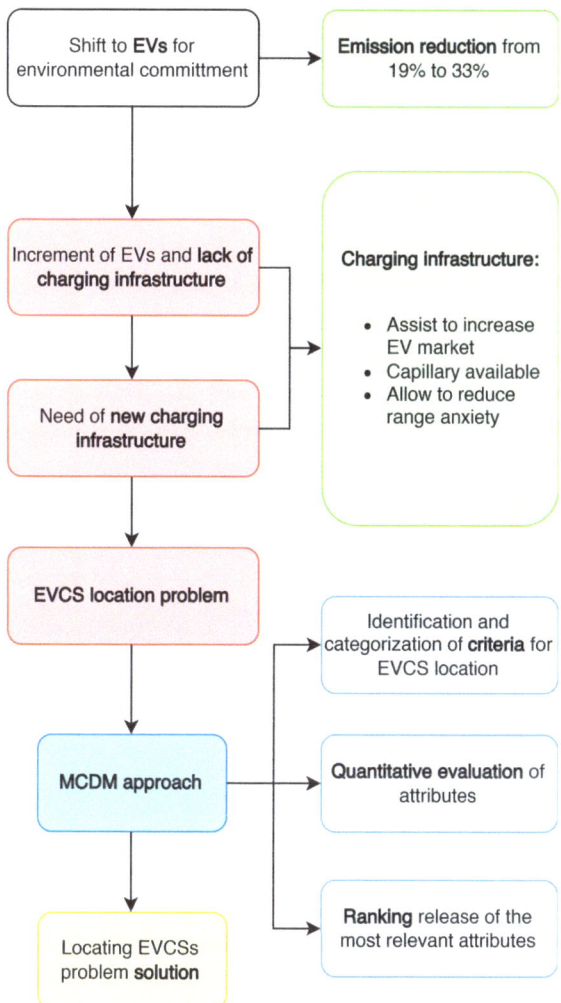

Figure 1. Locating EVCS problem roadmap.

2. Multi-Criteria Decision-Making Approaches

2.1. Overall View on Multi-Criteria Decision-Making

Multi-criteria decision-making problems have the general purpose of the identification of the preferred solution, which satisfies the decision maker's preferences. It is worth noting that in this conceptual framework, it is not possible to find an optimal solution, because this would imply that such a solution would present the best option in all the criteria considered and thus with no conflicts among them. Conversely, in the multi-criteria framework, this choice is made in the presence of conflicting information, which would lead to choosing a different solution depending on the prevalent feature considered. Hence, the choice of the preferred solution implies a comparison among different alternatives that can be either predetermined as the input of the problem or must be created from scratch by applying an appropriate methodology. This difference leads to the identification of two families of MCDM problems [10]:

- In the case of an initial set of a predefined number of alternatives, one has to "simply" select the most preferred one among those that compose an initial set; this case falls into a multi-attribute decision-making (MADM) problem.

- When MCDM methods are used to create the best solution (through, for example, an optimization method in a design process), the case falls into a multi-objective decision-making (MODM) problem.

Some common elements can be identified:

- Multiple attributes/objectives, representing the features that characterize the alternatives. In general terms, attributes/objectives can be called criteria. Relevant criteria must be adapted to the problem under analysis.
- Existence of conflict among the criteria. This means that no alternative is the best for all criteria.
- Different natures of the criteria: some of them can be numerical while some of them can be expressions (better, worse, higher, lower, and so on), which should eventually be translated into numerical terms. This aspect is much more relevant in MADM than in MODM, because the latter is usually based on a set of quantitative objectives and constraints rather than qualitative, as they better suit the design purpose.

Moreover, by definition, beyond the preferred and the optimal solutions, one can identify the following:

- *An ideal solution*: This is also called the *utopia point* and represents the solution characterized by the optimal values for all the objectives. This solution is unfeasible because of the conflicting nature of the criteria considered.
- *Non-dominated solutions*: These are also called *Pareto optimal* solutions. A solution α is non-dominated if and only if there is no other solution β improving at least one criterion with respect to the solution α without degrading at least another criterion. Being non-dominated is necessary (but not sufficient) for the preferred solution.
- *Satisfactory solutions*: Also called *compromise solutions*, these form a subset of the non-dominated solutions. They somehow exceed the acceptable level for all the criteria. The preferred solution is taken from this set of solutions.

The MCDM conceptual framework also allows us to include approaches based on decision theory: in this case, the criteria used to compare the alternatives are scenarios, and the application of decision theory approaches allows us to understand which alternative is more convenient. The scenarios are weighted with subjective weights or objective weights determined through the scenario occurrence probability or other mathematical elaborations based, for example, on the information entropy, or by combining subjective and objective weights [11]. The scenarios are built by including the potential evolution/modification of the boundary conditions (e.g., over a multi-year horizon) that can affect the values of attributes/objectives. Concerning the solution methods for MODM, depending on the nature of the problem and the variables involved, different approaches can be used, for example, as follows:

- *Multi-Objective Linear Programming* (MOLP): If the problem can be formulated in a linear optimization framework, the solution can be found by using linear programming, which guarantees convergence to the global optimum.
- *Evolutionary Multi-Objective Optimization* (EMO): when the computation times become prohibitive, the set of non-dominated solutions is approximated by using evolutionary algorithms that start with an initial set of solutions and improve these solutions iteratively until converging to a solution that becomes stable for a successive number of iterations.

2.2. Why Opt for Multi-Attribute Approaches?

The choice between MODM and MADM is essentially linked to the available information and the simulation approach, in particular, the presence of non-numerical attributes [12] and the identification of proper modelling approaches for social impact. In the presence of non-numerical attributes, the integration of the attributes within an optimization procedure is not straightforward: it is required to use proper scales to translate the non-numerical attributes into a quantitative numerical form. The evaluation of the impacts of some attributes (for example, the existence of a point of particular interest, such as malls or museums) may require complex simulation approaches referring to social aspects and human behaviour, which are only partially implementable and would require the creation of numerous customer profiles that can only be built with a large amount of detailed information.

In particular, the aspect of social simulation aspect constitutes the main obstacle to MODM implementation. Hence, the approach used by MADM essentially avoids providing a direct evaluation of the impact and allows for providing a relative comparison among different alternatives based on the available elements. For example, the presence of points of interest will not be evaluated by indicating the increase in number of accesses per hour, but instead, it will provide the information that the presence of a mall could impact much more than the presence of a museum because the use of cars is more common in the former case than in the latter one. Table 1 summarizes the elements to be considered when choosing the most appropriate approach, providing some brief notes for each element.

Table 1. Summary of MODM vs. MADM.

Characteristics	MODM	MADM	MADM Examples for EVCSs
Easy inclusion of non-numerical attributes	NO: A mathematical formulation is required	YES: Appropriate scales do exist	Providing the judgement by the DM in relation to the impact of different points of interest: the presence of a mall could have more of an impact compared to the presence of a museum because the use of cars is more common in the former case than in the latter one
Easy inclusion of potential mutual interactions of the features	NO: It is necessary for all the interactions in the model to be explicit	YES: If a feature has influence on another, it can be considered through appropriate weighting	Government support and installation permits are somehow linked together; they cover the question "how easy is it to do this business in this particular area of this particular country?"
Data required	Usually not negligible, either for validation of new model, or tuning of parameters	The amount of data required depends on the models developed to give the value of the attributes. In absence of data, the decision maker can make hypothesis to make a comparison among alternatives, enabling a successive sensitivity analysis	The evaluation of the impacts of some attributes (for example, the existence of point of particular interest, such as malls or museums) may require complex simulation approaches referring to social aspects and human behaviour, which are only partially implementable and would require the creation of numerous customer profiles that can be built only with a large amount of detailed information. The use of MADM would reduce the amount of data required (see the first item in this table)
Model updating	The update of the model is constrained by the number and types of state variables and on the optimization method	The framework is usually easy to modify, with some exceptions	The addition of one or more alternatives does not change the entire mathematical formulation (as instead may happen with optimization methods), even though the impact of the reversal ranking must be evaluated
Normalization	Depending on the method	Included as part of the procedures	-

2.3. Choice and Implementation of the MADM Method for EVCS Location

As shown in Table 1, the choice of MADM with respect to the MODM approach is ultimately linked to five elements, all of them favouring MADM compared to MODM. In fact, the following points were uncovered:

1. MADM methods allow the decision maker to use non-numerical attributes. Conversely, MODM approaches do not.

2. Mutual interactions among the features can be taken into account (even without formulating a model that links them together) through adequate weighting in MADM methods. As an example, government support and installation permits are somehow linked to each other (they are the features covering "how easy is it to do this business in this particular area of this particular country?"). The decision maker can provide weights whose sum represents how important the policy aspect is for him/her, without any model linking these two aspects. In MODM approaches, it would be quite complex to account for these interactions.
3. Without accurate and trustable data, MODM approaches are not suggested (because parameter tuning and the validation of new introduced models require a huge amount of "good enough" data).
4. The introduction of new features and the consequent updating of the model is usually simple with MADM approaches, while it is more difficult for MODM methods. In fact, introducing new features may involve the introduction of new state variables that must be included in the overall formulation. This limits the flexibility of use.
5. Data normalization is naturally included (and tested) in MADM methods, while it is truly "method-dependent" in the case of MODM optimization methods (i.e., it is an additional aspect to include).

In conclusion, (i) when all the features may be represented with a mathematical formulation, (ii) when it is not of interest to catch mutual interaction among features, and (iii) when the data (quantity and consistency) are enough for validating the model and tuning parameters, a MODM approach may be a viable option (even though the normalization and model update aspects must be carefully considered). Otherwise, MADM methods are the suggested choice. For EVCS location, there is a variety of features with possible mutual interactions, some of which are expressed in a categorical or qualitative way. Moreover, availability of enough data from the field cannot be guaranteed, and some choices need to be made by the data analyst. On these bases, adopting the MADM approach is suggested.

3. Proposed MADM Scheme

The MADM problem can be formalized in terms of the useful attributes found during the review of the scientific literature. To identify relevant case studies on the EVCS location problem, a systematic literature search was conducted using the main indexed databases. It was necessary to identify all studies that had as one of their objectives the EVCS location based on each attribute and characteristic. These studies were then combined with the results of the MCDM approach searches. Articles were considered based on their relevance and impact, excluding any publication that could not provide sufficient data on the EVCS location methodology. Publications that were outdated and those that were not peer-reviewed were also excluded. It was considered that studies more than 10 years old may not accurately represent the current problem. This filtering process resulted in a set of 43 scientific articles that helped to define categories, attributes and features. In this way, the selected studies were analysed to identify the most recurring attributes and features used to locate the EVCS. The attributes were coded and classified according to their frequency of occurrence and assigned weights. This analysis made it possible to determine the relative importance of each attribute and feature in the context of EVCS location. Since each scientific paper customizes the attribute classification (based on the purposes of their own analysis of the problem), it is worth reorganizing the attributes according to a novel scheme that better suits the purposes of the analysis. The new classification scheme is built according to a three-level arrangement, represented in a synoptic form in Table 2. The columns show the following data:

1. *Attribute category*: This identifies the macro-sector fields and includes all attribute subcategories. The most recurring and interesting attribute categories are the following:

    (a) Economic;
    (b) Territorial;

(c) Social;
(d) Technical.
2. *Attribute subcategory*: This highlights a particular aspect of the category to which it belongs. Each subcategory includes one or more attributes that complement and satisfy the meaning of the targeted aspect (for a total of 11 attribute subcategories).
3. *Output attributes* (forming the proposed classification): Starting from several basic attributes considered relevant with respect to the purposes of the EVCS location problem, along with other interesting attributes to be considered, the basic attributes have been grouped into 24 output attributes. Each category is described below by relating it to the basic attributes.

Table 2. Proposal of the new attribute framework.

Attribute Category	Attribute Subcategory	Basic Attributes	Output Attribute
Economic	Cost	Construction cost; Total Construction cost; Land occupation; Power grid connection costs; Equipment purchasing costs	Installation costs
		O&M costs	O&M costs
		Update/Removal costs	Update and removal costs
	Benefit	Annual profits; Solar energy potential/Renewable resources; Alternative revenue sources	Revenues
	Policy	Installation permits	Installation permits
		Incentives; Local government support; Maturity of the legal framework to implement tenders	Government support
Territorial	Traffic	Traffic convenience; Traffic condition	Traffic flow
		Road patency/topography; Slope; Number of roads; Main number of roads; Roads; Accessibility of the site	Road network characteristics
		Presence (and type) of EVCS (public/private); Public facilities; Coordination with the transportation network; Parking lots; Public transport; Hubs; More interaction with other infrastructures	Interactions with other infrastructures
	Geography	Service radius ("green" field)	Service radius
		Spatial coordination with urban development planning; Urban development	Urban development
		Terrain advantage; Heatwave zone; Flooding zone; Landslide zone; Earthquake zone; Forest; Soil type; Availability; Utilization	Land
	Environmental	Dismantling waste; Easiness of re-establishment in the future; Recycling	End of life management
		Sustainable development of charging station areas; Ecological influence; Destruction of soil, vegetation and landscape; Destruction of water resources	Territory sustainability
		Global emissions; Local pollutants/noise reduction; Air quality	Emissions
Social	Collective	Acceptability of new solutions; Adverse impact on people's lives; Improvement of employment; Benefits for people life	Impact on people's lives
		Population density; Population intensity; (Local) Number of vehicles; (Local) Number of EVs; (Local) EV sales; Residents' average income	Demographic information
		Social areas; Fuel station proximity	Points of interest
	Personal	Driver comfort; Home/private charging vs. public charging; ICE vs. BEV	User preferences

Table 2. *Cont.*

Attribute Category	Attribute Subcategory	Basic Attributes	Output Attribute
Technical	Grid side	Power and energy management; Power quality; Harmonic pollution on power grid; Impact on load levels of power grid; Impact on voltage; Power grid security implications; Consumption level; Electromagnetic interference; Level of penetration of RES	Grid operation
	Grid side	Power supply capacity of transmission and distribution systems; Distance to the substation; Substation; Substation capacity permits; Substation capacity; Power grid capacity	Grid planning
	User side	Further services to drivers; Charging services; Fast-charge ratio	Charging station services
	EVCS side	Possibility of EVCS capacity expansion in the future	EVCS planning
	EVCS side	Safety/Security and ability to tackle with the emergency; Reliability; Charging station capacity; Service capability/service capacity	EVCS operation and reliability

3.1. Category 1: Economic Attributes

The first category considers the economic aspects, directly or indirectly related to investments, construction and policy framework. It is divided into three subcategories: (i) costs, (ii) benefits, and (iii) policy.

3.1.1. Cost Subcategory

This subcategory takes into account some output attributes such as the installation costs, the operation and maintenance (O&M) costs, and the update and removal costs. The installation costs include the following basic attributes found in the literature:

- Construction cost: This includes land cost, demolition cost, equipment acquisition cost, and project investment cost [13–15]. In [16], the following items are listed: land lease or acquisition costs, survey and design costs, infrastructure construction costs, equipment and tool purchase costs, construction management and production costs, and project capital costs. Moreover, ref. [17] lists the following items: land acquisition costs, demolition costs, transportation costs, and auxiliary facilities costs.
- Equipment purchasing cost: In [18], this cost is reported with reference to a Battery Swapping Station (BSS) and is explained as the initial equipment acquisition cost during the construction of BSS. This concept is generalized for the equipment required for the EVCS construction.
- Land occupation cost: Considering BSS, it is described in [18] as the land that the Battery Swapping Station needs to occupy in order to store the battery, which will affect the cost and economic benefit.
- Power grid connection cost: The cost sustained for the connection of the EVCS to the power grid (Table 3). In [19], this cost depends on the distance of the charging station from the point of connection to the electric grid, as well as on the connection technology, assuming that the EVCS is directly connected to the electrical substation via a dedicated overhead line.
- Total construction cost: When no detailed description of the construction cost is available, often the total construction cost attribute is instead used, considering different aspects. These can refer to the equipment purchasing cost, land occupation cost, and power grid connection cost attributes explained above. The O&M costs include aspects such as the electricity charge, staff wages, financial expenses, taxes, battery depreciation, and so on [13–15]. The daily maintenance cost of machinery is also indicated in [15]. In [16], the operation and maintenance costs include personnel salaries, employee benefits, daily operation and maintenance, equipment depreciation, and business costs. The update and removal costs group the costs related to the

expected price of the surrounding land in the future and the fixed cost of the targeted EVCS site [17]. Higher update and removal costs mean that it would be more difficult to change the intended destination of use of the site.

Table 3. An example of investment and operating costs of EVCSs [20].

Economic Data	Unit	2 × 22 kW AC Charging Station	2 × 22 kW DC Charging Station
Equipment costs	[€]	5000	25,000
Grid connection costs	[€]	2000	5000
Authorization and planning costs	[€]	1000	1500
Installation and building costs	[€]	2000	3500
Total investment cost	[€]	10,000	35,000
Operating costs	[€/y]	1500	3000

3.1.2. Benefit Subcategory

The benefit subcategory considers all the possibilities to account for earnings and revenues related to the operation of the EVCS. It includes the revenues that can be broken down through the following basic attributes:

- *Annual profits*: Defined in [15] as the future revenues of the EVCS without an analytical expression, this basic attribute refers to the profits derived directly from charging operations.
- *Alternate revenue sources*: Proposed in [21], this is related to the capability of a location to profit from non-power sales such as advertising, participation in grid dispatching, and renewable energy generation. An additional example can be represented by the possibility to integrate different mobility solutions according to the needs to be charged, such as parking spot payment while charging the EV through a shared information technology platform. Another possible revenue source can be represented by solar energy potential related to Renewable Energy Sources (RESs). RESs can be exploited as an opportunity for implementing and feeding the power grid through a sustainable energy production network [22]. In particular, a practical example may indeed refer to the possibility of installing an RES production plant in areas suitable for selling the energy produced on the market.

3.1.3. Policy Subcategory

The policy subcategory considers all the issues that may arise facing the bureaucracy of a country to locate one or more EVCS. In particular, it reflects the actual legal conditions that may or may not allow for the installation of an EVCS in a given location. This subcategory encompasses two output attributes. The first one is the installation permit output attribute, including the necessary authorizations and approval procedures as strong factors for selecting a project. In addition to the licensing procedures for the charging station installation, construction approvals may be required depending on the space ownership and type [20]. This strongly depends on the legal framework of the country. The second output attribute is the government support, which is mostly related to the legal framework existing in the eligible location of installation for an EVCS and includes the following basic attributes:

- *Incentives* (or subsidies to increase the EV fleet): The adoption of measures, either financial incentives for EV purchase or non-financial traffic incentives for EVs, or tax exemptions and subsidies for charging infrastructure, all play a positive effect on the promotion of e-mobility, especially at the early stage of the market, when the economic viability of investments in charging infrastructure is uncertain [20].
- *Maturity of the legal framework to implement tenders*: In the case of developing public charging points through open tenders held by a municipality, the limited experience

for the implementation may adversely affect the interest in the charging infrastructure market [20].

- *Local government support*: This basic attribute includes the subsidy policy, favourable prices, and tax preferences, which are established to strongly promote the development of EVs [23]. Most of these aspects have already been reported in the attribute incentives and maturity of the legal framework. The EVCS project has a large initial investment cost and a long payback period, which is highly vulnerable to the influence of government policies [24]. Specifically speaking, the approval of construction land, the upgrading and transformation of the distribution network, the implementation of the subsidy policy, and the traffic planning in the vicinity of the EVCS all need government support. Currently, green policies are meant to be discussed and approved to push towards an electric conversion of mobility. Hence, the attitude of local government support is one of the indicators that must be considered.

3.1.4. Cost Functions

To provide some quantitative values, we refer to interesting analytical relations reported in [20] about the infrastructure costs; these are classified in investment costs, fixed operating costs, and variable operating costs. Investment costs concern equipment ownership and installation, grid connection, and licensing expenditures. In [20], the values reported in Table 8 of that paper are taken as reference, even though the authors stated that they *"are not precise cost figures but provide a clear picture of the economic parameters that serve this study to highlight the cost differences between the two technological options"*.

Almost all the papers contain a description of economic objectives without any analytical expression; an exception is [25], which reports an equation for the calculation of the costs of an EVCS. They consider V types of EV charging stations, Q cells and U EV charging units. An optimization problem is set up, in which the decision variables are x_{quv} (binary variables, equal to 1 if a charging station of type v is located in cell q with u charging units). The objective function is the total cost (intended as all the necessary costs of building a refuelling station). Let c_{qv} denote the cost of locating a new charging station of type v in cell q with one charging unit; the cost of constructing a new station with h charging points is $h^\delta c_{qv}$, where the exponent δ (with $0 < \delta < 1$) refers to the rate of cost increase as capacity options rise. The value of δ is less than unity because of the economies of scale for constructing a station. The construction cost saved by a gas station-based location is denoted by ϵ. The 0–1 parameter b_q is used to describe the existing gas station network. The total costs is thus determined:

$$C_T = \sum_{q=1}^{Q} \sum_{v=1}^{V} \sum_{u=1}^{U} (h^\delta c_{qv} - \epsilon b_q) x_{quv} \qquad (1)$$

Another reference that reports analytical expressions for cost determination is [26]. The total cost in [26] is the sum of the annual construction cost of the charging station, the annual O&M expense (including worker wage, maintenance expense, equipment depreciation expense, and electricity purchase expense) and operation expense of charging stations, and the wastage cost in the process of user charging (containing direct and indirect costs). Also, ref. [19] includes some analytical expressions, which are reported below. The "Station development cost" is the sum of the station equipment cost and land cost. The equipment cost is assumed to vary linearly with the station capacity, which is itself a function of the number and capacity of the connectors installed in the station [19]. For station e, the development cost DC_e is then calculated as

$$DC_e = C_{init} + 25 C_{land} NC_e + P_C C_{con}(NC_e - 1) \qquad (2)$$

where P_C is the single EVCS connection rate power (in kW). In the EVCS, more than one connection may exist. C_{con} is the connection development cost (in USD/kW or EUR/kW); C_{init} (in USD or EUR) is the EVCS fixed cost (i.e., the cost associated with basic equipment and facilities used to establish a charging station); C_{land} is the land rental cost (in this case, for 5 years); and NC_e is the number of connectors in the EVCS; hence, the capacity of the e-th EVCS is calculated as $P_{EVCS} = P_C NC_e$. The station electrification cost depends on the distance of the station from the point of connection to the electric grid, as well as the connection technology. It is assumed that the station is directly connected to the substation via a dedicated overhead line. The electrification cost of the e-th EVCS to the closest substation by a line with a given cross-section $A_e^{(line)}$ in [mm^2] is [19]

$$C_e^{(EL)} = \left(8000 + 65.7 A_e^{(line)}\right) l_e^{(line)} \tag{3}$$

where $l_e^{(line)}$ is the length of the line from the substation.

3.2. Category 2: Territorial Attributes

The second category of attributes considered is territorial. This category focuses on all aspects involving environmental variables, from the surrounding nature to the human activities. The identified subcategories are (i) traffic, (ii) geography, and (iii) environment.

3.2.1. Traffic Subcategory

It is strictly important to evaluate the traffic to decide where to locate an EVCS, since charging needs depend on traffic volumes. Furthermore, the physical characteristics of roads influence the traffic volumes—let us think about a large high-speed road rather than a narrow low-speed limited road. The location of the EVCS must also take into account the possibility of the potential interactions with other networks like public transport. In this way, intermodal e-mobility can be enhanced, accelerating the change in transport habits and mobility. Therefore, this subcategory considers the following output attributes: traffic flow, road network characteristics, and interactions with other infrastructures. Traffic flow consists of two very similar basic attributes:

- *Traffic convenience*: This refers to the number of main roads surrounding the targeted EVCS site, the level of traffic flow, and possibility of traffic jams. Convenient traffic implies that more consumers would be willing to use the targeted EVCS site and there would be higher potential customers [17]. In [27], this basic attribute is evaluated as the number of intersections within 5 km from site location.
- *Traffic conditions*: This is seldom defined as the actual distance between two adjacent EVCSs [15]. However, it can refer to the actual traffic criticalities being present in particular points or zones of the road network, thus giving a starting thumb-rule on identifying the critical points of traffic and hence concerning potential on-route charging demand.

Road network characteristics include all information regarding the roads. This output attribute is particularly interesting in terms of factors involved as basic attributes, since it declines different aspects, like the actual conditions of the road network, their topographic characteristics, and number. Below is the attributes in detail:

- *Road patency/topography*: The "patency" is defined as the average status of maintenance for the road surface. Sometimes, it is also meant to indicate road topography, with superimposition with the slope, the next basic attribute [28].
- *Slope*: It collects the slope of road sections considered within the area eligible to locate an EVCS. The location of an EVCS must avoid sites in which the road slope is high, and it is established that the maximum threshold slope is 7% [29]. Moreover, roads featured by high slopes offer a negative impact for construction and operations [22].
- *Number of roads*: This represents the total number of roads included within the eligible areas considered where to install an EVCS.
- *Main road number*: This defines the total number of main roads present within the considered area, thus neglecting roads of minor importance. It is closely related to the

previous basic attribute traffic conditions. The main difference is that here the number of main roads is taken into account.

- *Roads*: The meaning of this attribute seems to recall what was already seen for the previous road-related attributes. Here, the meaning is centred more on the energy demand depending on the vehicle mobility: the EVCS should be close to high-energy demand due to vehicle mobility [22]. The measure used is the Euclidean Distance.
- *Accessibility of site*: It is mentioned as an attribute in [30,31] without any definition published. It can be easily associated to guarantee an accessible EVCS location to allow for and facilitate charging operations.

Interactions with other infrastructures gather all information regarding the possible interchanges with every kind of transportation-oriented infrastructure. As previously exposed, the aim of this attribute is to create an inter-modal transportation system, thus pushing human behaviour to exploit inter-modality. This attribute is relevant in terms of the following:

- *Presence (and type) of EVCSs* (public/private): Since the location of alternative EVCSs should not be very close to existing EVCSs, the suitability of current EVCSs is examined and a comparison among current EVCSs is made [22]. No distinctions are made referring to EVCS ownership of competitors.
- *Public facilities*: This is mentioned in [28] with no definition given. According to the Collins dictionary, facilities are buildings, pieces of equipment, or services that are provided for a particular purpose. It can represent every kind of public infrastructure available in the eligible areas, i.e., mayor or other public institutions' offices, public network, etc.
- *Coordination with the transportation network*: This is an evaluation of the level of integration of EVCSs with the public transport network [32]. It is based on the availability of an already existing public transportation system near the EVCSs, which is essential when the EV user/driver intends to continue the journey by public transport [32]. Here, the drawback is represented by a transportation network that is too widespread and branched, since it would discourage the use of EVs—and the mobility of private vehicles in general—in favour of public transport.
- *Parking lots*: Since the EV charging time is long, parking lots are suitable EVCS locations [22]. The measure used is the Euclidean Distance. This attribute refers to the achievement of inter-modality in the transportation system. Parking lots are thus a very suitable area to install EVCSs since the vehicles can recharge when parked. Parking lots can also be managed by public transport operators themselves that are located and built in the neighborhood of a public transport line.
- *Public transport*: The measurement of the simplicity of accessing public transport [15]. It can be related to the ease of connection with the public transportation network. This attribute highlights that if the eligible area is close to a public transport service (line, terminal station or stop), the probability that customers will use the EVCS installed will be high. It is strictly linked with the previous attributes, parking lots, coordination with transportation network, and the following hubs basic attribute, since inter-modality is the main concept shared among them.
- *Hubs*: The EVCS should be close to a place with high-energy demand due to vehicle mobility [22]. The measure used is the Euclidean Distance. As previously recalled in the attribute roads, hubs (also called junctions) are meant as interchange spots with transportation services. This helps in increasing the potential charging demand. More interactions with other infrastructures are defined as the coordination with the main artery, inlet and outlet, residential areas, urban main functional areas, and a stable supply of electricity power [14]. This coordination is a benefit. It contributes to assign a high rate to the area considered if a high number of infrastructures of any type are present.

3.2.2. Geography Subcategory

Alongside traffic-related aspects, a relevant field to be analysed is the geography of the sites. This subcategory is more focused on examining the environmental and natural characteristics of the potential sites for locating EVCSs. The focus starts to move outside the urban area and evaluate the environmental impact of the EVCS location on the area. This subcategory considers the following attributes: (i) service radius, (ii) urban development, and (iii) land.

Service radius is expressed as the actual distance between two adjacent EVCSs [15]. This underlines the aspect already considered in the presence (and type) of EVCS (public/private) with an additional value. This attribute focuses on the aspect of the "green field", i.e., on the planning phase of new EVCSs to be added, and therefore, it focuses on areas that are not already reached by a capillary diffusion of EVCSs, thus contributing to increasing the diffusion of CS infrastructure.

Urban development gathers two basic attributes that results in a relationship between the EVCS infrastructure and the urban network. They are as follows:

- *Spatial coordination with urban development planning*: This highlights the integration of the EVCS infrastructure with the spatial development of urban pattern. Thus, the aspect highlighted by this basic attribute is the need of coordination between the charging needs and demand—that is expected to grow—with the expansion or improvement of urban areas [20].
- *Urban development* (or coordinated level of EVCS with urban development planning): This basic attribute gives the name to the corresponding output attribute and is defined as follows: It indicates if the targeted EVCS site satisfies the development planning for the urban electric grid and road network. If the targeted EVCS site is better coordinated with the urban development planning, there is less update and remove risk [17]. In this way, the meaning added by this last attribute goes to complete the global meaning of the output attribute. An EVCS plan coordinated with the urban development results in a less unpleasant impact on the urban pattern.

Land includes all information regarding the geographic characteristics of the areas considered. This output attribute aims to highlight the impact that the environment can have on the proposed location and also evaluates the possible produced drawbacks. Here, risks deriving from land characteristics are taken into account to assess if the location can be considered eligible for the installation of an EVCS. In detail, land is formed by the following basic attributes:

- *Terrain advantage*: It represents the eligibility of the area in terms of potential space to be used and traffic volumes. It is a general evaluation on the area.
- *Flooding risk*: This attribute was not found in the reviewed scientific literature. Since climate-related phenomena are becoming more and more destructive and aggressive on anthropic activities, it is reasonable to consider it. Flooding directly involves the EVCS infrastructure since its effects can heavily interfere with the electrical system. Historic and open-access data publicly available either from research institutes or released by public administration can be a good starting point to establish a rank of alternatives among the sites selected.
- *Heatwave risk*: Similar to the validity of the details for flooding risk, it is important to focus on heatwaves as well. Thermal phenomena can especially influence the underground distribution system, affecting the quality of the service.
- *Landslide risk*: Similar to the flooding risk, it is important to also consider the landslide attitude of the area within the process of selecting the appropriate location to install an EVCS. Landslide can compromise the availability of the EVCS and, in the worst case, can generate damages to the infrastructure. Therefore, the EVCS location must avoid sites in which the risk of landslide is high [29]. Also, here, open-access historic data can help in ranking the alternatives.

- *Earthquake risk*: Similar to the details for landslide risk, earthquakes can compromise the availability of an EVCS infrastructure as well. Therefore, the eligible locations for installing an EVCS must avoid sites in which earthquake events can downgrade the availability of EVCS [22] or damage it in an irreversible way.
- *Forest*: The presence of a forest surrounding the EVCS site location can represent a potential danger for the natural environment. Anthropic activities like construction works can interfere with wild fauna and vegetation and vice versa, undermining the full availability and operation of the EVCS infrastructure. Therefore, the potential location of an EVCS must be far from naturalistic areas, thus avoiding exploitation and interference with the surrounding environment of natural areas [22].
- *Soil type*: This strongly influences construction operations, since further technical aspects must be taken into account in the presence of a non-suitable soil (e.g., foundations, stability of soil type). Therefore, soil type influences the choice of the eligible location for the installation of EVCSs [23].
- *Availability*: With this basic attribute, a focus is set on the resources that are available for the construction phase of an EVCS once the location is selected. A site featured by the good availability of construction water and power should be given priority for the purpose of allowing for a fast construction schedule [23]. This is mainly determined by the nature of land use and intensity of land development [5,33]. Under the same conditions of residential land, different residential communities have different development intensities. With a larger intensity of land development, a greater charging demand is expected. An alternative name for this basic attribute could be a more generic resources distribution [31].
- *Utilization*: This attribute indicates aspects that are directly correlated to the previous attribute. In fact, it gives a measurement of the efficiency of resource utilization during the construction and operation of the EVCS, made by expert evaluation after discussions [16]. It can be classified as a preliminary evaluation of the potential eligible sites.

3.2.3. Environmental Subcategory

Once examined traffic-related and geographic-related aspects, the environmental characteristics of the sites must be considered. Here, the focus is on the environmental impact that all human activities connected to the installation of EVCSs can generate. The following output attributes are gathered: (i) end of life management, (ii) territory sustainability, and (iii) emissions.

End of life management groups all the basic attributes that focus on the future of the area selected to install the EVCS. In this way, the aim is to at most reduce the environmental impact of anthropic interventions. In particular, the basic attributes recurring here are as follows:

- *Dismantling waste*: This measures two fundamental aspects. The first is more related to the operative activities such as the construction garbage and sewage discharged during the EVCS construction, as well as battery disposal during the EVCS operation [14]. This is the most occurring definition given to waste problems. The second aspect that can be added is related to the waste that will be produced in case of dismantling the EVCS from the area. In this way, an accurate choice on the building materials can be set in advance during the preliminary design phase preferring eco-friendly or environmentally low-impacting materials, thus reducing the whole burden of environmental impact related to the dismantling phase.
- *Easiness of re-establishment in the future*: This gives a measurement of the simplicity of generalization and re-establishment of the area [15]. It completes the last aspect of the previous basic attribute since it focuses on the future destiny of the area selected. In this case, the post-business phase is considered.
- *Recycling*: With this basic attribute, the direct environmental impact of the EVCS installation is fully examined. Improving the recovery and utilization rate of resources

is crucial for achieving sustainable development [16]. This is a measure of the resources recovered during the construction and operation phases of the EVCS. It underlines the degree of recycling (or reuse) of the resources available in the area.

Territory sustainability focuses on all aspects that have a role on the destruction and ecological influence on the surrounding environment. Also, here, the aim is to at most reduce the environmental impact of anthropic interventions, joining and completing the target of the previous output attribute. Here, the following are considered:

- *Sustainable development of charging station areas*: This basic attribute focuses on the effects carried out by the presence of EVCSs on both the environment and humans. In particular, the benefits generated on e-mobility by the presence of EVCS infrastructure are reflected in exceeding the cost of financial incentives for new EV acquisition even in an adverse EV penetration scenario [20]. The EVCS infrastructure acts as a flywheel for EV penetration and plays a fundamental role in enhancing EV diffusion.
- *Ecological influence*: This prompts the measurement of "the influence on the flora and fauna surrounding the targeted EVCS site" [17], recalling the details marginally presented for the land attribute.
- *Destruction of soil, vegetation, and landscape*: This basic attribute is one of the most important, as it quantifies the measurement of "the vegetation deterioration due to the land development for building EVCSs" [14]. Sometimes, it is found to also be referred to as the water losses. For this peculiar aspect, it is better to reserve a dedicated basic attribute.
- *Destruction of water resources*: Similar to the previous one, it prompts the measurement of the damage to the surface flow and groundwater system [17].

Emissions is an output attribute that groups all the basic attributes that focus their attention on the future of the area selected for installing the EVCS. In this way, the aim is to at most reduce the environmental impact of anthropic interventions. In particular, the basic attributes recurring here are as follows:

- *Global emissions*: This attribute gives a measurement of the environmental pollutants' emission reduction by using EV rather than ICE vehicles [14]. In this case, the immediate effect carried out by the enhancement of EVs and EVCSs is evaluated as a benefit for citizens.
- *Local pollutant and noise reduction*: ICE vehicles cause significant noise pollution and have an adverse effect on community health. The enhancement of e-mobility contributes to a drastic reduction in noise pollution [20]. This basic attribute provides an additive part with respect to global emissions since it includes the noise reduction factor, which contributes to city life quality improvement.
- *Air quality*: Reducing air pollution is the biggest motivation for the use of EVs [22]. This basic attribute is defined in a very similar way to the two previous basic attributes. Moreover, here, it is seen from a social perspective, improving the effects on the use of EVs. It is evaluated as a benefit.

3.3. Category 3: Social Attributes

The third category is named "social". It regards all social factors that are involved in EVCS network expansion; these can positively (or negatively) influence or be influenced by the EVCS propagation. The subcategories identified here are (i) collective and (ii) personal.

3.3.1. Collective Subcategory

The collective social factors considered here are related to demographic, behavioural, and attitude aspects that can influence the location of EVCSs or that can be influenced by the chosen location of the EVCSs themselves. Here, the focus is on the environmental impact which can be generated by all human activities connected to the installation of EVCSs. The following output attributes are considered: (i) impact on people's lives, (ii) demographic information, and (iii) points of interest.

Impact on people's lives groups all basic attributes that are related to the influence that the operations of positioning the EVCS can have on the local people. This output attribute is better described by the following basic attributes:

- *Acceptability of new solutions*: Public awareness and support will affect the development of similar projects and the future development speed of EVs [24]. A diffused positive acceptance of EVCSs in the neighbourhood will increase the expansion of EVCS network, boosting the technical solutions offered. This can be achieved through social commitment in creating or developing new social areas capable of carrying forward the improvement of selected areas.
- *Adverse impact on people's lives*: This takes into account the adverse impacts of noise and electromagnetic field due to the construction and operation of EVCSs on the daily life of local residents [14]. An alternative approach is to account in advance for the local resident attitude and opinion on the EVCS construction and operation. This enables to find out in advance whether the local population is inclined to tolerate noise and electromagnetic field due to the construction and operation of the charging station [23].
- *Improvement on employment*: The construction and maintenance of EVCSs can provide more job opportunities, including for local people in different fields. In this way, if the employment rates of the local territory are low, it can offer work opportunities; therefore, employment rates can be boosted up [5]. This can become an important aspect regarding the social well-being of the local areas.
- *Benefits for people's lives*: The difference compared to the previous basic attribute is that, here, it is defined in a more general way and can also consider positive effects, i.e., improving the quality of life of the residents, in people's opinion, which are underlined here [15]. An alternative point of view is given considering that the construction and operation of EVCSs may generate poor acceptance among the local population due to the negative effects of noise and electromagnetic radiation. This can lead to forcing the shutdown of the project even at the very beginning, particularly in residential communities. Therefore, efforts must be put into practice by investors to change the level of acceptance of residents to reduce investment losses at most [5]. For example, if the local area sees a contextual improvement of the residential zone through the construction of new social areas or the redevelopment of the same neglected areas, this can lead to changing the mentality of local residents, pushing them to accept rather than refuse the presence of EVCSs.

Demographic information includes all the basic attributes that can address the needed information related to EV diffusion. These can be summed up as follows:

- *Population density*: This attribute indicates that the need for charging stations is higher in areas where EVs are frequently used. Population density can be used as an indicator to determine which regions are best suited to see the location of one or more EVCSs, since population density may represent a potential ideal charging request. If the location is characterized by a high population density, it will be more suitable [34]. The information suggested here needs to be strengthened by considering further information given by the next basic attributes listed here; otherwise, it will have no meaning when considered alone.
- *Population intensity*: This is defined in the same way as population density, but it seldom appears to be called with a different denomination.
- *Number of vehicles (local)*: This considers the total number of vehicles (of all types) in the local area selected. It represents an additive information with respect to population density, since it prompts the indication of high vehicle potential and the transformation of conventional vehicles into EVs [22]. This information must be associated with the next basic attribute: the number of EVs (local).
- *Number of EVs (local)*: This considers the actual number of EVs being present in the local area considered as eligible to locate EVCSs. It is important to be considered because it addresses the relation between the charging demand and EV ownership.

The former is meant to increase if the latter increases. It gives the estimated potential charging demand at the beginning [22].
- *EV sales (local)*: This basic attribute addresses the projected number of EVs that the EVCS site is called to serve. When the number of EV sales in the area surrounding the targeted EVCS site is higher, a higher number of the EVCS is needed [17].
- *Residents' average income*: The consumption characteristics and income levels of residents in different residential communities are diverse, which depend on the employment level, the consumption structure, the growth of consumer expenditure, and the cost of living [5]. High-income-rate districts are meant to be suitable to locate EVCSs [34].

Points of interest includes all the basic attributes that focus on locating EVCS in correspondence of nodes important for what concerns the public utility, seen by the user perspective. These can be summed up by the following:
- *Social areas*: EVCS locations should be close to popular centres like shopping malls, stadiums, universities, public buildings, hospitals, due to merging the needs of mobility, sociality, and public services [22]. Also, working areas can represent a potential location in terms of charging demand.
- *Fuel station proximity*: This basic attribute takes into account two aspects, given by the variety of EV typologies. PHEVs need both fuel products and electricity, while BEVs require longer charging times. Therefore, the proximity of fuel stations can represent a constraint for EVCS location [22].

3.3.2. Personal Subcategory

Personal social factors considered here are related to behaviours and attitude aspects seen by the user perspectives. Hence, the only output attribute is user preferences. This is described by the following basic attributes:
- *Driver's comfort*: This refers to whether the driver can immediately start charging operations and avoid waiting times due to queuing. If the EVCS is located in a place featured by heavy traffic and large charging volumes, it may generate longer waiting times, thus reducing the drivers' comfort [18]. This last concept is defined for the location of Battery Swapping Stations, but it can be easily applied to the location of EVCS cases.
- *Home/private charging vs. public charging*: Since charging needs for EV owners is becoming more and more urgent, the balance between public and private infrastructure must be accounted for, since home charging can show "high rates of preference by EV users" [20].
- *ICE vs. EV*: This aspect was not found in the literature review, but it constitutes a threshold attitude for users. Even though the available EVs ensure a relatively long duration of a fully charged battery, the users can still prefer to travel by using an ICE for covering long hauls rather than using an EV. In addition, waste management for existing vehicles replaced by EVs could impact the possibility to purchase an EV by benefiting from dedicated incentives for vehicle replacement or fiscal discounts applied to the use of EVs.

3.3.3. Social Category: Analytical Expressions

Analytical expressions are provided in [32] for evaluating the level of integration of EVCSs with the public transport network, and it is divided into two terms, i.e., the intensity of integration of primary and secondary transport networks with a considered EVCS. The values considered are the number of stops of the public transport system located at a distance not exceeding a given threshold:

$$\max \sum_{l=1}^{L} \sum_{p=1}^{P} x_l (B_l(\bar{R}) \cap S_S) \tag{4}$$

where L indicates the number of potential EVCS locations; $x_l \in [0;1]$ is a binary variable, i.e., 1 if EVCS is located in the l-th alternative location and 0 otherwise; B_l is a zone around l potential location of the ECVS with radius $\bar{R}$; and S_S is a set of s stops that belong to the public transport system. The reference [32] also provides similar analytical expressions for evaluating the integration of EVCSs with the main roads of the city system in terms of the number of EVCSs located no more than a threshold distance from the main roads of the city. Adequate indicators to represent the integration of EVCSs with points of interest are also reported.

3.4. Category 4: Technical Attributes

The last category is named "technical". It regards all technical factors and engineering aspects that are involved in both EVCS and grid planning and operation. The interaction between an EVCS and an electrical grid can represent an obstacle to the physical integration of the EVCS infrastructure. The identified subcategories are (i) grid side, (ii) user side, and (iii) EVCS side.

3.4.1. Grid Side Subcategory

Grid planning gathers several technical aspects that can impact the electrical grid transmission and distribution once the suitable location of EVCS is chosen. This is described by the following output attributes: (i) grid operation and (ii) grid planning.

Grid operation focuses on all aspects concerning the operability of the EVCS infrastructure. Here, several technical issues are considered and explained:

- *Power and energy management*: This aspect involves the effects of the EVCS operation on the actual balance of electric loads influencing the power stability of the grid. EVCS constitutes a non-negligible component of medium- and low-voltage distribution systems. As an immediate consequence, the EVCS should be located in an area that is sufficiently far from the heavy loaded electric lines to ensure a stable operation of the distribution network [23,35,36]. Moreover, to improve the stability of the grid, energy storage systems can be installed to increase the reliability of the grid and its response to extended overloads.
- *Power quality*: This is defined with the same meaning for power and energy management, but with the focus that is put on an EVCS, from an opposite perspective. As already mentioned in the previous attribute, the quality for EVCS-delivered electric power can be improved with the installation of energy storage systems, thus contributing also to stabilize the network against unforeseen overloads or voltage drops.
- *Harmonic pollution in the power grid*: This basic attribute focuses on the harmonic distortion of the EVCS. This is due to a large amount of charging demand that generates harmonics injection in the power grid. If it cannot be effectively compensated and filtered, it will seriously affect the power supply quality, damage the already existent capacitors, and threaten the safety of the whole power grid [18].
- *Impact on the load levels of power grid*: This basic attribute is associated with an aspect that is becoming more relevant in the last period, that is, vehicle-to-grid (V2G). It is assumed that the battery can also serve as an energy storage system for the grid while satisfying the charging needs of EVs. In order to ensure the stability of the power grid and to avoid the rising of huge impacts on the power grid, the real-time load levels of the power grid itself should be taken into account, and the charging and discharging threshold of the battery should be reasonably selected, ending in a good compromise [18].
- *Impact on voltage*: This is defined as the quality of the electricity supplied to the targeted EVCS site that determines the service quality of targeted EVCSs. Since the EVCS is usually planned within an electric power distribution network, when charging operations start, a higher power load will be generated, which will cause the voltage to drop by seriously endangering the safe and stable operation of the power grid [17,18].

- *Power grid security implications*: This basic attribute refers immediately to the previous one, since it quantifies through a significant indicator the measurement of the influence of the targeted EVCS site on power grid the [17]. A higher score of this index indicates a greater threat to the local power grid security.
- *Consumption level*: This refers to an energy efficiency measure that can be seen either under an energy point of view, i.e., if the EVCS shows high efficiency with minimized energy and thermal losses, or from an economic perspective in terms of missing cash flows [28,37]. Although it is not occurring with the following meaning within the scientific literature examined, it can also refer to the difference between the potential demand initially estimated and the actual charging demand.
- *Electromagnetic interference*: This can be wrongly misunderstood and confused with the effects of electromagnetic fields on the natural environment. Conversely, it identifies the interference produced by electromagnetic fields generated by large radio transmitters and industrial electromagnetic fields on the site location of EVCSs. Therefore, it measures the influence of an electromagnetic interference on the power supply stability of the EVCS. It is assumed that at a longer distance, a weaker electromagnetic interference on the targeted EVCS will be observed, ensuring a stable feeding of charging power [17].
- *Level of penetration of RES*: This aspect is not adequately pointed out from the literature review. Despite this, it can represent an important aspect due to the following reasons. First, a high-RES penetration can enable us to dedicate an RES production that is able to reinforce the actual electric/power grid distribution network feeding the EVCS, and thus increase the responsiveness of the EVCS infrastructure against the overloads and unforeseen peak demands. Secondly, it can represent a huge potential for what concerns an increment of the capacity of EVCS sites.

Grid planning accounts for all the technical issues that can influence the supply capacity and thus the technical problems that can potentially rise in the presence of EVCSs. In particular, it is described by the following basic attributes:

- *Power supply capacity of transmission and distribution systems*: This is defined as the amount of electric power that must be delivered by the grid when the EVCS operates and electricity loads are supplied. It strongly depends on the charging services that the EVCS will provide that must show compatibility with the actual state of the grid. Therefore, it should be adapted to the power supply capacity of the local transmission and distribution systems [17,18,35].
- *Distance to the substation*: This is defined as the distance between the EVCS infrastructure and the first useful substation, which should be close enough to areas characterized by high energy demand [22]. In a few cases, this aspect is included in a cost item [19]. In fact, the farther away the substation is, the longer the wiring will be; therefore, the higher the power losses will be. This can be related with O&M costs output attribute.
- *Substation*: This refers to the concept of substation proximity, with a very close meaning already reported by the previous basic attribute [22].
- *Substation capacity permits*: This is defined as a measure of the integration degree between the electricity demand of the targeted EVCS site and the substation capacity of the located area [17]. It can also indicate the level of overloading of the substation and its attitude to sustain these conditions. A higher score to this index indicates that the site is more suitable and can obtain permits for its installation.
- *Substation capacity*: This is used with the same reference of the previous definition [28]. Here, it indicates the power capacity of the substation.
- *Power grid capacity*: This basic attribute focuses on an important aspect that must not be overlooked when defining the planning phase. The power grid capacity is an important factor for the integration of the charging infrastructure. Major technical work may occur due to a strengthening of the existing network or the need for transformer installations to enable a full operability of EVCSs in the area [20].

3.4.2. User Side Subcategory

The focus is now progressively set on all aspects concerning the services and options offered by the EVCS infrastructure. Here, only the charging station services output attribute exists, whose basic attributes are described as follows:

- *Charging services*: This basic attribute refers to the service level offered by the EVCS. This is defined as the EV number and service radius that the EVCS can serve [16]. This basic attribute can consider the different possibilities of charging the EV offered by EVCS, like, for instance, DC/AC sockets, and the related maximum capacity.
- *Further services to the drivers*: Although in the scientific literature the services are limited to the charging services offered to the drivers—indicated at the previous point—the services can be extended by also referring to different additional services that can be offered to the drivers while charging. This basic attribute is the opportunity to offer appropriate services to the drivers in correspondence to the EVCS, meant as a benefit indicator. Often, the notion of Electricity Accessibility (EA) is introduced, aimed at measuring the service quality of a charging station network. EA is measured by the average time spent by a random driver to complete charging [15,25]. The analytical formulation of the EA used in [25] is represented with t_{qz}, i.e., the travel time from cell q to cell z, and with t_v the service time of charging stations of type v. The objective function to be minimized is the average EA, where F is the total number of charging demand in the network; D_q is the demand in cell q; and y_{qzv} is the fraction of vehicles in cell q that is served by charging station of type v in cell z, as reported in (5). The perspective here is seen as the opposite of the point of interest attribute, where the EVCS is located depending on an already existent service of public interest. The difference here is that an additional point of interest can be created, with paybacks that could also directly involve the local population.

$$EA = \sum_{q=1}^{Q} \sum_{z=1}^{Z} \sum_{v=1}^{V} D_q (t_{qz} + t_v) \frac{y_{qzv}}{F} \quad (5)$$

- *Fast-charge ratio*: This is defined in the literature as the ratio of the number of fast-charging stations to the total number of EVCSs. EV users can prefer using fast-charging facilities to save time rather than conventional charging. Therefore, the location served by EVCS infrastructure with a higher fast-charge ratio is thus more likely to provide efficient charging services and to attract more customers to charge, thereby exploiting fast-charging solutions [21].

3.4.3. EVCS Side Subcategory

The EVCS infrastructure is now focused on the planning phase. This subcategory is described by the two last output attributes, concerning several aspects: (i) EVCS planning and (ii) EVCS operation and reliability.

EVCS planning accounts for the possibility of capacity expansion in the future, since the expansion of the capacity of EVCSs needs to consider the number of charging users in the future, the projected new EV sales in the area, land resources nearby, government policies, and the upgrade of the distribution network [24].

EVCS operation and reliability focuses on the aspects of safety and reliability of the EVCS. In particular, here, several basic attributes are grouped, each described with a technical or safety issue that must be considered:

- *Safety/security and ability to tackle the emergency*: This refers to the capability to sustain emergency conditions and also evaluate the protection of the EVCS. It can consider the security of the EVCS in an emergency situation, including grid safety, fire protection facilities, and the resilience properties of the EVCS site, i.e., the ability to resist natural disasters [15,16].
- *Reliability*: This evaluates the reliability as the resistance and durability of the EVCS with respect to many external conditions. It is measured as the stability of alternative

EVCS sites to future changes in external conditions. It sometimes accounts for the reliability of the power supply located near the site locations, meant as time to failure. It is often defined as derived from the concepts of Mean Time To Failure or Mean Time Between Failures [15,16,27]. A high score means high reliability.
- *Charging station capacity*: The power capacity of the EVCS determines the maximum number of daily charging sessions. These are essentially the "sales units" of the investment. A high-power 50 kW charging station can serve up to 60 charging sessions per 24 h, while the maximum capacity of a normal-power 22 kW station is limited to 26 charging sessions [20]. During the operation phase, an increased number of EVCS units available on the same charging site could emerge as needed to satisfy the demand.
- *Service capability/service capacity*: This is defined as the number of EVs that can obtain access to the charging service provided by the EVCS, the daily charging volume, and the maximum charging volume. It can also be defined as the daily service volume and the maximum number of EVs that could obtain access to the charging service provided by the charging station [14,23].

4. Importance of the Attributes

This section provides a review of the attribute appearances and weights as they are used in the literature, organized in the form of matrices. In this context, absolute and relative weight values can be distinguished throughout the literature examined. In particular, the following can be noted:
- An absolute weight provides the importance of a single attribute compared to the total attributes considered in all categories. To be as clear as possible, the absolute weight value defines the global influence of one specific attribute on the rest of the attributes considered.
- A relative weight defines the importance of one attribute in comparison to the others within the same attribute category. It defines the local influence of the single attribute among the others that belong to the same category of attributes.

The evaluation of absolute and relative weights requires a broad and in-depth literature review, so that it is possible to extrapolate the weights of each attribute from every single study contribution and then evaluate the impact of the weight of each attribute by considering the whole set of attributes. According to this rationale, the sum of the weights for the single literature contribution will be unitary. A practical example, with values referring to the contents of one of the papers considered [17], is shown in Table 4. The output attributes mentioned in this paper are marked with one, while at least one of the basic attributes included in the output attribute appears. Conversely, the output attributes that are not included are marked with null value. Furthermore, the corresponding weights assigned by the authors are reported in the "Weights" column. The total number of the appearances reported at the bottom of Table 4 indicates the number of attributes appearing in [17], while the sum of all weights assigned reaches unity (i.e., 100%). This operation has been repeated for covering all the works selected and reviewed from the scientific literature.

It is important to highlight that the weight values found in the literature are strictly connected to the total number of attributes considered in the single literature contribution. Moreover, the same attribute may be calculated differently in different papers, because of the different attributes considered within the same paper. In fact, as indicated in Section 3, the literature contributions examined were retrieved to deal with different attributes among them, but that showed similar semantics. And seven significant papers are taken as examples here. With a focus on the overall multiplicity of the criteria, only 10 attributes are considered in [34], rising up to 13 in [20] and 15 in [15,29]. Higher numbers are found in [22,30], with 19 and 21 total attributes, respectively, while [36] considers up to 45 attributes jointly. Moreover, despite the variability encountered, a further difference is identified in the way of proceeding the aggregation for the attributes considered. For example, three categories are exploited in [29], four categories in [15,30], five in [20,22], and nine are used to group attributes in [36]. Conversely, [34] does not use any category to enclose attributes. Treating the attributes recalled in those contributions semantically, it is possible to observe that the same meaning is not always reflected uniformly across the examples with the same undertone. If only the *economic* category is isolated, no attributes are considered by the authors of [34], which represents an outlier in this sense, while [29] focuses on land cost, thus discarding construction costs and O&M costs as [15,20,22,30,36] have reportedly performed.

The approach chosen by the authors of [20] is noticeable since it considers O&M costs and equipment cost rather than construction costs. This comparison can also be extensively repeated for other categories (environmental, social, and technical), where basically each paper proposes its own framework of attributes, thus increasing the multiplicity of criteria classification and hence the sparsity of approaches, given that the topic covered is common. Therefore, a re-ordered categorization of attributes aimed at constituting a common practice needs to be provided to be adopted in future research, allowing for direct comparisons between different strategies and consequently increasing uniformity.

Table 4. Example of appearances and weights calculated from the contents of [17].

Attribute Category	Attribute Subcategory	Output Attribute	Index (row,col)	From Xu et al. (2018) [17]	
				Appearance	Weight
Economic	Cost	Installation costs	1	1	4.50%
		O&M costs	2	1	4.30%
		Update and removal costs	3	1	3.40%
	Benefit	Revenues	4	1	5.50%
	Policy	Installation permits	5	0	0
		Government support	6	0	0
Territorial	Traffic	Traffic flow	7	1	6.40%
		Road network characteristics	8	0	0
		Interactions with other infrastructures	9	0	0
	Geography	Service radius	10	0	0
		Urban development	11	1	3.70%
		Land	12	0	0
	Environmental	End of life management	13	0	0
		Territory sustainability	14	1	24.80%
		Emissions	15	0	0
Social	Collective	Impact on people life	16	0	0
		Demographic information	17	1	12.10%
		Points of interest	18	0	0
	Personal	User preferences	19	0	0
Technical	Grid side	Grid operation	20	1	29.70%
		Grid planning	21	1	5.40%
	User side	Charging station services	22	0	0
	EVCS side	EVCS planning	23	0	0
		EVCS operation and reliability	24	0	0
		Total		10	100.00%

Starting from the information relating to each paper, the basic attributes of the entire sets were jointly considered, with the aim of providing the outcome regarding the relative importance of the output attributes to which they refer. The relative importance of each attribute is provided by considering different aspects: (i) the occurrence of the output attribute in the literature and (ii) its weight. The generalized scheme of attributes described in Section 3 is validated using numerical analysis to motivate generalized considerations, in order to orient the decision maker about the importance that each attribute has in the literature. Therefore, to provide a synthetic analysis of the two aspects mentioned above, two matrices have been built and calculated as illustrated below, considering the general case with N_T output attributes and N_P papers analysed:

$$\mathbf{A}_O \in \mathbb{N}^{N_T, N_T}, \ a_o(g,j) = \begin{cases} \sum_{p=1}^{N_P} x_p^{(j)} & g = j \\ \sum_{p=1}^{N_P} x_p^{(g)} & g \neq j \end{cases} \quad (6)$$

where $x_p^{(j)} = 1$ if the output attribute j exists in the literature contribution p, and $x_p^{(g)} = 1$ if both output attributes j and g are mentioned in the literature contribution p, and

$$\mathbf{A}_W \in \mathbb{R}^{N_T, N_T}, \ a_w(g,j) = \begin{cases} \sum_{p=1}^{N_P} w_p^{(j)} & g = j \\ \sum_{p=1}^{N_P} w_p^{(g)} & g \neq j \end{cases} \quad (7)$$

where $w_p^{(j)}$ is the weight of attribute j in the literature contribution p, and $w_p^{(g)}$ is the weight of attribute g in the literature contribution p when attribute j is also included in the literature contribution p. These two matrices have been computed and graphically rearranged in heatmap form, as shown in Figures 2 and 3, considering the correspondence shown in the column named "Index (row, column)" of Table 4. The rearrangement in the heatmap of the matrix $\mathbf{A}_O$ shows that the higher the number of papers citing the output attribute j, the darker the colour of the element $a_o(j,j)$. Moreover, it is worth to note that matrix $\mathbf{A}_O$ is symmetric due to its construction. In fact, let us take as an example the values contained in $a_o(1,3)$ and $a_o(3,1)$. In $a_o(1,3)$, the output attribute of row 3 (update/removal costs) appears 13 times together with the output attribute of column 1 (installation costs). The dual condition is represented by the element $a_o(3,1)$, which contains the number of appearances of the output attribute in row 1 (installation costs) when the output attribute of column 3 (update/removal costs) appears, and it is already known that it appears 13 times.

Normalization of the matrix $\mathbf{A}_O$ could lead to the notion that a relative importance with respect to appearances of attributes can emerge. This is misleading, since the symmetry of the matrix does not prompt reciprocity relationships between the elements below and above the main diagonal. With reference to the matrix $\mathbf{A}_W$, shown in Figure 3, the higher the value of $a_w(g,j)$, the higher the relevance of the g-th output attribute with respect to the j-th output attribute. Conversely, the lower the value of $a_w(g,j)$, the lower the relevance of the g-th output attribute with respect to j-th output attribute. This provides an immediate comparison among the different output attributes. In addition, if $a_w(g,j) > a_w(j,g)$, the g-th output attribute takes more importance than the j-th output attribute. When $a_w(g,j) = 0$, no weights were assigned in the literature for the given association of the g-th and the j-th output attribute, i.e., the two attributes were never considered together. Also in this case, the normalization of the matrix $\mathbf{A}_W$ would lead to misleading results. An immediate example is given considering column 5—the one featured by the lowest value on the main diagonal. Normalizing the values of column 5 with respect to cell $a_w(5,5)$ will result in having values higher than 1, which are difficult to interpret, as the matrix $\mathbf{A}_W$ is non-symmetric and no reciprocity relation exists with the corresponding elements on row 5.

The matrix of occurrence for the output attributes shown in Figure 2 provides a detailed view of the frequency with which each attribute is mentioned in the literature. This specific visual representation provides an immediate understanding of the absolute importance of the attributes. On the other hand, the matrix of weights for the output attributes, proposed in Figure 3, provides an immediate comparison between the different attributes, showing the relative importance of each one. This systematic, visual approach makes it possible to identify meaningful associations between attributes, providing new insights that may influence future EVCS siting decisions.

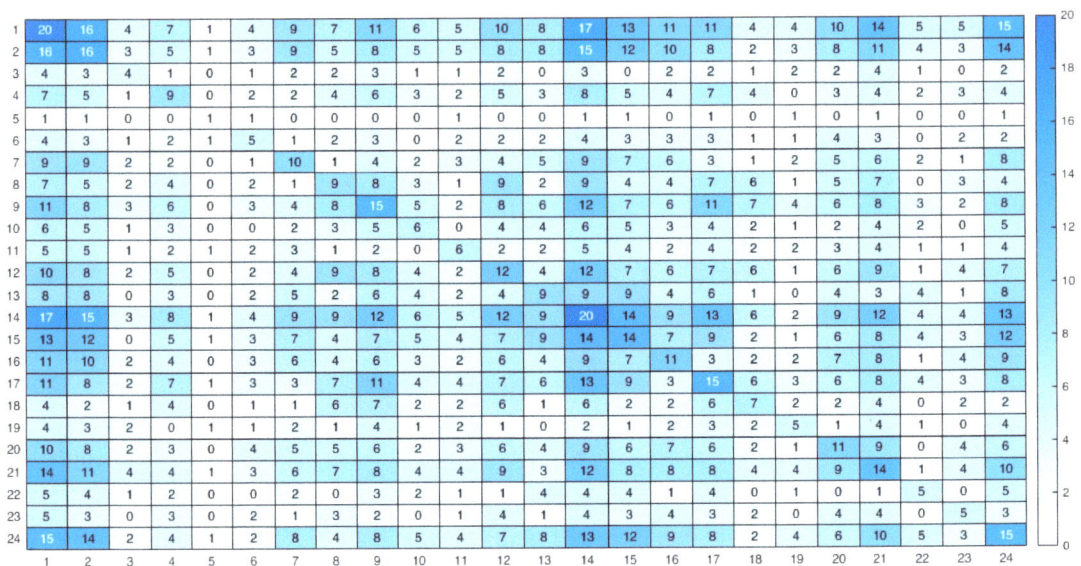

Figure 2. Matrix of occurrence for the output attributes.

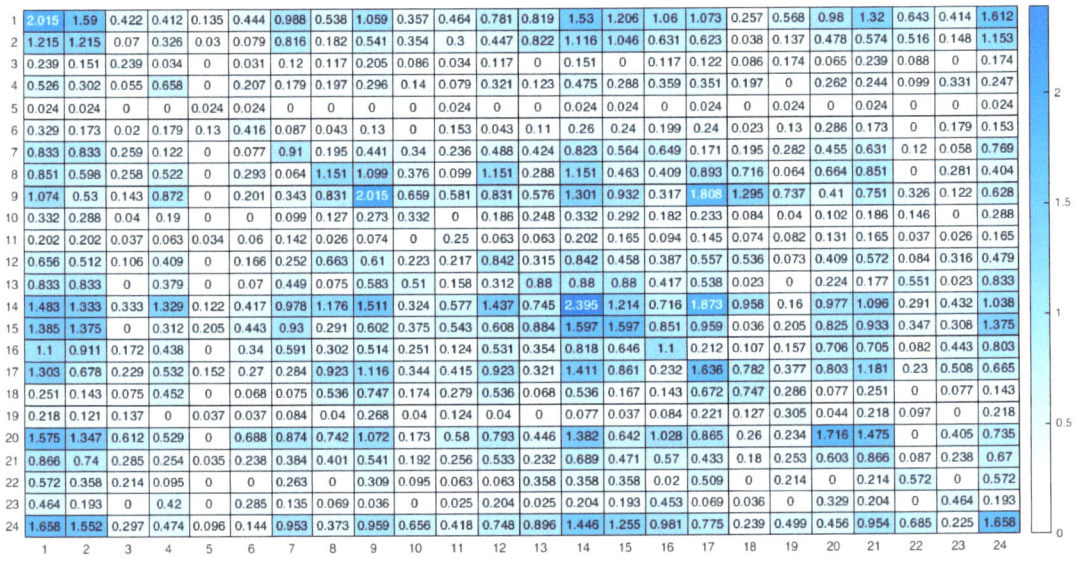

Figure 3. Matrix of weights for the output attributes.

By combining the information from both matrices, it is possible to list the ten most relevant output attributes:

- 14—Territory sustainability;
- 1—Installation cost;
- 9—Interactions with other infrastructures;
- 20—Grid operation;
- 24—EVCS operation and reliability;
- 17—Demographic information;

- 15—Emissions;
- 2—O&M costs;
- 8—Road network characteristics;
- 16—Impact on people's lives.

The same procedure has been repeated considering the subcategories of attributes, thus aggregating the original starting data according to the scheme reported in Table 5. Thus, the results are constituted by a pair of $[11 \times 11]$ matrices for subcategories of attributes, i.e., $\mathbf{A}_O^{(SC)}$ and $\mathbf{A}_W^{(SC)}$, with the first for appearances and the second for weights assigned (shown in Figure 4). If only subcategories of attributes are considered, the reduction in the matrix dimensions is observed with the corresponding increase in the values contained in the cells. In the presence of one or more recurring attributes belonging to the same subcategory in the literature, the appearance will be set as equal to 1 within the same paper considered. For instance, the attributes belonging to the *economic costs* subcategory in Table 5 all appear in [17], but the appearance of the *cost* subcategory for that particular paper remains to be equal to unity.

Here, the aggregation of attributes belonging to the same subcategory is performed. The matrix $\mathbf{A}_O^{(SC)}$ has been computed and reported in Figure 4a. As was previously reported for the matrix $\mathbf{A}_O$, here, the rearrangement of the matrix into a heatmap also points out which subcategories are considered more relevant. It appears that rows and columns 3, 8, and 10 are less considered in the literature, corresponding to the *economic–policy*, *social–personal* and *technical–user side* attribute subcategories. The matrix $\mathbf{A}_W^{(SC)}$ instead considers in each matrix element the corresponding sum of weights assigned in the literature. The matrix $\mathbf{A}_W^{(SC)}$ remarks the distinction pointed out by the matrix $\mathbf{A}_O^{(SC)}$, considering the aggregation of weights. Hence, weights of attributes included within the same subcategory are summed up for each paper, thus resulting in the matrix reported in Figure 4b. It is possible to note that weights are now taking a very high value with respect to the values that appear in the matrix $\mathbf{A}_W$, because of the aggregating procedure of weights coming from the first step. It is possible to perform a subsequent step towards aggregating all subcategories of attributes in their corresponding categories; thus, a pair of $[4 \times 4]$ matrices for categories of attributes, i.e., $\mathbf{A}_O^{(C)}$ and $\mathbf{A}_W^{(C)}$ shown in Figure 5, can be computed. These will deliver the idea of which category is predominant among the others. The indices of matrices are reported in Table 6 for the sake of simplicity. As already presented for the subcategories, all the attributes belonging to the *economic* category in Table 6 appear in [17], but the appearance of the *economic* category for that paper is always equal to unity.

Table 5. Indices of attribute subcategories.

Attribute Category	Attribute Subcategory	Index (row,col)
Economic	Cost	1
	Benefit	2
	Policy	3
Environmental	Traffic	4
	Geography	5
	Environmental	6
Social	Collective	7
	Personal	8
Technical	Grid side	9
	User side	10
	EVCS side	11

The attribute subcategory matrices shown in Figure 4 aggregate the data to clearly show which subcategories are most highly regarded in the literature. As for the matrices for the output attributes, the occurrence matrix representation defines the relevance of the subcategory in the research, while the weight matrices define its relative importance among them. This representation offers a level of detail and clarity that helps us to better understand current trends in EVCS research. From Figure 5a, it is possible to appreciate the appearance distribution of the four categories of attributes in matrix $\mathbf{A}_O^{(C)}$. The total number of occurrences of each category compared to the other categories is very close, thus pointing out that all attribute categories are considered with equal importance. Only the associations of the *technical* category with the *social* and *territorial* categories show a slightly lower number of appearances. This means that the scientific literature reviewed focuses mainly on *economic* with *social* and *territorial* aspects, giving relatively less importance to *technical* aspects. The matrix $\mathbf{A}_W^{(C)}$ in Figure 5b instead points out the aggregation of weights assigned by the different authors to these four categories. From this distribution, it is possible to note that the aspects related to the *territorial* category and linked with the *economic* and *social* categories are featured by a higher aggregated weight assigned, thus considered relevant within the literature examined. *Social* seems to be less important than the other attribute categories. This can be explained with the fact that, in general, lower weight is assigned to this specific category. Finally, as far as the *technical* category is concerned, the results shown in the matrix $\mathbf{A}_W^{(C)}$ allow us to classify this category as having a relevance similar to the *economic* one.

Table 6. Indices of attribute categories.

Attribute Category	Index (row,col)
Economic	1
Environmental	2
Social	3
Technical	4

The analysis carried out herein allows us to formulate further considerations, contributing to set a path for future research related to the EVCS location problem. Keeping in mind the aforementioned information that is valid as general considerations and mediated by numerical analysis—i.e., the inter-relationships discovered between categories and subcategories of attributes—it is worth noting that the analysed literature contributions have mainly focused on solving the EVCS location problem under technical and environmental aspects, since those attribute categories are retrieved to have high aggregated weight compared to the rest. This is what Figure 4b points out, where the aggregated weight provides a hierarchical order of attribute categories: *territorial*, *technical*, *economic*, and *social*. Motivations can be extrapolated from this, extending the attention from attribute categories to output attributes as follows. The urgency of pushing towards the widespread use of EVs with a charging infrastructure aware of the surrounding environment and harmonized with existing electrical grids and loads should be realized.

The *territorial* aspect is strongly considered, even much more than *technical*, in order to propose environmentally sustainable solutions oriented towards providing less impact on the natural environment and a satisfactory level of integration with the surrounding environment. The *economic* category is therefore only considered afterwards, but the higher aggregated weights set a priority in considering this category with respect to *social*. These statements are reflected in the analysis of how the aggregated weighing is distributed across the subcategories and, furthermore, across the output attributes. In fact, within the *territorial* category, all subcategories are considered (4—*traffic*; 5—*geography*; 6—*environmental*), with 4 and 6 predominantly weighted, while *technical* presents high weights for 9 and 11 (*grid side* and *EVCS side*, respectively). Then, both *economic* and *social* are represented with only one subcategory, having high weight among the others belonging to the same attribute category (i.e., 1—*cost* and 7—*collective*, respectively).

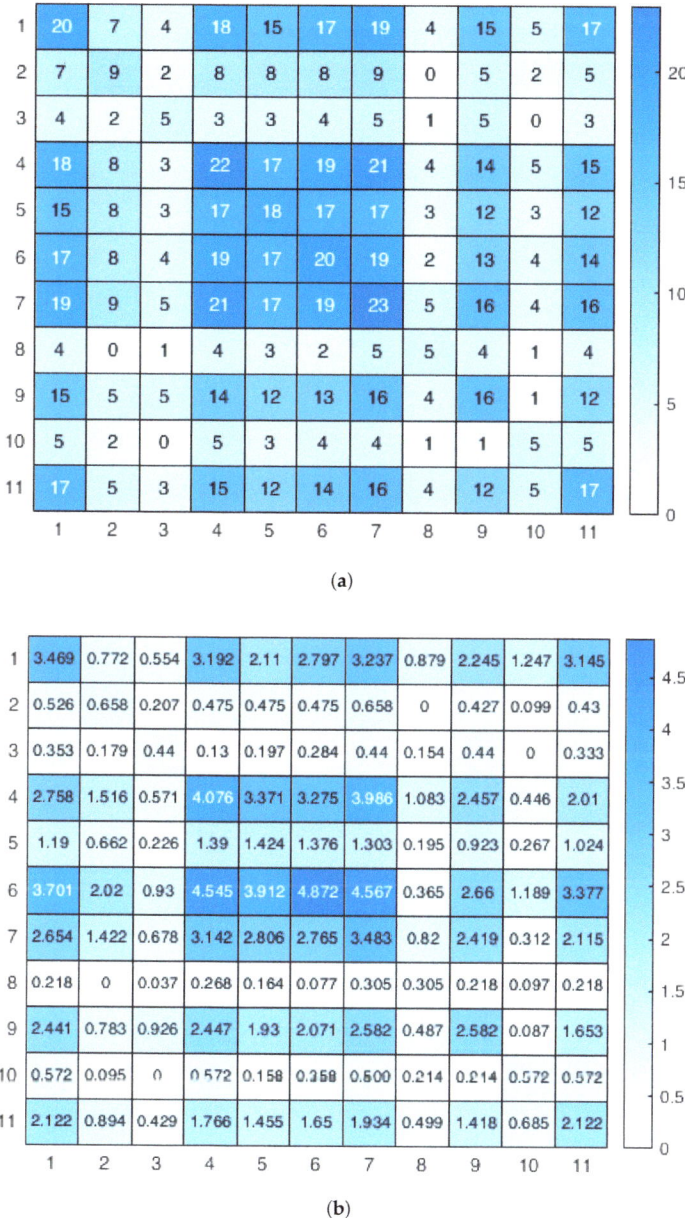

Figure 4. Matrices for attribute subcategories: (**a**) occurrence, (**b**) weights.

Other subcategories that are less considered are, for instance, 2 (*benefits*) and 3 (*policy*) in *economic*, 5 (*geography*) in *territorial*, and 8 (*personal*) in *social*. These last considerations can be analysed item by item considering the weighing of output attributes. As recalled before, aspects related to the *territorial* category are targeted by the existing research, with some valuable differences among them. Output attributes no. 8 (*road network characteristics*), 15 (*emissions*), 9 (*interactions with other infrastructures*), and 14 (*territory sustainability*) are the most predominantly weighted; this implies a selection of EVCS locations able to increase the level of integration into the citizens' pattern and facilities in order to potentially increase the future use of EVCSs from EV users.

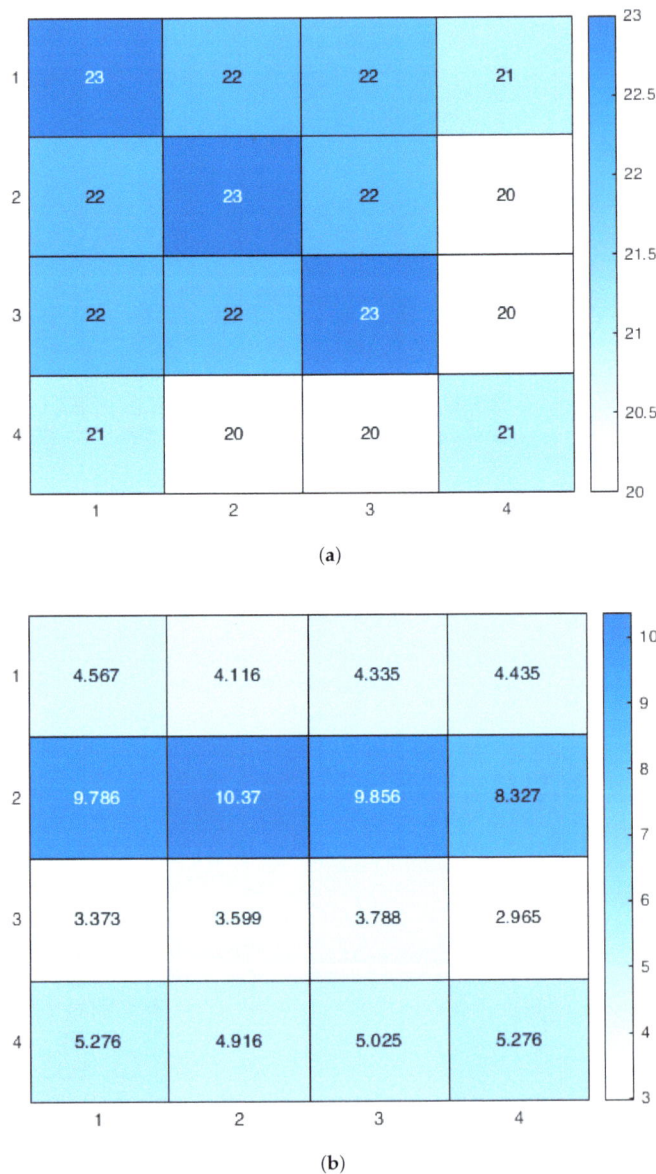

Figure 5. Matrices for attribute categories: (**a**) occurrence, (**b**) weights.

As far as *technical* aspects are concerned, a strong attention is set on 20 (*grid operation*) and 24 (*EVCS operations and reliability*) to ensure that the locations of EVCSs are fully integrated in the existing infrastructure for electric distribution, minimizing the disturbances induced on the grid operation and risk of outages. In the *economic* category, the majority of the literature contributions addressed 1 (*installation costs*) and 2 (*O&M costs*), while lower weights are assigned to other attributes. An outlier here is represented by 6 (*government support*), which is considered in only a few papers (5). For *social*, the two most relevant attributes are 17 (*demographic information*) and 16 (*impact on people's lives*), while less relevance is assigned in general to 18 and 19 (*points of interest* and *user preferences*).

The discussion proposed here strongly depends on the point of view considered, i.e., of the different stakeholders involved, in the process and on the targeted focus preliminarily set by researchers. In fact, the recently developed research is strongly dedicated to reducing invasive impacts or interferences arising from the installation of a new infrastructure in an already-existing context, i.e., grids and roads. Furthermore, the location of EVCSs must address social aspects in an important way, aiming to maximize the future exploitation of the charging infrastructure from the user side, and thus also increasing economic incomes. Here, the capillary presence of EVCSs on a given area must be intended as strategic to allow for the increase in EV penetration within the private vehicle fleet, in terms of being diffused based not only on traffic volumes, but also on social activities, and thus reaching the highest number of EV users and capturing their need to charge. Moreover, the policies being progressively approved in Europe to increase the number of EVs in the private sector is enhancing the local administrations to concede more areas to be included in public tenders to be assigned for the installation of EVCS infrastructure. Therefore, this last point is expected to be considered with increasing weight with respect to the past, as stakeholders will orient their business strategies where public governments are supporting this change. This analysis offers a valuable overview of the broad set of key factors influencing the EVCS infrastructure's location selection process and a practical approach to systematically categorize and weigh the main attributes to support the deployment strategy of a CPO. The framework aids in making informed decisions that balance technical, economic, and environmental factors, facilitating a more streamlined and effective roll-out of charging stations. The emphasis on attributes such as territory sustainability, grid operation, and installation cost is in line with practical considerations in infrastructure planning. Furthermore, the inclusion of both numerical and non-numerical attributes in the MADM approach allows us to consider a very broad spectrum of factors, thereby enhancing the robustness of the decision-making process. Overall, this study provides a comprehensive basis that can support a CPO's approach to strategic EVCS location planning, fostering the development of a more efficient and user-oriented charging network.

5. Conclusions

This paper has presented a structured categorization of the attributes considered within a multi-attribute analysis that addresses the EVCS location. The analysis of the literature highlighted the existence of different nomenclatures for similar aspects, introducing difficulties in defining the multi-criteria problem, as well as confusion in choosing the most convenient attributes. Therefore, the proposed categorization started from the information found in the literature and introduced a novel structure with basic attributes, categories, and subcategories. In this way, similar aspects have been methodically merged, allowing the decision makers to easily find high-level aspects to be included in the analysis. In particular, the new framework is composed of input attributes obtained from the literature review, combined with additional items relevant for the actual application, shaping the features contained as categories and subcategories. The new framework is defined according to four main categories (economic, territorial, social, and technical), and the features are defined by tailoring the solution on real cases, which represents the novelty of this work.

The new feature framework also focuses on the relative importance among groups of features, evaluating both the occurrences and the assigned weights based on the literature outcomes. A numerical assessment was carried out through the computation of aggregated appearances and weights organized into two matrices for all the literature contributions considered. Through this analysis, it was possible to examine both the distribution of the criteria and their relative importance, finally establishing a hierarchical order based on the actual literature background retrieved. The proposed categorization is also convenient for better understanding the most common attributes used in the literature and their relative importance. This provides suggestions to the decision maker on the choice of weights.

As a future development, the analysis can be extended by setting up a comparison among the different MCDM methods mainly exploited for EVCS location, evaluating their performance, limitations, and strengths. Moreover, an interesting insight is to examine the relevant points of view of the different actors involved in the EVCS location problem. Based on this last point, it can be observed how the categorization of the attributes will change, highlighting which output attributes will be considered or excluded in the analysis.

The application of the new criteria presented here offers a conventional basis for future research works in the field. The adoption of this framework can allow for a direct comparison among different works and proposed solutions. This makes it possible to partially attempt to resolve the inhomogeneity of the retrieved attributes, which constitutes the main challenging aspect in the application of the proposed framework.

Author Contributions: The work has been carried out with similar contributions by all authors. All authors have read and agreed to the published version of the manuscript.

Funding: This research received no external funding.

Data Availability Statement: Data is contained within the article.

Conflicts of Interest: This paper has been originated by a common activity of the authors, initially supported by Atlante Srl under the contract "Metodi decisionali per la localizzazione ottimale di infrastrutture di ricarica di autoveicoli" (in Italian, November 2022–February 2023), followed by unsupported common activity carried out by the authors to continue the work for the preparation of the paper. The authors Paolo Ciliento, Marco De Donno and Francesco Lamberti are employed by the company Atlante Srl. The author Francesca Mapelli is employed by the company Atlante Italia Srl (100% controlled by Atlante Srl).

Nomenclature

BSS	Battery Swapping Station
CPO	Charging Point Operator
EA	Electricity Accessibility
EMO	Evolutionary Multi-objective Optimization
EV	Electric Vehicle
EVCS	Electric Vehicle Charging Station
ICE	Internal Combustion Engine
MADM	Multi-Attribute Decision-Making
MCDM	Multi-Criteria Decision-Making
MOCO	Multi-Objective Combinatorial Optimization
MODM	Multi-Objective Decision-Making
MOLP	Multi-Objective Linear Programming
O&M	Operation and Maintenance
RESs	Renewable Energy Sources

References

1. Cazzola, P.; Gorner, M.; Tattini, J.; Schuitmaker, R.; Scheffer, S.; D'Amore, L.; Signollet, H.; Paoli, L.; Teter, J.; Bunsen, T. *Global EV Outlook 2019—Scaling Up the Transition to Electric Mobility*; International Energy Agency: Paris, France, 2019. [CrossRef]
2. Ahmad, F.; Iqbal, A.; Ashraf, I.; Marzband, M.; Khan, I. Optimal location of electric vehicle charging station and its impact on distribution network: A review. *Energy Rep.* **2022**, *8*, 2314–2333. [CrossRef]
3. Deeum, S.; Charoenchan, T.; Janjamraj, N.; Romphochai, S.; Baum, S.; Ohgaki, H.; Mithulananthan, N.; Bhumkittipich, K. Optimal Placement of Electric Vehicle Charging Stations in an Active Distribution Grid with Photovoltaic and Battery Energy Storage System Integration. *Energies* **2023**, *16*, 7628. [CrossRef]
4. Stanko, P.; Tkac, M.; Kajanova, M.; Roch, M. Impacts of Electric Vehicle Charging Station with Photovoltaic System and Battery Energy Storage System on Power Quality in Microgrid. *Energies* **2024**, *17*, 371. [CrossRef]

5. Wu, Y.; Xie, C.; Xu, C.; Li, F. A Decision Framework for Electric Vehicle Charging Station Site Selection for Residential Communities under an Intuitionistic Fuzzy Environment: A Case of Beijing. *Energies* **2017**, *10*, 1270. [CrossRef]
6. Mhana, K.H.; Awad, H.A. An ideal location selection of electric vehicle charging stations: Employment of integrated analytical hierarchy process with geographical information system. *Sustain. Cities Soc.* **2024**, *107*, 105456. [CrossRef]
7. Elomiya, A.; Křupka, J.; Jovčić, S.; Simic, V.; Švadlenka, L.; Pamucar, D. A hybrid suitability mapping model integrating GIS, machine learning, and multi-criteria decision analytics for optimizing service quality of electric vehicle charging stations. *Sustain. Cities Soc.* **2024**, *106*, 105397. [CrossRef]
8. Gulbahar, I.T.; Sutcu, M.; Almomany, A.; Ibrahim, B.S.K.K. Optimizing Electric Vehicle Charging Station Location on Highways: A Decision Model for Meeting Intercity Travel Demand. *Sustainability* **2023**, *15*, 16716. [CrossRef]
9. Ademulegun, O.O.; MacArtain, P.; Oni, B.; Hewitt, N.J. Multi-Stage Multi-Criteria Decision Analysis for Siting Electric Vehicle Charging Stations within and across Border Regions. *Energies* **2022**, *15*, 9396. [CrossRef]
10. Hwang, C.L.; Yoon, K. Methods for Multiple Attribute Decision Making. In *Multiple Attribute Decision Making: Methods and Applications A State-of-the-Art Survey*; Springer: Berlin/Heidelberg, Germany, 1981; pp. 58–191. [CrossRef]
11. Zhao, H.; Hao, X. Location decision of electric vehicle charging station based on a novel grey correlation comprehensive evaluation multi-criteria decision method. *Energy* **2024**, *299*, 131356. [CrossRef]
12. Wang, X.; Xia, W.; Yao, L.; Zhao, X. Improved Bayesian Best-Worst Networks with Geographic Information System for Electric Vehicle Charging Station Selection. *IEEE Access* **2024**, *12*, 758–771. [CrossRef]
13. Guo, S.; Zhao, H. Optimal site selection of electric vehicle charging station by using fuzzy TOPSIS based on sustainability perspective. *Appl. Energy* **2015**, *158*, 390–402. [CrossRef]
14. Liu, H.C.; Yang, M.; Zhou, M.; Tian, G. An Integrated Multi-Criteria Decision Making Approach to Location Planning of Electric Vehicle Charging Stations. *IEEE Trans. Intell. Transp. Syst.* **2019**, *20*, 362–373. [CrossRef]
15. Rani, P.; Mishra, A.R. Fermatean fuzzy Einstein aggregation operators-based MULTIMOORA method for electric vehicle charging station selection. *Expert Syst. Appl.* **2021**, *182*, 115267. [CrossRef]
16. Feng, J.; Xu, S.X.; Li, M. A novel multi-criteria decision-making method for selecting the site of an electric-vehicle charging station from a sustainable perspective. *Sustain. Cities Soc.* **2021**, *65*, 102623. [CrossRef]
17. Xu, J.; Zhong, L.; Yao, L.; Wu, Z. An interval type-2 fuzzy analysis towards electric vehicle charging station allocation from a sustainable perspective. *Sustain. Cities Soc.* **2018**, *40*, 335–351. [CrossRef]
18. Wang, R.; Li, X.; Xu, C.; Li, F. Study on location decision framework of electric vehicle battery swapping station: Using a hybrid MCDM method. *Sustain. Cities Soc.* **2020**, *61*, 102149. [CrossRef]
19. Sadeghi-Barzani, P.; Rajabi-Ghahnavieh, A.; Kazemi-Karegar, H. Optimal fast charging station placing and sizing. *Appl. Energy* **2014**, *125*, 289–299. [CrossRef]
20. Anthopoulos, L.; Kolovou, P. A Multi-Criteria Decision Process for EV Charging Stations' Deployment: Findings from Greece. *Energies* **2021**, *14*, 5441. [CrossRef]
21. Zhang, L.; Zhao, Z.; Yang, M.; Li, S. A multi-criteria decision method for performance evaluation of public charging service quality. *Energy* **2020**, *195*, 116958. [CrossRef]
22. Kaya, Ö.; Tortum, A.; Alemdar, K.D.; Çodur, M.Y. Site selection for EVCS in Istanbul by GIS and multi-criteria decision-making. *Transp. Res. Part D Transp. Environ.* **2020**, *80*, 102271. [CrossRef]
23. Wu, Y.; Yang, M.; Zhang, H.; Chen, K.; Wang, Y. Optimal Site Selection of Electric Vehicle Charging Stations Based on a Cloud Model and the PROMETHEE Method. *Energies* **2016**, *9*, 157. [CrossRef]
24. Zhou, J.; Wu, Y.; Wu, C.; He, F.; Zhang, B.; Liu, F. A geographical information system based multi-criteria decision-making approach for location analysis and evaluation of urban photovoltaic charging station: A case study in Beijing. *Energy Convers. Manag.* **2020**, *205*, 112340. [CrossRef]
25. Bai, X.; Chin, K.S.; Zhou, Z. A bi-objective model for location planning of electric vehicle charging stations with GPS trajectory data. *Comput. Ind. Eng.* **2019**, *128*, 591–604. [CrossRef]
26. Ren, X.; Zhang, H.; Hu, R.; Qiu, Y. Location of electric vehicle charging stations: A perspective using the grey decision-making model. *Energy* **2019**, *173*, 548–553. [CrossRef]
27. Hosseini, S.; Sarder, M. Development of a Bayesian network model for optimal site selection of electric vehicle charging station. *Int. J. Electr. Power Energy Syst.* **2019**, *105*, 110–122. [CrossRef]
28. Liu, A.; Zhao, Y.; Meng, X.; Zhang, Y. A three-phase fuzzy multi-criteria decision model for charging station location of the sharing electric vehicle. *Int. J. Prod. Econ.* **2020**, *225*, 107572. [CrossRef]
29. Erbaş, M.; Kabak, M.; Özceylan, E.; Çetinkaya, C. Optimal siting of electric vehicle charging stations: A GIS-based fuzzy Multi-Criteria Decision Analysis. *Energy* **2018**, *163*, 1017–1031. [CrossRef]
30. Karaşan, A.; Kaya, İ.; Erdoğan, M. Location selection of electric vehicles charging stations by using a fuzzy MCDM method: A case study in Turkey. *Neural Comput. Appl.* **2018**, *32*, 4553–4574. [CrossRef]
31. Tang, Z.C.; Guo, C.; Hou, P.X.; Fan, Y. Optimal Siting of Electric Vehicle Charging Stations Based on Voronoi Diagram and FAHP Method. *Energy Power Eng.* **2013**, *5*, 1404–1409. [CrossRef]
32. Schmidt, M.; Zmuda-Trzebiatowski, P.; Kiciński, M.; Sawicki, P.; Lasak, K. Multiple-Criteria-Based Electric Vehicle Charging Infrastructure Design Problem. *Energies* **2021**, *14*, 3214. [CrossRef]

33. Wu, F.; Sioshansi, R. A stochastic flow-capturing model to optimize the location of fast-charging stations with uncertain electric vehicle flows. *Transp. Res. Part D Transp. Environ.* **2017**, *53*, 354–376. [CrossRef]
34. Guler, D.; Yomralioglu, T. Suitable location selection for the electric vehicle fast charging station with AHP and fuzzy AHP methods using GIS. *Ann. GIS* **2020**, *26*, 169–189. [CrossRef]
35. Zhao, H.; Li, N. Optimal Siting of Charging Stations for Electric Vehicles Based on Fuzzy Delphi and Hybrid Multi-Criteria Decision Making Approaches from an Extended Sustainability Perspective. *Energies* **2016**, *9*, 270. [CrossRef]
36. Ayyildiz, E. A novel pythagorean fuzzy multi-criteria decision-making methodology for e-scooter charging station location-selection. *Transp. Res. Part D Transp. Environ.* **2022**, *111*, 103459. [CrossRef]
37. Bahaj, A.S.; Turner, P.; Mahdy, M.; Leggett, S.; Wise, N.; Alghamdi, A. Environmental assessment platform for cities racing to net zero. *J. Phys. Conf. Ser.* **2021**, *2042*, 012140. [CrossRef]

Disclaimer/Publisher's Note: The statements, opinions and data contained in all publications are solely those of the individual author(s) and contributor(s) and not of MDPI and/or the editor(s). MDPI and/or the editor(s) disclaim responsibility for any injury to people or property resulting from any ideas, methods, instructions or products referred to in the content.

MDPI AG
Grosspeteranlage 5
4052 Basel
Switzerland
Tel.: +41 61 683 77 34

Energies Editorial Office
E-mail: energies@mdpi.com
www.mdpi.com/journal/energies

Disclaimer/Publisher's Note: The title and front matter of this reprint are at the discretion of the Guest Editors. The publisher is not responsible for their content or any associated concerns. The statements, opinions and data contained in all individual articles are solely those of the individual Editors and contributors and not of MDPI. MDPI disclaims responsibility for any injury to people or property resulting from any ideas, methods, instructions or products referred to in the content.

www.ingramcontent.com/pod-product-compliance
Lightning Source LLC
LaVergne TN
LVHW073334090526
838202LV00019B/2419